Research Series on the Chinese Dream and China's Development Path

Project Director
Xie Shouguang, President, Social Sciences Academic Press

Series editors
Li Yang, Vice president, Chinese Academy of Social Sciences, Beijing, China
Li Peilin, Vice president, Chinese Academy of Social Sciences, Beijing, China

Academic Advisors
Cai Fang, Gao Peiyong, Li Lin, Li Qiang, Ma Huaide, Pan Jiahua, Pei Changhong, Qi Ye, Wang Lei, Wang Ming, Zhang Yuyan, Zheng Yongnian, Zhou Hong

Drawing on a large body of empirical studies done over the last two decades, this Series provides its readers with in-depth analyses of the past and present and forecasts for the future course of China's development. It contains the latest research results made by members of the Chinese Academy of Social Sciences. This series is an invaluable companion to every researcher who is trying to gain a deeper understanding of the development model, path and experience unique to China. Thanks to the adoption of Socialism with Chinese characteristics, and the implementation of comprehensive reform and opening-up, China has made tremendous achievements in areas such as political reform, economic development, and social construction, and is making great strides towards the realization of the Chinese dream of national rejuvenation. In addition to presenting a detailed account of many of these achievements, the authors also discuss what lessons other countries can learn from China's experience.

More information about this series at http://www.springer.com/series/13571

Xinli Zhao · Minrong Li
Maoxing Huang · Alexander Sokolov
Editors

BRICS Innovative Competitiveness Report 2017

SSAP

Springer

Editors
Xinli Zhao
China Science and Technology Exchange Center
Beijing
China

Minrong Li
China Institute of Science and Technology Evaluation
Beijing
China

Maoxing Huang
School of Economics
Fujian Normal University
Fuzhou, Fujian
China

Alexander Sokolov
Institute for Statistical Studies and Economics of Knowledge
National Research University Higher School of Economics
Moscow
Russia

ISSN 2363-6866 ISSN 2363-6874 (electronic)
Research Series on the Chinese Dream and China's Development Path
ISBN 978-981-10-8077-7 ISBN 978-981-10-8078-4 (eBook)
https://doi.org/10.1007/978-981-10-8078-4

Jointly published with Social Sciences Academic Press

The print edition is not for sale in China Mainland. Customers from China Mainland please order the print book from: Social Sciences Academic Press.

Library of Congress Control Number: 2018930513

This Springer imprint is published by the registered company Springer Nature Singapore Pte Ltd.
The registered company address is: 152 Beach Road, #21-01/04 Gateway East, Singapore 189721, Singapore

Foreword

Remarks at the Opening Ceremony of the 5th BRICS Science, Technology and Innovation Ministerial Meeting

Honourable ministers of BRICS countries,

Ladies and Gentlemen,

Dear Friends,

Good morning! It is my great pleasure to gather with you in Hangzhou, one of the most innovative cities in China, for the Fifth BRICS Science, Technology and Innovation Ministerial Meeting by the beautiful West Lake. First of all, I would like to extend, on behalf of the Chinese government, sincere welcome and best wishes to all the ministers, delegates and guests coming from afar.

Last September, leaders of the Group of Twenty (G20), which includes the BRICS countries, gathered in Hangzhou for a Summit meeting on the theme of "Towards an innovative, invigorated, interconnected and inclusive world economy" and achieved fruitful results. For the first time, the Summit listed "Innovating upon growth patterns" as a key topic. On 4 November 2016, the G20 Science, Technology and Innovation Ministers Meeting was held in Beijing, and attended by BRICS science, technology and innovation ministers. During the event, we worked together and reached a *Statement of the G20 Science, Technology and Innovation Ministers Meeting*. All these have pointed to a new direction, planned a new path and injected new impetus to the social and economic development of not only G20 countries but also our BRICS countries and even the whole world.

Today, we, the ministers of science and technology of the BRICS countries, gather here to further implement the outcome of the G20 Hangzhou Summit. It is of great significance for putting forward a "BRICS Solution" for and contributing "BRICS Wisdom" to innovative growth.

As the first echelon of emerging markets of the G20 countries, the BRICS countries play an irreplaceable role in South–South cooperation in science and technology and in North–South dialogues on innovation. The BRICS countries are the main group that leads science and technology innovation in developing countries, as well as an important force of science and technology innovation in the world. As for science and technology innovation, the BRICS countries make up approximately 17% of the world's gross annual R&D expenditure, have a high-tech product export volume of nearly $6 trillion or approximately 28% of the world's total, and boast 590,000 published papers in science and technology journals, accounting for approximately 27% of the world's total. There has been a gradual increase in BRICS' contribution to science, technology and innovation in the world, and the international influence of the BRICS countries is increasing. The BRICS countries have become a "bellwether" in their respective regions, leading the development of science and technology, economy and society in neighbouring countries.

Looking back, the First BRICS Science, Technology and Innovation Ministerial Meeting on the theme of "A Strategic Partnership for Equitable Growth and Sustainable Development" was held in South Africa in February 2014. At the meeting, we signed the *Cape Town Declaration* together, marking the official establishment of the mechanism of cooperation in science, technology and innovation between the BRICS countries.

In March 2015, the Second BRICS Science, Technology and Innovation Ministerial Meeting was held in Brazil. We adopted the *Brasilia Declaration* and signed the *Memorandum of Understanding Between BRICS Countries on Intergovernmental Cooperation in Science, Technology and Innovation*, identifying 19 priority fields of cooperation and specifying new directions for cooperation in science, technology and innovation.

In October 2015, the Third BRICS Science, Technology and Innovation Ministerial Meeting was held in Russia, carrying a theme of "BRICS Partnership—a Powerful Factor of Global Development". The meeting issued the *Moscow Declaration* and agreed on BRICS cooperation within large research infrastructures, coordination of the existing large-scale national programs of the BRICS countries, development and implementation of a BRICS Framework Program for funding multilateral joint research projects, technology commercialization and innovation, as well as establishment of BRICS Research and Innovation Networking Platform. The meeting also approved the *2015–2018 BRICS Work Plan for Science, Technology and Innovation.*

In October 2016, the Fourth BRICS Science, Technology and Innovation Ministerial Meeting was held in India. Carrying a theme of "Building Responsive Inclusive and Collective Solutions", the meeting adopted the *Jaipur Declaration*. The First BRICS Young Scientists Forum was also held in India.

The current BRICS Science, Technology and Innovation Ministerial Meeting has "Leading through innovation & deepening cooperation" as its theme and will focus on exchange of views concerning BRICS STI policies, cooperation in thematic fields, joint sponsorship of multilateral R&D projects, youth innovation and entrepreneurship, science park cooperation and other important topics. Thanks to this meeting, we hope that we can continue to strengthen innovation and entrepreneurship policy and conduct exchanges between BRICS countries, promote

S&T innovation and entrepreneurship platform building in the BRICS countries, and deepen technology cooperation among enterprises, technology transfer and commercialization, science park cooperation and youth innovation and entrepreneurship cooperation between the BRICS countries. This meeting intends to issue a *Hangzhou Declaration* and will deliberate on and adopt a *BRICS Innovation Cooperation Action Plan* to contribute STI plans to the outcomes of the BRICS Leaders Meeting in Xiamen in September.

Ladies and Gentlemen, Friends,

Science and technology is the foundation of national prosperity, and innovation is the soul of national progress. The Chinese government attaches great importance to science, technology and innovation, regards it as a strategic support for raising productivity and improving comprehensive national strength and places it at the centre of overall national development. Currently, China is thoroughly implementing an innovation-driven development strategy. In recent years, China has made a series of breakthroughs in science, technology and innovation: national technological strengths and innovation abilities have been further improved, and various science, technology and innovation achievements have been made; science, technology and innovation has been integrated into overall economic and social development, speeding up new driving forces for growth and playing a markedly improved role in supporting and leading the supply-side structural reform; mass innovation and entrepreneurship is thriving and the whole society is ever more enthusiastic about supporting and participating in innovation; a main science and technology system reform framework has been basically established, substantive breakthroughs have been made in the key fields of enterprise innovation policy, planned funds management, science and technology results industrialization and income distribution system reform, and there has been a further strengthening of research personnel's sense of gain; and China's international standing in science, technology and innovation has been rising continuously.

In 2016, China's gross R&D expenditure reached RMB1.55 trillion, accounting for 2.1% of its GDP, with enterprises contributing to 78% of the total R&D expenditure. For six consecutive years, China has ranked second in the world in terms of international research papers published and has rose to fourth place in the world in terms of SCI-indexed paper citations. So far, China has approved a total of 17 national independent innovation demonstration zones and 156 national high-tech development zones, giving full play to their role in leading and driving regional economic and social development.

China attaches great importance to international cooperation in science, technology and innovation. In May this year, the Chinese Ministry of Science and Technology issued the *Thirteenth Five-Year Special Plan for International Cooperation in Science, Technology and Innovation*, setting the development goal of further deepening cooperation in science, technology and innovation with other countries and helping to build new international relations centring on win–win cooperation. In the next 5 years, China will continue to step up its opening up to and cooperation with the outside world and build wider partnerships for science, technology and innovation with the rest of the world, including other BRICS countries.

Ladies and Gentlemen, Friends,

The world economy is now on a tortuous path to recovery through deep adjustment. It is in a critical period of functionals change from the old to the new. A new round of science and technology revolution and industrial change is poised to take place. Mankind is entering a period of active and intensive science, technology and innovation. This provides us with rare opportunities and challenges for deepening cooperation in science, technology and innovation. The BRIC countries have their respective advantages in the areas of accumulated talents, theories and practices of science, technology and innovation, material and financial resources for science, technology and innovation, and geographical distributions across the globe. There is huge potential for their linked development. Overall planning and consultations in science, technology and innovation, cooperation and co-implementation of planned projects, and prompt sharing of project results between the BRICS countries will boost the growth mode transformation and upgrading, support and lead economic and social development in all five countries and provide new impetus for global economic growth. Meanwhile, these will also give a greater say to the BRICS countries and even all developing countries in global science, technology and innovation, as well as in political, economic, cultural and many other fields.

Innovation drives development, and cooperation leads the future. The BRICS countries share the same fate with each other and are both a community of shared interests whose members go forward hand in hand and a closely connected community of common destiny. In May this year, China successfully held the Belt and Road Forum for International Cooperation. In his keynote speech, Chinese President Xi Jinping stressed the need to adhere to the Silk Road spirit which centres on peaceful cooperation, openness and inclusiveness, mutual learning, mutual benefit and win–win cooperation, to work together to promote the construction of the Belt and Road, build the Belt and Road into a belt and road of peace, prosperity, openness, innovation and civilization Road, and march towards a better tomorrow. President Xi Jinping pointed out that it is necessary to adhere to innovation-driven development and build a digital Silk Road of the twenty-first century. He called for pushing forward the deep integration of science and technology with industry and with finance so as to create entrepreneurship space and workshops for young people of all countries in the Internet age, as well as for efforts to put the new concept of green development into practice and achieve the 2030 Sustainable Development Goals together. At the recently concluded 2017 G20 Summit and the BRICS Leader's Informal Meeting held during the Summit, the BRICS countries reached important consensuses on strengthening unity and cooperation, working together to build an open world economy, improving global economic governance and promoting sustainable development. All these are highly consistent with the idea of cooperation under the BRICS mechanism. China will uphold the BRICS spirit of openness, inclusiveness, cooperation and win–win cooperation, work with other countries and make joint efforts to plan a new BRICS development blueprint and write a new chapter of cooperation in science, technology and innovation between the BRICS countries.

As an old Chinese proverb puts it, “Even mountains and seas cannot distance people with common aspirations”. Though far apart, the BRICS countries share the same aspirations and no mountains or seas can limit their cooperation. As long as we think about and work on the same goal, BRICS cooperation in science, technology and innovation will surely open a better tomorrow and usher in the next “golden three years”.

Finally, I wish the meeting a complete success, and all our friends smooth work, happy living and good health during your stay in Hangzhou! Thank you!

July 2017

Wan Gang
Vice Chairman of the CPPCC National Committee
Minister of Science and Technology of China
and President of China Association for Science
and Technology

Editorial Board

Translators
Li Zhiqiang
Zheng Wei
Pan Hua
Chen Yuheng
Dai Le
Xia Huanhuan

Remarks by Minister Naledi Pandor, Minister of Science and Technology, at the Fifth BRICS Science, Technology And Innovation Ministerial Meeting, Hangzhou, China: 18 July 2017

Session on Policies on Innovation among BRICS countries

Minister Wan Gang, Minister of Science and Technology of the People's Republic of China

Deputy Ministers and Senior Officials of the BRICS partners,

Ladies and Gentlemen,

I am truly delighted to be back in the People's Republic of China, for this Fifth BRICS Science, Technology and Innovation Ministerial Meeting. I would in the first instance like to express my sincere appreciation to our Chair and Host, Minister Wan Gang, for the very efficient organization of our meeting, and the warm and gracious hospitality afforded to all of us.

There cannot be a more appropriate theme for our meeting than that chosen by the Chinese Presidency: "Leading through innovation and deepening cooperation". It is indeed our mission to strengthen our partnership to ensure science, technology and innovation play an optimal role in further enhancing the competitiveness of the BRICS economies and in improving the quality of living of all our citizens. Let us not forget that this critical task was entrusted to us by our Heads of State when at the G20 Summit held here in Hangzhou last year, they explicitly recognized innovation as a key driver for growth and sustainable development.

As policy-makers for our governments, the BRICS Ministerial Meeting provides a valuable opportunity for us to share within our partnership, our respective experiences and expertise, in formulating and implementing policies and strategies, which will ensure innovation-driven growth. This will be especially useful for us in South Africa, as my Ministry is currently preparing a new White Paper, the highest level policy document in our Government, on science, technology and innovation. The White Paper is intended to guide our efforts to ensure we are successful in promoting inclusive development in South Africa through science and innovation.

Our new policy document will have an important focus on the role Government should play in enabling innovation, notably by instilling a national innovation culture across all spheres of our government. This is certainly an area where we can learn much from our BRICS partners. Also receiving attention will be ensuring that appropriate supply-side measures such as funding and other incentives fulfil their role as drivers for innovation performance. Equally important will be the role of demand-side measure to promote the so-called innovation pull, through, for example, the leveraging of public procurement and specific support for small and medium enterprise development. Another priority will be the promotion of grass-roots innovation, ensuring all South Africans enjoy the opportunity to create and exploit innovation opportunities. Grassroots innovation is notably a strategic focus area in our bilateral cooperation with India.

Of course we will not achieve our vision, without successful policies and practices to facilitate the contribution of industry and business to innovation-driven growth. Without the participation of our enterprises, the so-called innovation chasm, which hampers our economies, will continue to persist. Unless our enterprises assist us to bridge this divide between research and the marketplace, we will be faced with the continued inability to translate a significant proportion of our research and development results into socio-economically useful products and services. Through science, technology and innovation, the business sector also has a critical role to play in diversifying our economy, helping us to develop a competitive knowledge economy, not dependent on raw materials and other commodities.

For South Africa, cooperation with and learning from our BRICS partners is crucial also in this domain. We, thus, greatly value the strategic partnership with China on Science Park Cooperation, launched during Her Excellency, Vice Premier Liu Yandong's visit to South Africa, last April. We, for example, admire the success of the Skolkovo Innovation Centre in the Russian Federation, a prime example of concerted investment in a public–private partnership to boost innovation and the development of high-technology enterprises.

We should also spare no effort to promote youth innovation and entrepreneurship. In this regard, I would like to congratulate the Government of China on hosting the very successful BRICS Young Scientist Forum last week. I look forward to the report on the Forum to be presented later today, because as put simply, the youth is our future. A few weeks ago, I had convened South Africa's inaugural Youth in Science, Technology and Innovation Summit, to provide an opportunity for South Africa's young innovators to communicate to policy-makers their concerns but above all their ambitions. It was a truly inspiring event, which left me more convinced than ever, that eliminating poverty, unemployed and inequality, starts by investing in our youth. The immense potential of their ingenuity, drive and commitments are our most precious assets for the future, and these should be permitted to blossom.

Dear colleagues, in South Africa we live by the credo that science knows no borders. It is only by sharing our resources, experience and expertise, that the global community will effectively put research and innovation at the service of our societies, achieving the Sustainable Development Goals. The Chinese Presidency's chosen theme for the Xiamen Summit in September—Stronger Partnership for a Brighter Future—captures this imperative most eloquently. We should therefore concertedly invest in and further develop our BRICS science, technology and innovation partnership, including within the framework of the BRICS Action Plan for Innovation, we will endorse later today.

It is also my hope that we will be able to launch as soon as possible the next call under the BRICS framework programme for multilateral research and innovation cooperation. Our BRICS partnership has become a recognized force in the global science area. Successful cooperation between South Africa, China and India is, for example, one of the pillars underpinning the global partnership to advance the Square Kilometre Array (SKA) global radio telescope project. We hope the Russian Federation and Brazil will join soon.

Furthermore, in another example of the impact of our collaboration, South Africa and Brazil had launched last week a science plan for South–South research cooperation in the South Atlantic Ocean. I do also hope that we will be able support the building of science, technology and innovation capacities elsewhere in Africa through our BRICS partnership and in this regard would like to acknowledge the efforts of China under the Forum for China–Africa Cooperation (FOCAC.)

Dear colleagues, in conclusion, I pledge that under the forthcoming South African Presidency, building on the foundation we have laid here in Hangzhou, and our work in Brasilia, Moscow and Jaipur, we will work concertedly with all of you to advance, strengthen and deepen our strategic and privileged partnership.

Permit me to finally share, that today, 18 July, marks the birthday of democratic South Africa's first President, our beloved and iconic leader, Nelson Mandela. The United Nations have asked that on this day, around the world, Nelson Mandela International Day is observed, by recognizing that all individuals have the ability and the responsibility to change the world for the better. Mandela Day is an occasion for everyone, including us gathered here in Hangzhou, to take action and inspire change.

In this spirit, my final remark will be to quote one of the other historic leaders of South Africa's liberation struggle Chief Albert Luthuli, who in 1961 stated that "Scientific inventions, at all conceivable levels should enrich human life, not threaten existence. Science should be the greatest ally, not the worst enemy of mankind". I have no doubt this is the sentiment, which will also guide our work here in Hangzhou.

I thank you.

Dr. Naledi Pandor
South African Minister of Science and Technology

Memorandum of Understanding on Cooperation in Science, Technology and Innovation Between the Governments of the Federative Republic of Brazil, the Russian Federation, the Republic of India, the People's Republic of China and the Republic of South Africa

Preamble

The Government of the Federative Republic of Brazil, The Government of the Russian Federation, The Government of the Republic of India, The Government of the People's Republic of China, and The Government of the Republic of South Africa (hereinafter referred to as the "Parties");

Reaffirming the overarching vision embodied in the BRICS Summit Declarations, including the 2011 BRICS Sanya Declaration which identified the need "to explore cooperation in the sphere of science, technology and innovation, including the peaceful use of space";

Noting the recommendations of the First, Second and Third BRICS Science, Technology and Innovation Senior Officials' Meetings, held respectively in Dalian, China, in September 2011 Pretoria, South Africa, in November 2012, and New Delhi, India, in December 2013;

Harnessing potential bilateral synergies and other forms of multicountry frameworks of cooperation among Brazil, Russia, India, China and South Africa in science, technology and innovation;

Desirous to further strengthen cooperation in the fields of science, technology and innovation for accelerated and sustainable socio-economic development among the five countries;

Recognizing the importance of cooperation based on the principles of voluntary participation, equality, mutual benefit, reciprocity and subject to the availability of earmarked resources for collaboration by each country;

Recognizing the variable geometry of the research and development systems of the BRICS member countries;

Hereby Agree as follows:

Article 1

Competent Authorities

The competent authorities responsible for the implementation of this Memorandum of Understanding will be the following designated organizations:

(a) For the Federative Republic of Brazil, the Ministry of Science, Technology and Innovation (MCTI);
(b) For the Russian Federation, the Ministry of Education and Science (MES);
(c) For the Republic of India, the Department of Science and Technology (DST, India);
(d) For the People's Republic of China, the Ministry of Science and Technology (MOST);
(e) For the Republic of South Africa, the Department of Science and Technology (DST, South Africa).

Article 2

Objectives

The main objectives of this Memorandum of Understanding are:

(a) To establish a strategic framework for cooperation in science, technology and innovation among the BRICS member countries;
(b) To address common global and regional socio-economic challenges in the BRICS member countries utilizing shared experiences and complementarities in science, technology and innovation;
(c) To co-generate new knowledge and innovative products, services and processes in the BRICS member countries utilizing appropriate funding and investment instruments;
(d) To promote, where appropriate, joint BRICS science, technology and innovation partnerships with other strategic actors in the developing world.

Article 3

Areas of Cooperation

The main areas of cooperation under this Memorandum of Understanding shall include but not be confined to:

(a) Exchange of information on policies and programmes and promotion of innovation and technology transfer;
(b) Food security and sustainable agriculture;

(c) Natural disasters;
(d) New and renewable energy, energy efficiency;
(e) Nanotechnology;
(f) High-performance computing;
(g) Basic research;
(h) Space research and exploration, aeronautics, astronomy and earth observation;
(i) Medicine and biotechnology;
(j) Biomedicine and life sciences (biomedical engineering, bioinformatics, biomaterials);
(k) Water resources and pollution treatment;
(l) High-tech zones/science parks and incubators;
(m) Technology transfer;
(n) Science popularization;
(o) Information and communication technology;
(p) Clean coal technologies;
(q) Natural gas and non-conventional gases;
(r) Ocean and polar sciences;
(s) Geospatial technologies and its applications.

Article 4

Mechanisms and Modalities of Cooperation

The principal mechanism for cooperation shall be this Memorandum of Understanding. The Parties or their designated institutions may enter into sub-agreements which shall be governed by the terms of this Memorandum of Understanding.

The modalities of cooperation under this Memorandum of Understanding and sub-agreements arising therefrom between the Parties in the fields of science, technology and innovation shall take the following forms:

(a) Short-term exchange of scientists, researchers, technical experts and scholars;
(b) Dedicated training programmes to support human capital development in science, technology and innovation;
(c) Organization of science, technology and innovation workshops, seminars and conferences in areas of mutual interest;
(d) Exchange of science, technology and innovation information;
(e) Formulation and implementation of collaborative research and development programmes and projects;
(f) Establishment of joint funding mechanisms to support BRICS research programmes and large-scale research infrastructure projects;
(g) Facilitated access to science and technology infrastructure among BRICS member countries;

(h) Announcement of simultaneous calls for proposals in BRICS member countries;
(i) Cooperation of national science and engineering academies and research agencies.

Article 5

Governing Structures

The main structures governing cooperation under this Memorandum of Understanding shall include:

1. BRICS Science, Technology and Innovation Ministerial Meeting
2. BRICS Science, Technology and Innovation Senior Officials' Meeting
3. BRICS Science, Technology and Innovation Working Group

1. The BRICS Science, Technology and Innovation Ministerial Meeting (comprising Ministers responsible for science, technology and innovation in Brazil, Russia, India, China and South Africa) shall convene at least once every year during the presidency of a member country. The main responsibilities of the BRICS Science, Technology and Innovation Ministerial Meeting will include:
   (a) Providing an overarching vision and advice on institutional and financial frameworks for major BRICS science, technology and innovation programmes and initiatives;
   (b) Facilitating linkages between the BRICS Science, Technology and Innovation Working Group and other BRICS sectoral working groups or BRICS expert groups to ensure the effective implementation and realization of the objectives of this Memorandum of Understanding;
   (c) Setting priorities for cooperation and joint action in science, technology and innovation among BRICS member countries for a given period of time, taking into account the priority areas indicated in Article (3) above.
2. The BRICS Science, Technology and Innovation Senior Officials' Meeting will constitute Directors-General (or equivalent) of BRICS member countries as the leaders of delegation, BRICS science, technology and innovation country coordinators, focal points, scientists, experts and other relevant officials.
   The BRICS Science, Technology and Innovation Senior Officials' Meeting will meet annually in the country where the BRICS Summit is hosted.
   Responsibilities of the BRICS Science, Technology and Innovation Senior Officials' Meeting will include:

(a) Exchanging information on recent science, technology and innovation developments as well as identifying common policy challenges in BRICS member countries;
(b) Supporting the implementation of strategic decisions related to science, technology and innovation taken by the BRICS Summits, as well the high-level decisions emanating from BRICS Science, Technology and Innovation Ministerial Meetings;
(c) Facilitating BRICS science, technology and innovation cooperation mainly through the prioritization of the thematic areas identified in Article (3) of this Memorandum of Understanding;
(d) Configuring appropriate funding mechanisms and instruments to support BRICS science, technology and innovation cooperation;
(e) Harnessing synergies in respect of science, technology and innovation priority directions at bilateral, multilateral and poly-lateral levels within BRICS;
(f) Approving 3- to 5-year cycles for BRICS science, technology and innovation initiatives and programmes;
(g) Reviewing periodically progress in terms of implementation with respect to science, technology and innovation cooperation under this Memorandum of Understanding, as well as identifying new areas, activities and cooperation modalities of mutual interest;
(h) Providing recommendations for consideration by the BRICS Science, Technology and Innovation Ministerial Meeting to enhance effective implementation of this Memorandum of Understanding;
(i) Considering other agenda matters deemed appropriate by the BRICS member countries.

3. The BRICS Science, Technology and Innovation Working Group will constitute the five BRICS science, technology and innovation country coordinators whose responsibilities will include:

(a) Fulfilling the function of Secretariat for the BRICS Science, Technology and Innovation SOM (developing the agenda and annotations for the BRICS science, technology and innovation SOM; recording proceedings of the SOM, etc.);
(b) Convening Science, Technology and Innovation Working Group meetings between sessions of the Science, Technology and Innovation SOM.

Article 6

Funding Mechanisms and Instruments

Science, technology and innovation cooperation under this Memorandum of Understanding will be supported by appropriate BRICS country funding mechanisms, instruments and national rules.

The key objectives of the BRICS science, technology and innovation funding mechanisms and instruments shall be:

(a) To establish R&D programmes in frontier and priority research areas in support of sustainable development in BRICS member countries;
(b) To promote the co-generation of new knowledge and innovative products, services and processes;
(c) To co-invest in large-scale research infrastructure projects;
(d) To facilitate technology and knowledge transfer and implementation;
(e) To facilitate policy development in science, technology and innovation;
(f) To facilitate linkages with various forums dealing with business, academia, research and development centres, government agencies and institutions.

Article 7

Management of Intellectual Property Rights

1. The parties will ensure adequate and effective protection and fair allocation of intellectual property rights of a proprietary nature that may result from the cooperative activities under this Memorandum of Understanding, according to their respective national laws and regulations and their international obligations.
2. The condition for the acquisition, maintenance and commercial exploitation of intellectual property rights over possible products and/or processes that might be obtained under this Memorandum of Understanding will be defined in the specific programmes, contracts or working plans of the activities of cooperation.
3. The specific programmes, contracts or working plans relating to the activities of cooperation mentioned in Paragraph 2 of this Article will set out the conditions regarding the confidentiality of information whose publication and/or disclosure might jeopardize the acquisition, maintenance and commercial exploitation of intellectual property rights obtained under this Memorandum of Understanding. Such specific programmes, contracts or working plans related to the activities of cooperation will establish, where applicable, the rule and procedures concerning the settlement of disputes on intellectual property matters under this Memorandum of Understanding.

Article 8

Final Dispositions

1. This Memorandum of Understanding will come into force on the date of signature and will remain valid for five (5) years. Thereafter, this Memorandum of Understanding shall be renewed automatically for successive equal periods, unless one of the Parties notifies the others in writing its intention to terminate this Memorandum of Understanding.
2. The present Memorandum of Understanding may be amended at any time, by mutual consent of the Parties, through diplomatic channels.
3. Any Party may, at any time, notify the others of its intention to terminate the present Memorandum of Understanding. Termination will be effective six (6) months after the date of the notification and will not affect the ongoing activities of cooperation, unless otherwise agreed by the Parties.
4. Any dispute related to the interpretation or implementation of the present Memorandum of Understanding will be settled by direct negotiations between the Parties, through diplomatic channels.

In Witness Whereof the undersigned, being duly authorized thereto by their respective Governments, have signed this Memorandum of Understanding in five originals, in Portuguese, Russian, Hindi, Chinese and English languages, all texts being equally authentic. In case of any divergence of interpretation, the English text will prevail.

BRICS Action Plan for Innovation Cooperation (2017–2020)

I. Foreword

We, BRICS countries,

1. With 42% of the world population, contribute 18% of global GDP, 17% of global R&D investment and 27% of science papers published on international journals, as an important force of international economic cooperation and one of the most dynamic and promising emerging economies, BRICS countries are major representatives of emerging economies in the world. Our collective efforts are to undertake innovation and cooperation and facilitate innovation-driven development for sustainable development of the world economy.
2. Reaffirm that innovation refers to the embodiment of an idea in a technology, product or process that is new and creates productive value. An innovation is the implementation of a new or significantly improved product (good or service), or process which derives from creative ideas, technological progress, a new marketing method or a new organizational method in business practices, workplace organization or external relations. Innovation covers a wide range of domains with science, technology and innovation (STI) as the core.
3. We will actively promote cooperation in STI under bilateral and multilateral frameworks in accordance with the *MoU on Cooperation in STI between the Governments of BRICS Countries*, *Jaipur Declaration*, and the theme of the 5th BRICS STI Ministerial Meeting, thus drive rapid and sustainable economic growth and social progress in the BRICS countries.
4. Stress that innovation is one of the key driving forces of global sustainable development, playing a fundamental role in promoting economic growth, supporting job creation, entrepreneurship and structural reform, enhancing productivity and competitiveness, providing better services for the citizens and addressing global challenges. The BRICS countries aim to encourage innovation through practical actions to promote sustainable economic growth today and lay a solid foundation for tomorrow.

II. Action Plan

BRICS countries are facing new challenges in economic development though our economic prospects and growth momentum remain unchanged. In this context, we are committed to the following steps:

1. Promoting exchanges and good practices among the BRICS countries on innovation strategies and policies; enhancing mutual understanding, complementarity and coordination for the BRICS cooperation in innovation, and in particular, for the attainment of socio-economic progress driven by scientific, technological and social innovation, for the building of a BRICS community of shared values and common future, and for the realization of sustainable development goals.
2. Strengthening cooperation in scientific and research activities, enhancing cooperation in innovation based on existing mechanisms and joint research programmes including such cooperation conducted through public–private partnerships; fostering strategic and long-term university–industry partnerships to address the needs of industry and contributing directly to economic growth and development; continuing to encourage and support research and development projects in the areas of fundamental and applied research and innovation within bilateral and multilateral frameworks and continuing to carry out joint calls for STI projects; understanding the importance of implementing BRICS initiatives related to research and innovation; promoting open science and the sharing of research infrastructure; developing and initiating international mega science programmes.
3. Organizing joint activities on identifying priorities for STI cooperation of BRICS countries based on foresight and monitoring of global STI development.
4. In view of the importance of science and technology parks for regional economic development, encouraging cooperation among science parks including supporting the transnational establishment of BRICS high-tech enterprises in S&T parks. We welcome the establishment of exchange mechanisms for science parks and expanding areas of cooperation in these domains.
5. Encouraging technology transfer among the BRICS countries, strengthening training of technology transfer professionals, developing platforms for collaboration among businesses and academia, enabling extensive and orderly transfer and translation of innovation achievements in the BRICS countries. Utilizing existing technological network platforms as instruments of search for foreign partners for technological collaboration and initiation of joint STI projects.
6. Promoting BRICS Partnerships on Youth Innovation and Entrepreneurship to carry out pragmatic cooperation, advocating the entrepreneurial spirit of encouraging innovation and tolerating failure, and to create a favourable ecosystem for innovation and entrepreneurship among the younger generation.
7. Acknowledging the importance of supporting STI investment and the need to establish inter-BRICS investment instruments, we support explore the possibilities of driving BRICS cooperation on innovation and entrepreneurship through the National Development Banks, New Development Bank and other existing financing institutions.

8. Supporting the mobility of STI human resources, especially exchanges among young scientists and entrepreneurs, supporting efforts to help address the future demand for new skills, sharing best practices on enhancing skills training for innovation and entrepreneurship, including improving access to Science, Technology, Engineering and Mathematics (STEM) education, creating jobs through joint research and collaboration in innovation and entrepreneurship, and stressing the role of youth in innovation. Stressing the role of women in science, technology and innovation activities as one of the key priorities of the BRICS STI Agenda.

III. Implementation

The BRICS Science Technology Innovation and Entrepreneurship Partnership (STIEP) Working Group will be responsible for the development of mechanisms and opportunities to implement the Action Plan, which will in the first period focus on the following deliverables:

1. Creation of networks of science parks, technology business incubators and SMEs, where the innovation actually happens.
2. Creation of cross-cultural talent pools for converting ideas into solution in domains of ICT, materials, water, health, energy, natural disaster risk reduction and resilience, etc.

July 2017
Hangzhou, China

Contents

Introduction

Strengthen Innovation Cooperation to Shape the Future

Science, technology and innovation are crucial driving forces in the development of a country and a nation and of the entire human society at large. The competition in comprehensive national strength, in essence, is the competition in science, technology and innovation. In the backdrop of globalization, a country which has strong science, technology and innovation capabilities is more advantageously positioned in the division of labour in industries and better able to create new industries and can own more advanced intellectual properties needed to achieve further development. Science, technology and innovation hold the golden key to discovering new fountainheads of growth and unlocking dormant growth potential. Although the global economy remains sluggish overall, a new round of scientific, technological and industrial revolution is creating new historic opportunities as new concepts and new sectors such as "Internet+", 3D printing and smart manufacturing emerge and new technologies keep coming up, especially in artificial intelligence, information technology, life science and biotechnology, opening up unprecedented opportunities and development impetus, also with a massive potential of transforming traditional industries. In addition, science, technology and innovation play an irreplaceable basic role in the effort to respond to global challenges and can not only effectively promote the addressing of global challenges such as climate change, food shortage, resource depletion and poverty but also accelerate the achievement of the goals set forth in the 2030 Agenda for Sustainable Development for the benefit of the entire humankind.

The formation and development of the BRICS group of countries (Brazil, Russia, India, China and South Africa) have reflected the process from quantitative to qualitative change in the evolving of the international economic and political landscape, adapted to the trends of the times and advanced the establishment of a fairer and more rational international order. The BRICS cooperation mechanism has become a role model of cooperation between emerging economies and developing countries and will continue to generate benefits for the peoples of its member countries and make important contributions to the effort of promoting global economic growth, driving science, technology and innovation, and achieving sustainable development. The BRICS countries represent approximately 42% of the world population and occupy 30% of the earth's territory with a combined nominal GDP of approximately 23% of the world GDP and a combined trade volume of approximately 16% of the world trade. Over the past decade, the BRICS countries have contributed over half of the global economic growth. As the leading group of emerging market countries in the G20, the BRICS countries play an irreplaceable role in South–South cooperation in science and technology and South–North dialogue on innovation, serving as a principal group in leading science, technology and innovation in developing countries and an important force in global science, technology and innovation. BRICS countries invest heavily in research and development, with the annual R&D expenditures accounting for approximately 17% of the world's total, high-tech exports reaching nearly USD6 trillion or approximately 28% of the world's total, and publications of science papers totalling 590,000, approximately 27% of the world's total. As their contribution to global science, technology and innovation steadily increases, the international influence of the BRICS countries has been improving as well. The BRICS countries are pacesetters and leaders in their respective regions and lead regional countries in scientific, technological, economic and social development.

The world economy is experiencing a zigzag recovery from a deep decline and going through a crucial period of transition with traditional drivers being replaced by new ones. The new round of scientific and industrial revolution and industrial transformation is gathering momentum as the world enters a period of active and intensive innovation. The BRICS countries have their respective strengths in extensive areas including talent, science and technology, and resources and have a huge potential of achieving interconnected development. The close cooperation of the BRICS countries will increase the say of developing countries on international political, economic and science and technology affairs, promote timely sharing of their respective experience, accelerate their economic transformation, and provide new drivers to the global economic growth.

The year 2017 has commenced the second decade of the BRICS cooperation and will see the convening of the Fifth BRICS STI Ministerial Meeting in Hangzhou, China, in July, and the Ninth BRICS Summit in Xiamen, China, in September. Developing from "BRIC" to today's "BRICS", the BRICS mechanism of science and technology cooperation has become increasingly mature. As the influence of the BRICS continues to increase, the BRICS countries will have even better development prospects. The steadily growing science, technology and innovation force of the BRICS countries will also strengthen the BRICS cooperation mechanism and its influence. Let's act in unison to respond to the four expectations put forward by Chinese President Xi Jinping of the Ninth BRICS Summit: deepen pragmatic cooperation to achieve mutual benefit; strengthen global governance to address challenges together; carry out people-to-people exchanges to solidify popular support; and promote institutionalization to build partnership in wider areas.

Innovation drives development, and cooperation creates a bright future for all. The BRICS countries, who share the same destiny, are a community of shared interests and benefit from acting in concert. We believe that, through the concerted efforts of the BRICS countries, the BRICS science, technology and innovation cooperation will open up an even better future. The future is for us to create together!

People's Republic of China
July 2017

Dr. Huang Wei, Vice Minister
Ministry of Science and Technology

Part I
General Reports

Chapter 1
Forecast and Evaluation of Innovation Capabilities and Review of STI Cooperation of BRICS

Xinli Zhao, Dan Wang, Yi Xiao, Quanchao Dong, Hongwei Huo, Zongwen Ma and Bingqing Xin

Innovation is the primary driving force of human development and plays a critical role in promoting healthy social-economic development, accelerating institutional reform, upgrading both productivity and competitiveness, and addressing global challenges. It is the foundation that enables more benefits to human society as well as sustainable social development.

BRICS countries are major countries leading STI development among developing countries as well as a globally important STI force. BRICS countries account for 30% of global territory (5/6 latitude and 2/3 time zones covered by BRICS countries shown in Fig. 1), 42% world population, 23 and 16% global GDP and trade respectively. In terms of STI level in the world as a whole, BRICS countries shared 17% of global R&D input, 28% (totaling around 6 trillion dollars) of total export value of Hi-tech products, 27% (around 590,000 papers) of published S&T papers and journals in the world. These five countries are making increasing more contribution to the global economic growth and having more international influence in the world.

Recently, BRICS Countries have seen flourishing multilateral STI collaboration, and China also witnessed increasingly in-depth bilateral STI cooperation with other four countries. At a critical conjuncture of economic transformation for BRICS countries, it is particularly important to analyze and forecast the development of comprehensive innovation capabilities of BRICS countries.

With a view to better playing out the guiding and leading role of STI cooperation among BRICS countries for other developing countries and supporting 2017 BRICS Summit in Xiamen and BRICS STI Ministers' Meeting, this report, based on the evaluation and forecast of comprehensive STI capabilities and competi-

D. Wang
Northeast Agricultural University, Harbin, China

X. Zhao (✉) · Y. Xiao · Q. Dong · H. Huo · Z. Ma · B. Xin
China Science and Technology Exchange Center, Beijing, China
e-mail: Zhaoxl@cestc.org.cn

X. Zhao et al. (eds.), *BRICS Innovative Competitiveness Report 2017*,
Research Series on the Chinese Dream and China's Development Path,
https://doi.org/10.1007/978-981-10-8078-4_1

Fig. 1 Geographic location of BRICS Countries

tiveness of G20 countries previously done, studies comprehensive STI competitiveness of BRICS countries and forecasts their future STI development; moreover, the report studies the status-quo of China's STI collaboration with other four BRICS countries, based on which the report analyze the problems and come up some proposals.

1 Evaluation of BRICS Countries' Comprehensive Innovation Competitiveness

Supported by the Chinese Ministry of Science and Technology and the Chinese Academy of Social Sciences, Fujian Normal University, S&T Section of Chinese Permanent Mission to UN and China Science and Technology Exchange Center took the lead to evaluate innovation competitiveness since 2001 of G20 countries including BRICS countries and global 100 powers. As of now, acknowledged by the international community, the *Report on the Group of Twenty (G20) National Innovation Competitiveness Development (Yellow Book)* has been published consecutively 5 times and *the Report on World Innovation Competitiveness Development (Yellow Book)* 1 time. Additionally, other four categories of reports, including *Global Competitiveness Report* by the World Economic Forum, *Global Innovation Index* by INSEAD, *Bloomberg Innovation Index* and *Global Innovation Index* jointly published by WIPO and Cornell University, etc., also evaluate the innovation competitiveness of major economic powers in the world. These assessment, well acknowledged internationally as five major reports on STI capabilities, vary from each other in terms of different focuses, features, and thus with different conclusions being reached. In order to more soundly and inclusively reflect the results of different evaluation systems, during 2016 G20 session when China was then rotating presidency of G20, led by the China Science and Technology Exchange Center, the research team, based on the statics offered by abovementioned five categories of reports, adopted DJI like (Dow Jones Indexes) calculation to rank the comprehensive STI competitiveness of G20 members and forecast their future STI development through fitting and extrapolating the historic data. Such research methods have been extensively recognized by 2016 and 2017 T20 Summits as well as the first G20 STI Ministers Meeting in 2016. The corresponding papers developed have been included as important contents into the book *Science, Technology and Innovation for Development of G20*, which was recognized as one of top 50 *Best Policy Study/Reports Produced by a Think Tank* according to *2016 Global Go To Think Tank Index Report.*

This chapter evaluates the comprehensive innovation competitiveness of BRICS countries for 2001–2016 according to the model of national comprehensive innovation competitiveness indexes so to better inform 2017 BRICS Summit when China being as the rotating president of this years' summit.

1.1 Changes in the Rankings

For a comparison at a larger scale, we ranked BRICS countries by comparing them among G20 countries rather than just among five countries themselves.

The ranking of BRICS countries among G20 countries in 2016 in terms of comprehensive innovation competitiveness was China, Russia, South Africa, Brazil and India; while such ranking in 2001 was Russia, China, South Africa, Brazil and India.

Shown by Fig. 2, from 2001 to 2016, Brazil, India and China have risen in their ranking in terms of national comprehensive innovation competitiveness; the ranking of Russia and South Africa remained same.

The report also referred to *the Global Competitive Report*, *Global Innovation Index (GII)* and the ranking of BRICS countries among G20 members in terms of their national innovation competitiveness, shown in Table 1.

The Global Competitive Report shows in 2016 among BRICS countries China stood 28th place, India 39th, Russia 43rd, South Africa 47th and Brazil 81st[1]; while the ranking for 2001 was South Africa 34th place, China 39th, Brazil 44th, India 57th, and Russia 63rd.[2]

[1]World Economic Forum (2017).

[2]World Economic Forum (2002).

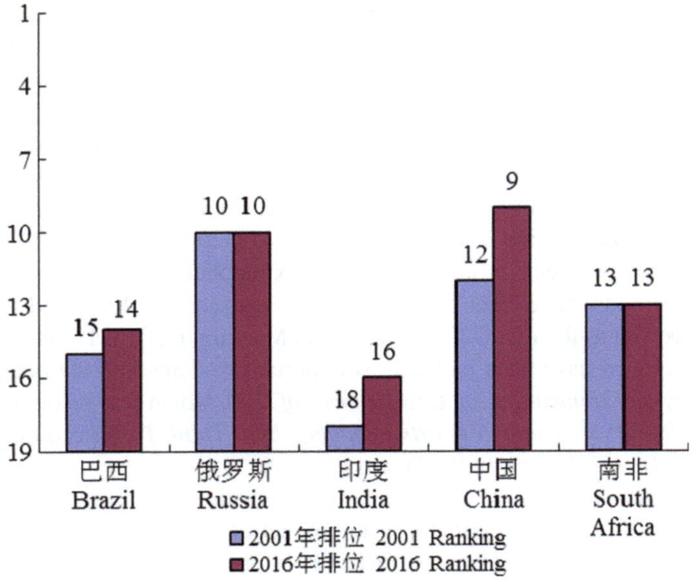

Fig. 2 Changes in the ranking of BRICS countries in terms of their national comprehensive innovation competitiveness (2001–2016)

According to 2016 *GII*, China ranked 22nd place, Russia 4th, South Africa 57th, India 60th, and Brazil 69th[3]; the ranking in 2006 was as follows: India ranked 23rd place, China 29th, South Africa 38th, Brazil 40th and Russia 54th.[4]

In line with the ranking developed by the *Report on the Group of Twenty (G20) National Innovation Competitiveness Development (Yellow Book)* among BRICS countries, China in 2015 ranked 8th place, Russia 11th, Brazil 15th, South Africa 17th and India 18th[5]; The ranking in 2001 was Russia 10th, China 12th, South Africa 13th, Brazil 15th and India 18th.[6] Additionally, *Bloomberg Innovation Index* ranked BRICKS countries in 2017 as follows: China ranked 21st place, Russia 26th, Brazil 46th and with South Africa and India not being included in the list.

1.2 Changes in the Scores

From 2001 to 2016, scores of BRICS countries in terms of their comprehensive innovation competitiveness evaluation indexes, though fluctuated in some year, averagely was on the rise, increasing by 17.86 points (Fig. 3).

[3]Dutta et al. (2017).

[4]Dutta and Calkin (2007).

[5]Li et al. (2017).

[6]Li et al. (2011).

Table 1 The rankings of BRICS countries listed by major global innovation evaluation reports

Items		Countries				
		Brazil	Russia	India	China	South Africa
Global Competitiveness Report	Rankings in 2001	44	63	57	39	34
	Rankings in 2016	81	43	39	28	47
	G20 Rankings in 2001	13	17	16	11	10
	G20 Rankings in 2016	18	13	11	9	15
Global Innovation Index	Rankings in 2006	40	54	23	29	38
	Rankings in 2016	69	45	60	22	57
	Rankings in 2006	14	17	9	11	13
	G20 Rankings in 2016	16	12	15	8	13
Report on the Group of Twenty (G20) Nationals Innovation Competitiveness Development (Yellow Book)	Rankings in 2001	15	10	18	12	13
	Rankings in 2015	15	11	18	8	17
Bloomberg Innovation Index	Rankings in 2017	46	26	–	21	–
	G20 Rankings in 2016	13	11	–	9	–

Table 2 testifies that the overall level of BRICS countries' comprehensive innovation competitiveness has been rising over past sixteen years. Firstly, China's ranking has risen quickly and constantly topped other BRICS countries. Among G20 members, China has risen from 12th place in 2001 to 9th place in 2016, the only one developing country that has risen to be among top 10 in the world. Secondly, the comprehensive innovation competitiveness of BRICS countries has seen rapid increase. The past sixteen years have seen over 10% increase in the comprehensive innovation competitiveness index of BRICS countries. Thirdly, the growth rate of comprehensive innovation competitiveness of each BRICS varies. China and Russia led the increase in growth rate while India saw a medium growth rate; the growth rate of Brazil and South Africa has been relatively slow.

Figures 4 and 5 respectively show the change in scores ranked by GII and G20 Innovation Evaluation Index.

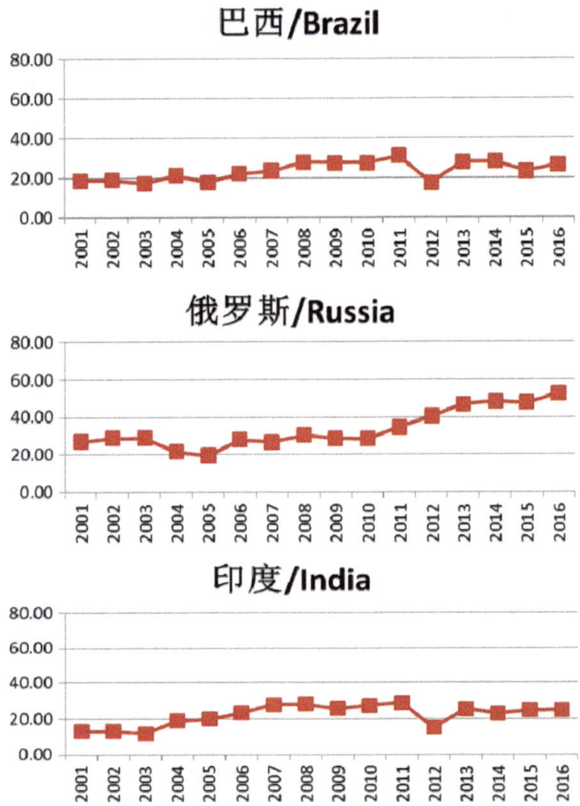

Fig. 3 Changes in the score of BRICS countries in terms of comprehensive innovation competitiveness evaluation indexes (2001–2016)

1.3 Analysis of the Changes in the Score

The Report on the Group of Twenty (G20) Nationals Innovation Competitiveness Development (Yellow Book) evaluated national innovation competitiveness by five secondary indicators, including the competeveness of the innovation foundation, innovation environment, innovation output, and sustained innovation.

Table 3 shows the scores and rankings according to the secondary indicators of national innovation competitiveness of BRICKS respectively in 2001 and 2015. Figures 6, 7, 8, 9 and 10 present radar charts of the changes in rankings of BRICS countries in terms of secondary indicators for national innovation competitiveness. Over past 15 years, Brazil's rankings slightly rose in terms of innovation environment and sustained innovation, but the ranking in innovation output has dropped significantly. Thus Brazil saw slow growth in comprehensive innovation capability. In case of Russia, it has seen remarkable achievements in innovation environment and innovation output with the ranking for innovation environment jumping to the 6th place and the rapidly strengthened comprehensive innovation capability. India maintained a medium growth rate in its comprehensive innovation capability with

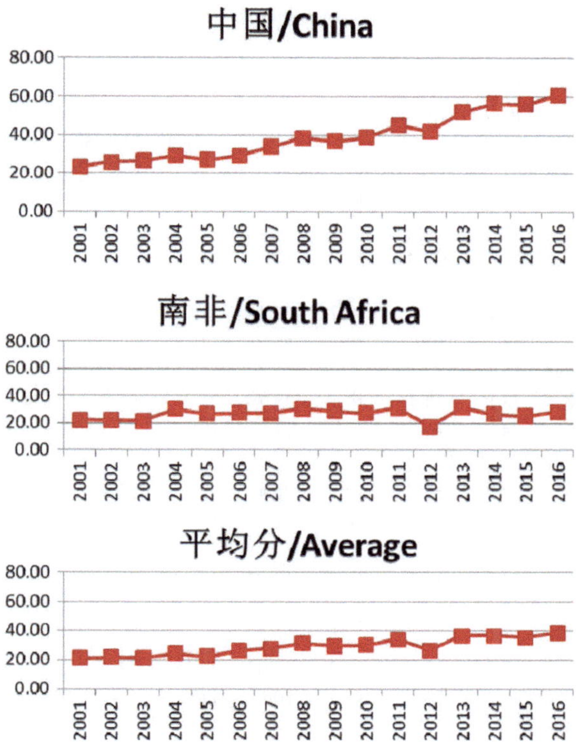

Fig. 3 (continued)

significant rise in the rankings for innovation output and input, the rankings for other indicators actually remaining stable. China have seen great rise in all five secondary indicators with the ranking for innovation environment remaining 13th place unchanged, thus the comprehensive innovation capability also saw rapid development. South Africa saw relatively slow growth of innovation capability with rankings of five secondary indicators all slightly dropping.

In line with the national innovation competiveness evaluation indexes of BRICS countries (2014–2015) by the *Report* shown in Table 4, China and Russia are with evenly high and strong scores for all these five secondary indicators, thus both saw rapid growth in comprehensive innovation capabilities. India though with one strong and one medium weighted indicator, the shortfalls in innovation environment, sustained innovation and innovation foundation undermined the growth rate of its comprehensive innovation capabilities. Brazil and South Africa have performed weakly in all five indicators and some indicators even turned out negative, resulting in slow growth rate of national innovation competitiveness of these two countries.

Table 2 The comparison of comprehensive innovation competitiveness evaluation index of BRICS countries (2001–2016)

Items		Country							
		Brazil	Russia	India	China	South Africa	G20 highest score	G20 lowest score	G20 average score
Overall changes	Scores	7.99	25.66	11.77	37.56	6.32	−7.86	14.62	14.18
	G20 Rankings	+1	0	+2	+3	0	–	–	–
2001	Scores	18.6	27	13.2	23.3	21.9	88.8	7.7	34.97
	G20 Rankings	15	10	18	12	13	–	–	–
2002	Scores	18.8	28.9	13.1	25.7	21.8	89.6	8	35.79
	G20 Rankings	15	10	18	11	13	–	–	–
2003	Scores	17.3	29	12	26.4	21.2	88.8	8.3	35.33
	G20 Rankings	15	10	18	11	13	–	–	–
2004	Scores	21.23	21.46	18.88	29.03	29.98	78.31	12.06	36.19
	G20 Rankings	14	13	16	11	10	–	–	–
2005	Scores	17.78	19.68	19.96	27.03	26.58	77.28	11.43	34.82
	G20 Rankings	17	14	13	10	11	–	–	–
2006	Scores	22.08	28.01	23.36	29.21	27.46	75.38	20.66	38.9
	G20 Rankings	17	11	15	10	12	–	–	–
2007	Scores	23.5	26.68	27.83	33.68	27.14	74.61	19.54	38.43
	G20 Rankings	16	13	11	10	12	–	–	–
2008	Scores	27.99	30.41	27.97	38.52	30.51	72.08	22.36	41.01
	G20 Rankings	14	12	15	9	11	–	–	–
2009	Scores	27.46	28.65	25.57	36.85	28.62	67.17	23.17	38.82
	G20 Rankings	14	11	17	9	12	–	–	–
2010	Scores	27.48	28.43	27.22	38.71	27.32	60.74	24.55	37.51
	G20 Rankings	12	11	14	9	13	–	–	–
2011	Scores	30.99	34.59	28.77	44.97	31.05	65.87	24.15	41.95
	G20 Rankings	13	11	16	9	12	–	–	–
2012	Scores	17.17	40.29	15.01	41.9	17.08	83.91	11.37	40.24
	G20 Rankings	13	10	18	9	14	–	–	–

(continued)

Table 2 (continued)

Items		Country							
		Brazil	Russia	India	China	South Africa	G20 highest score	G20 lowest score	G20 average score
2013	Scores	28.04	46.91	25.43	51.85	31.37	85.6	23.88	49.94
	G20 Rankings	14	11	17	9	13	–	–	–
2014	Scores	28.19	48.29	23.03	56.43	26.54	84.39	21.98	49.78
	G20 Rankings	13	11	17	9	14	–	–	–
2015	Scores	23.43	47.57	24.71	55.99	25.38	81.85	20.83	47.59
	G20 Rankings	17	10	14	9	13	–	–	–
2016	Scores	26.59	52.66	24.97	60.86	28.22	80.94	22.32	49.15
	G20 Rankings	14	10	16	9	13	–	–	–

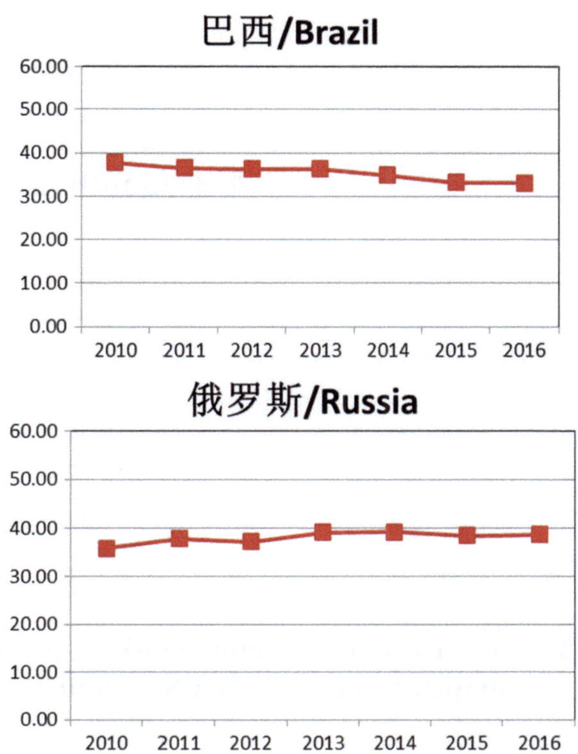

Fig. 4 Changes in the scores of BRICS according to GII (2010–2016)

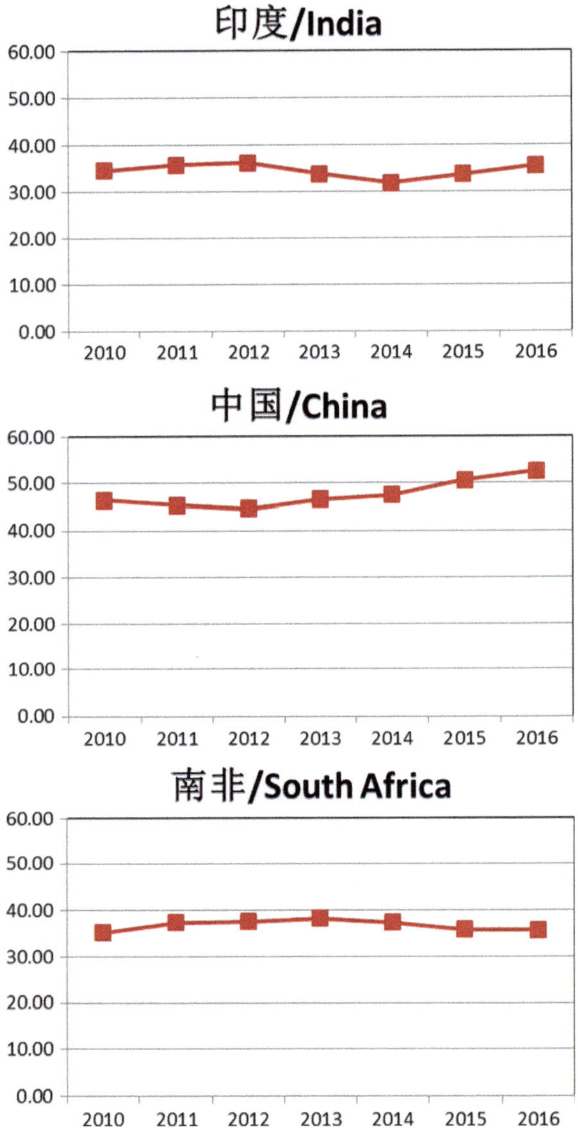

Fig. 4 (continued)

2 Forecast on the Comprehensive Innovation Competitiveness of BRICS Countries

This report calculates innovation competitiveness index of BRICS countries for future 5 years and analyzes their rankings among G20 members based on the score of comprehensive innovation competitiveness indexes from 2001 to 2016.

2.1 Overall Forecast

Statistics shows the changes in comprehensive innovation indexes of BRICS countries vary from each other.

Compared with the average score of BRICS' innovation indexes for 2001–2016, the average score of innovation index of each BRICS country would be on the rise for coming 5 years, reflecting the national comprehensive innovation competitiveness of BRICS countries constantly being strengthened over times. China and Russia would see fast growth. The innovation indexes of other BRICS countries and major developing economies would see relatively slight increase as their innovation competitiveness has been growing comparatively slowly. India maintains a medium growth rate while Brazil and South Africa grows at a slow speed in the hope of rising from the trough soon. The compared scores of comprehensive innovation competitiveness of BRICS countries were shown in Table 5. In future 5 years, comprehensive innovation indexes of BRICS countries could expect an annually average increase around 1.5% on the premises of no significant financial fluctuation, domestic turmoil and big natural disasters happening in these countries, even though this figure is still lower than the annual average growth rate of 2.8% registered by developed countries.

The forecast shows in terms of innovation competitiveness, the USA would continue to be No. 1 and Japan would see its ranking rise over time. China, after the USA, Japan, Germany, would rank among top 5 from the 9th place in 2016. Among BRICS countries, the rankings of Brazil and South Africa would remain comparatively stable. Table 3 shows the changes of national innovation competitiveness.

It is expected by 2030 comprehensive STI competiveness of BRICS countries would be continuously strengthened. First, China would see steady growth in national comprehensive STI competitiveness, leading other BRICS countries and becoming top 3 among G20 members around 2030. Secondly, the overall innovation capabilities of BRICS countries would be upgraded. With the prerequisite of no big economic and financial fluctuation, domestic turmoil and serious natural disasters in BRICS countries, the comprehensive innovation competitiveness indexes would see an average of more than 1.5% growth annually in each BRICS country though this figure is still lower than annual growth rate of G20 members. Thirdly, the growth rate of national innovation competiveness would continue vary from each other. It is predicted that the comprehensive innovation competitiveness of India would see a significant rise with its growth rate probably surpassing China between 2025 and 2030; the growth rate of Russia would fall and India would probably take over Russia in terms of its comprehensive STI competitiveness by 2030. Other BRICS countries would see steady increase in their STI competitiveness with the growth rate of Brazil and South Africa lower than the average figure of BRICS countries even deficit growth rate in South Africa in some year.

Fig. 5 Changes in the scores of BRICS by G2O innovation evaluation index (2001–2015)

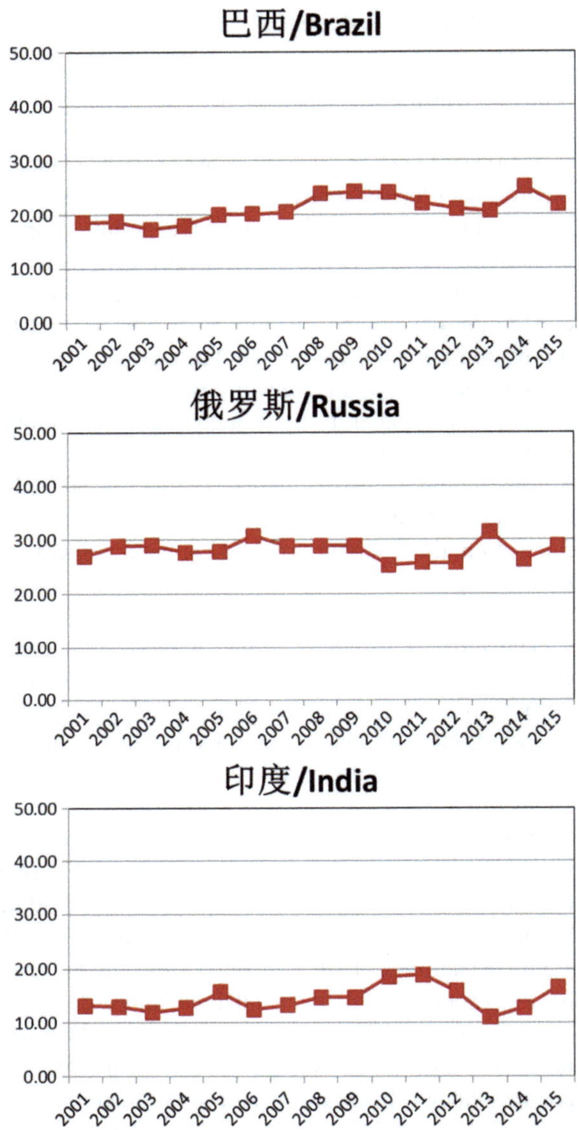

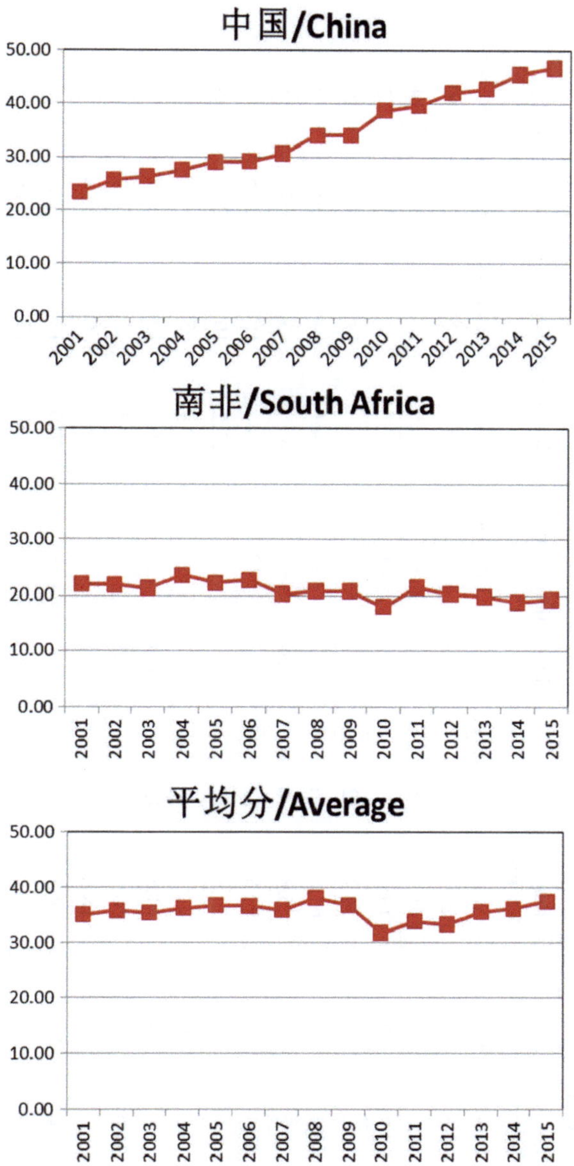

Fig. 5 (continued)

Table 3 Scores and rankings according to the secondary indictors for national innovation competiveness of BRICS countries

Countries		Indicators									
		Innovation foundation		Innovation environment		Innovation input		Innovation output		Sustained innovation	
		Scores	Rankings	Scores	Rankings	Scores	Rankings	Scores	Rankings	Scores	Rankings
Brazil	2001	8.3	14	23	16	22.2	12	13	12	26.5	16
	2015	11.6	16	41.1	15	14.1	13	10.5	16	32.4	14
	Changes	3.3	−2	18.1	1	−8.1	−1	−2.5	−4	5.9	2
Russia	2001	13.2	12	36.2	12	35.7	9	8	17	41.9	10
	2015	16.4	14	61.5	6	21.2	11	14.1	14	30.8	15
	Changes	3.2	−2	25.3	6	−14.5	−2	6.1	3	−11.1	−5
India	2001	1.4	18	22.6	17	7.7	18	9.8	15	24.5	17
	2015	5.4	18	24.8	19	7.6	15	22.2	10	23.3	18
	Changes	4	0	2.2	−2	−0.1	3	12.4	5	−1.2	−1
China	2001	6.9	15	31.5	13	30.4	10	16.1	10	31.7	14
	2015	31.9	9	50.1	13	48.5	7	53.3	2	50.1	5
	Changes	25	6	18.6	0	18.1	3	37.2	8	18.4	9
South Africa	2001	4.6	17	44.5	10	19	13	10.9	13	30.3	15
	2015	4.9	19	51.5	12	8.8	14	6.8	17	24.6	17
	Changes	0.3	−2	7	−2	−10.2	−1	−4.1	−4	−5.7	−2

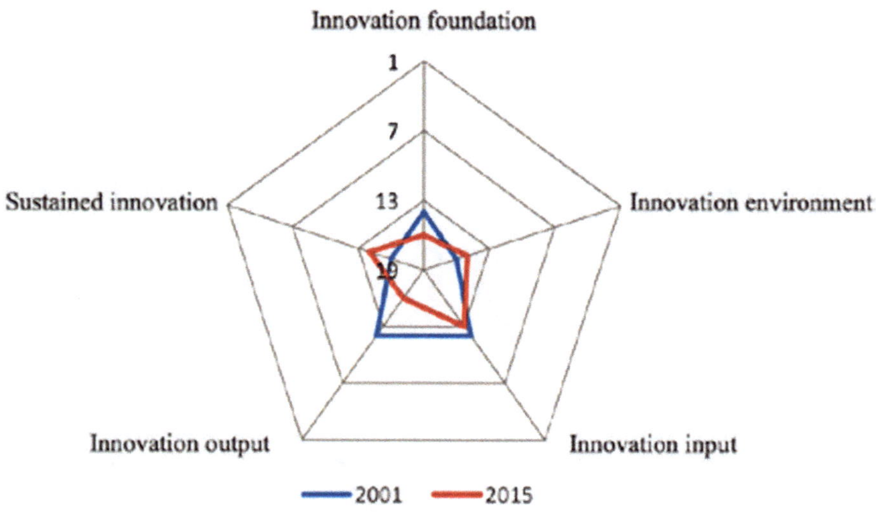

Fig. 6 Changes in secondary indicators for Brazil's national innovation competitiveness

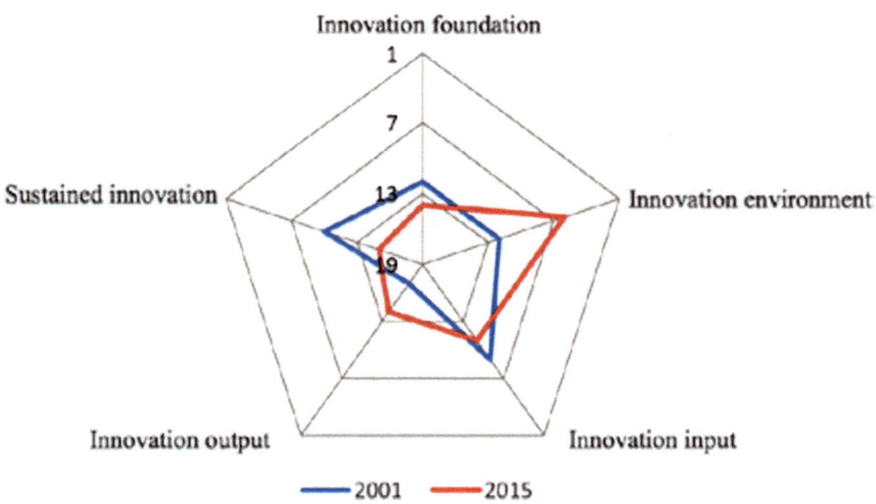

Fig. 7 Changes in secondary indicators for Russia's national innovation competitiveness

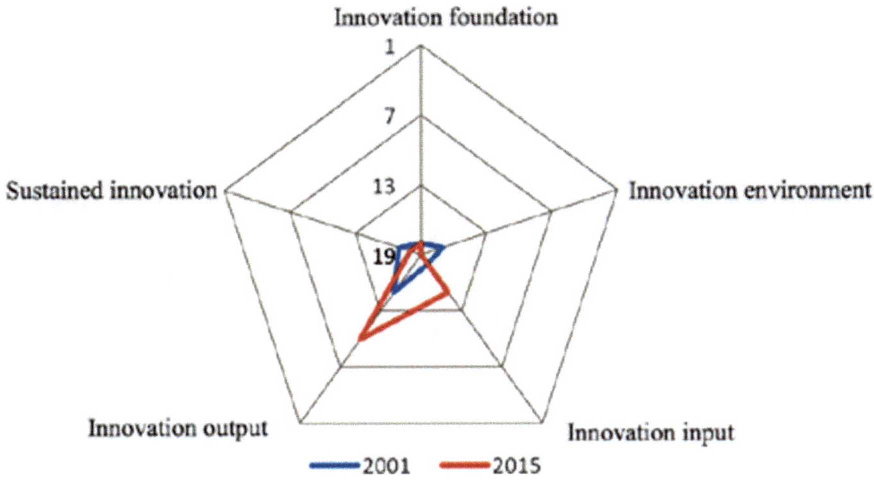

Fig. 8 Changes in secondary indicators for India's national innovation competitiveness

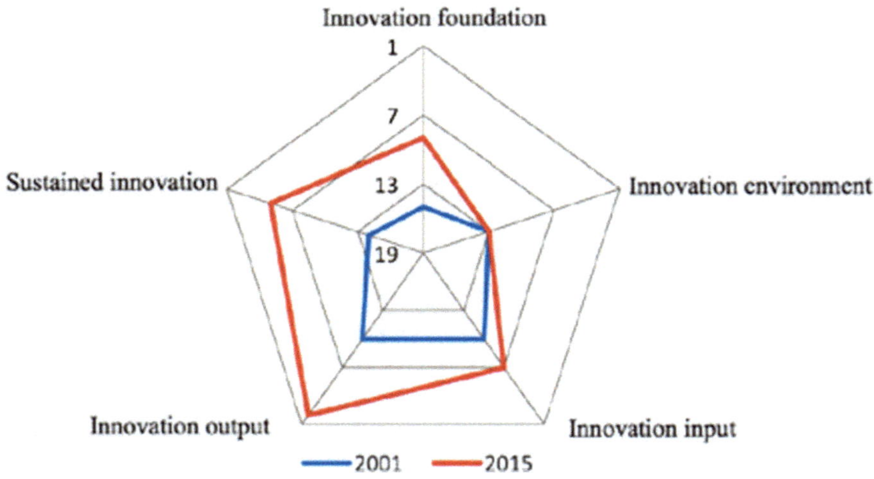

Fig. 9 Changes in secondary indicators for China's national innovation competitiveness

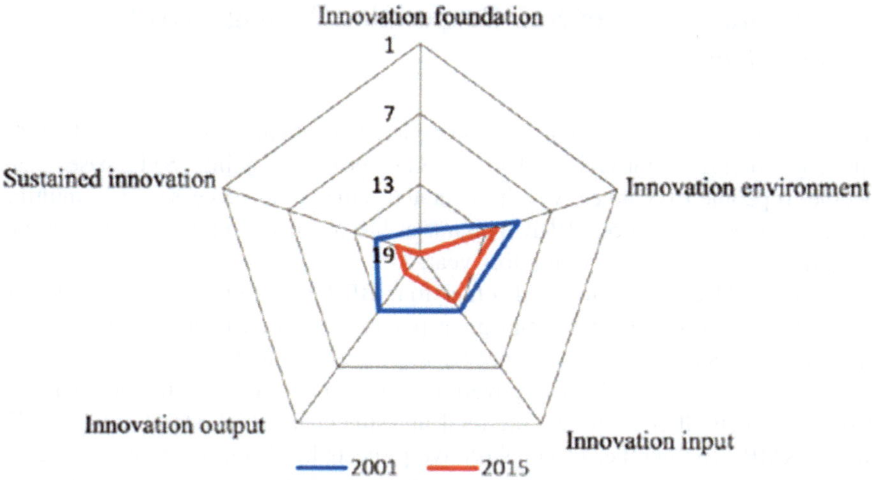

Fig. 10 Changes in secondary indicators for South Africa's national innovation competitiveness

Table 4 Comparison of indicators for national innovation competitiveness of BRICS countries

Country	Strong indicators	Positive indicators	Medium weighted indicators	Negative indicators
Brazil	–	–	Innovation environment, innovation input, sustained innovation	Innovation foundation, innovation output
Russia	–	Innovation environment	Innovation output, sustained innovation, innovation foundation, innovation input	–
India		Innovation output	Innovation input	Innovation environment, sustained innovation, innovation foundation
China	Innovation output, sustained innovation	Innovation foundation, innovation input	Innovation environment	–
South Africa	–	–	Innovation environment, innovation input	Sustained innovation, innovation foundation, innovation output

3 The Status-Quo of STI Cooperation Among BRICS Countries

Recent years China has extensively conducted practical cooperation with other BRICS countries through bilateral STI projects, joint STI bases and people-to-people exchanges, which generated remarkable successes. The multilateral STI cooperation among BRICS countries has just been in place and is expected to start substantive collaboration this year.

China's STI collaboration with other four BRICS countries features rich contents, significant achievements and great potential through international STI joint projects, joint STI bases and STI people to people exchanges.

For 2007–2015, the Chinese government has supported 665 international STI projects with BRICS countries, the total investment in which stood around 2.729 billion RMB. The number of collaborative projects has been increasing as well as the rising investment from the Chinese government. The collaboration mainly occur in the areas of materials, engineering, information and life sciences. China has jointly established 190 nation-level international STI collaboration bases related to BRICS countries in 7 areas of life sciences, advanced manufacturing, information technology, material sciences, earth sciences, energy and environment, agriculture. The number of collaboration bases has evenly been dispensed in each area.

STI people to people exchanges mainly include the technology training offered as an aid by the Chinese government to developing countries and the Program of attracting Young STI talents to work in China. Since 2006, the Ministry of Science and Technology itself has already organized 411 technology training sessions for 7885 participants from developing countries, among which 566 participants from BRICS countries attended 221 sessions, accounting 7.18% of total number of participants. The average number of participants from BRICS country is twice that of other developing countries. The technology training covers areas of agriculture, information, manufacturing, solutions to climate changes, resources and environment, new energy, healthcare, medicine and STI policy and management.

(1) **China-Russia STI Collaboration**

China-Russia STI collaboration is a typical example of collaborative successes compared with China's STI cooperation with other BRICS countries. Through joint STI projects and bases, China has together worked with Russia in more than 600 projects. Both in terms of the number of joint projects and investment, China-Russia collaboration account 90% of China's joint STI efforts with BRICS countries. Major collaboration is in the areas of materials, engineering and technology, information, etc. 157 international bases in China engage the collaboration with Russia, leading China's cooperation with other BRICS countries.

(2) **China-India STI Collaboration**

People to people exchange is the mainstream of China-India STI cooperation. The number of participants from India who attended the technology training sessions

Table 5 Forecast on the ranking of STI competitiveness of BRICS countries (2016–2022)

Year/Countries		China	Russia	South Africa	India	Brazil
2016	Innovation indicators	60.86	52.66	28.22	24.97	26.59
	Rankings	9	10	13	16	14
	Changes compared with the rankings in 2016	0	0	0	0	0
2017	Innovation indicators	64.5	56.06	27.23	24.9	27.01
	Rankings	9	10	13	15	14
	Changes compared with the rankings in 2016	0	0	0	−1	0
2018	Innovation indicators	68.38	59.66	27.29	25.4	27.4
	Rankings	8	10	14	15	13
	Changes compared with the rankings in 2016	1	0	−1	−1	1
2019	Innovation indicators	72.41	63.42	27.36	26.02	27.8
	Rankings	5	10	14	15	13
	Changes compared with the rankings in 2016	4	0	−1	1	1
2020	Innovation indicators	76.59	67.37	27.42	26.79	28.22
	Rankings	4	10	14	15	13
	Changes compared with the rankings in 2016	4	0	−1	1	1
2021	Innovation indicators	76.99	68.22	25.38	27.71	28.95
	Rankings	4	10	15	14	13
	Changes compared with the rankings in 2016	5	0	−2	2	1
2022	Innovation indicators	77.23	68.34	25.95	28.74	29.05
	Rankings	4	10	15	14	13
	Changes compared with the rankings in 2016	5	0	−2	2	1

China organized for the developing countries is the largest among BRICS countries. There has been totally 256 participants since 2006 accounting 45% of total participants attending the training sessions. The number of China-India project and investment are both relatively small. From 2007 to 2015, there has been around 9 China-India projects with total funding around 22.67 million RMB, among which 5 were China-Pakistan projects with the funding around 15.01 million RMB. There are 18 international bases that engage collaboration with India, the number of which is still relative low.

(3) **China-South Africa STI Collaboration**

China joined hands with South Africa mainly through STI projects. There has been totally 89 China-South Africa projects of which 30 are joint research project with the total funding around 63.03 million RMB. Except Russia, China and South Africa maintained a good momentum of STI collaboration both in terms of the number of projects and the scale of project investment.

In the meantime, joint projects between China and other African countries totaled around 27 with the funding around 1070 million RMB, among which 3 projects were with Egypt with the funding of 15.7 million RMB, 2 more respectively with Kenya and Algeria with the funding around 11.53 million RMB and 9.94 million RMB. There has been 14 international STI bases that engage the collaboration with South Africa, the number of which is the least among BRICS countries.

(4) **China-Brazil STI Collaboration**

The STI collaboration is relatively weak with the people-to-people exchange at an initial stage. From 2017 to 2015, China has conducted 17 joint projects with Brazil with the total project funding around 31.21 million RMB. At the same period, China has carried out 9 projects with other South American countries with the total funding around 91.16 million RMB, among which 8 projects were with Argentina with the total funding around 89.16 million RMB. There has been 15 international bases that engage the collaboration with Brazil.

3.1 Intergovernmental Collaborative Network Between China and BRICS Countries

With a better understanding of current status of STI collaboration between China and BRICS countries, the report studies the international STI cooperation network among BRICS countries by analyzing S&T reports submitted by Chinese project carriers via *national science and technology report system*[7] whose projects involve STI collaboration between China and BRICS countries. The quantitative analysis was done from the perspective of social network[8] in consideration of the nature of network structure on the collaboration realities and features of intergovernmental projects between China and other BRICS countries.

Figure 11 visualizes the structure of the project collaboration network between China and BRICS countries. It is clear that collaborative scale in terms of joint projects between China and BRICS countries is still small and the collaborative network has not yet been well linked. The China- BRICS collaborative network is

[7]National Science and Technology Report System (website: http://www.nstrs.cn/).

[8]Social network refers to a relatively stable relation network developed through interaction among individuals of the society. It focuses on the interaction and connection among social members. It is a social structure composed of many knots which connect an individual or an organization. The social network represents various social relations.

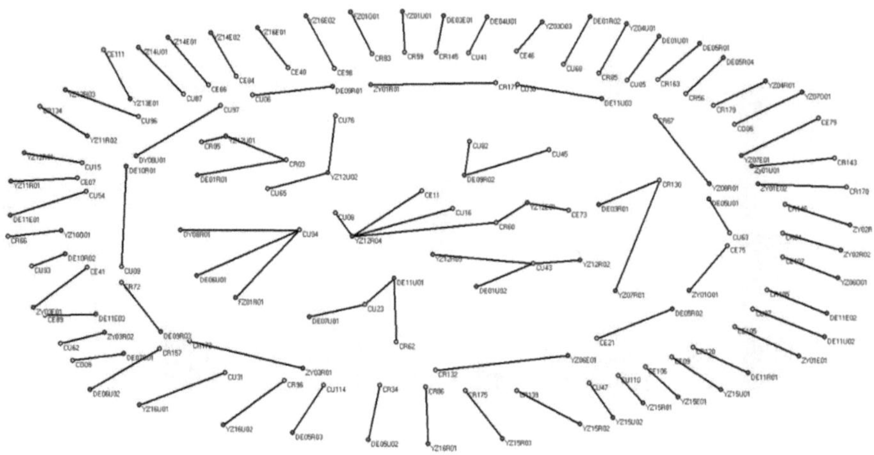

Fig. 11 The network of intergovernmental projects between China and BRICS countries

composed of some 30 poorly linked small components, calling for more collaboration into future.

Notes:

1. C + one letter + two digits, for this set of signifiers, C stands for a Chinese organization, R, U, E, O for the one letter respectively stands for a research institute, a university, an enterprises and other organizations, and two digits specify different entities falling into different categories of organizations.
2. Two letters + two digits + one letter + two digits, for this set of signifiers, first two letters means names of geological regions, such as YZ for Asia, FZ for Africa, SM for South Africa; first two digits specify different countries in respective regions; for the one letter, R, U, E, O still respectively stand for research institutes, universities, enterprises, and other organization; two digits means different entities falling into different categories of organizations (Fig. 12).

The report adopts the same approach to analyze the data for the features of international STI collaboration between China and key regional countries. People-to-people exchanges and joint research still remain as the major collaborative modes of STI cooperation between China and USA, China and Europe, China and BRICS countries, and China and "B&R" countries. 80% collaborative projects involve people-to-people exchanges and 60% collaboration are joint research projects. Additionally, the establishment of joint platforms is another major topic for collaboration between China and Europe, China and "B&R" countries, which the STI collaboration between China and BRICS countries falls short of. More collaboration on technology standards, accounting for 40% of collaboration modes, occur between China and BRICS countries as well as China and "B&R" countries. Such modes of collaboration happen less between China-USA and China-EU, which accounts for only 20% of total collaboration modes

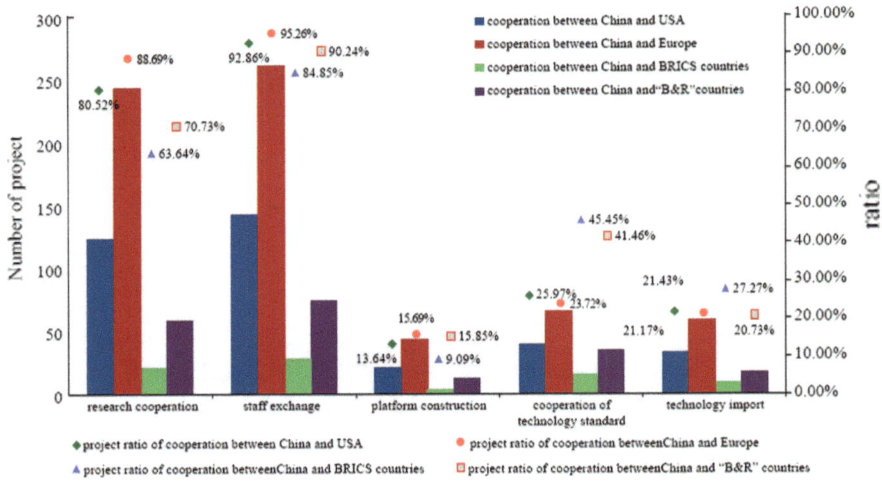

Fig. 12 Features of the international collaboration between China and key regional countries

respectively with them. In terms of the number of technology transfer cases, China transferred more technologies from USA and Europe, and particularly more from European countries. Statistics on technology transfer and collaboration of technology standards shows China has strong complementarities and great potential for win-win results with relevant STI partners in technology collaboration.

3.2 *The Output of STI Collaboration Between China and BRICS Countries*

As we will analyze the output of STI collaboration between China and BRICS countries in terms of jointly authored papers in the second Chapter, in this part we will mainly discuss the joint technology output out of collaboration based on the number of joint patents. Table 6 shows the number of patents jointly applied by China and BRICS countries.

The overall number of patents jointly applied by BRICS countries is still small. According to PCT[9] statistics, as of the end of 2014, the total PCT number jointly applied by BRICS countries was only around 56. China has the largest number of 46 PCT application with other BRICS countries. In the meantime, the number of China-Japan PCT application, China-EU and China USA PCT application

[9]PCT (Patent Cooperation Treaty, PCT) was an agreement on international cooperation of patents which entered into effect in 1978. The treaty stipulates a unified procedure for its member countries to follow when they file a patent application. The patents applied in line with PCT are regarded as international patent application, abbreviated as PCT application.

Table 6 The number of patents applied jointly by China and BRICS countries[a]

Dataset: International co-operation in patents										
Patent office	Patent applications filed under the PCT									
Partner country	Total patents	Total co-operation with abroad	Japan	United States	European Union (28 countries)	Brazil	China	India	Russian Federation	South Africa
Priority date	2014									
Country										
Japan	43,064.0	869.0	–	354.0	241.0	2.0	143.0	33.0	5.0	0.0
United States	56,702.0	7259.0	354.0	–	3333.0	85.0	1140.0	528.0	106.0	17.0
European Union (28 countries)	54,454.0	5961.0	241.0	3333.0	3108.0	61.0	477.0	252.0	60.0	32.0
World	206,187.0	13,347.0	869.0	7259.0	8485.0	153.0	1911.0	858.0	211.0	61.0
Brazil	746.0	153.0	2.0	85.0	61.0	–	5.0	3.0	2.0	1.0
China (People's Republic of)	26,778.0	1911.0	143.0	1140.0	477.0	5.0	–	31.0	9.0	1.0
India	2480.0	858.0	33.0	528.0	252.0	3.0	31.0	–	2.0	2.0
Russia	1148.0	211.0	5.0	106.0	60.0	2.0	9.0	2.0	–	0.0
South Africa	315.0	61.0	0.0	17.0	32.0	1.0	1.0	2.0	0.0	–

[a]OCED database

Data extracted on Jul 2017 05:36 UTC (GMT) from OECD.Stat

respectively stood around 143,477 and 1140. The number of PCT application between India and other BRICS Countries stood 38, ranking the 2nd place in terms of PCT number among BRICS countries. This figure for Russia, Brazil and South Africa respectively is 13, 11 and 4. Moreover, there was no PCT jointly applied by South Africa and Russia.

3.3 Achievements of STI Cooperation Among China and BRICS Countries

(1) **Benefiting All Participating Parties Through Advanced Applicable Technologies Cooperation**

STI cooperation of BRICS countries constitutes an important part of bilateral and multilateral cooperation within BRICS mechanism. China presents a good international image as a responsible countries in the world through the collaboration with BRICS countries on advanced applicable technologies, which benefit both the people and all participating nations.

China University of Geosciences (Wuhan) carried out the project, namely, "*The utilization of underground water resources contaminated by arsenic in India and arsenic removal research*" for the arsenic contaminated underground water in West Bengal of India. The project utilized natural geological material to purify the water with high arsenic, effectively solving the local problem for drinkable water by reducing volume of the arsenic in the water from 800 to 5 μg/L.

(2) **Deepening Partnership with BRICS Countries Through People-to-People Exchanges**

Under the framework of training and cultivating leading talents from developing countries, China helps developing countries, including BRICS countries cultivate leading STI talents and build up long-term cooperation among research entities, universities and enterprises, consolidating foundation for further STI cooperation with strengthened STI people-to-people exchanges, understandings and improved capabilities of participating entities.

In recent years, Institute of Geographic Sciences and Natural Resources, CAS organized the training sessions on the technologies that enable the sharing of environmental and resources data of silk road economic belt. 9 Russian students have been enrolled to the training sessions and the STI cooperation agreement has been signed with the Institute of Natural Resources, Ecology and Cryology, SB, RAS, which makes possible the international scientific investigation along China-Mongolia-Russia economic Corridor.

(3) **Sharing S&T Resources for Common Development**

Sharing S&T resources constitutes an important content of S&T collaboration among BRICS. BRICS countries would benefit from S&T progress achieved through strengthening the sharing of S&T resources particularly in areas of basic and frontier research and for people's livelihood, open access to important S&T infrastructures and sharing best STI experiences and practice. With the permission of the governments of Brazil, China and South Africa, China Center for Resources Satellite Data and Application, by carrying out the project named *the construction of ground system in South Africa for CBERS-02B*, enables 13 countries in Southern Africa to access data provided by CBERS-02B, benefiting local social and economic development with the application of data in agriculture, environmental protection and disaster prevention and reduction.

(4) **Jointly Participating in International Mega Scientific and Engineering Projects for Bigger Say in International S&T Arena**

BRICS should actively participate in international mega scientific programs and engineering projects taking account of the strategic development needs, domestic strengths and development realities of BRICS countries so to increase their say on multilateral S&T issues and help address important global challenges as well as problems in basic research.
India, China and South Africa are major member countries of SKA. Within SKA, three countries have already conducted a series of collaboration, including joint research, personnel exchanges, joint astronomic observation and data analysis. Such cooperation helps improve capabilities of BRICS countries in basic research and plays an important role for BRICS countries to have a better say on international issues.

4 Challenges and Suggestions

Presently, the international community lacks an in-depth strategic study on international collaboration among BRICS countries; moreover, due to historical bias, different domestic/political system and STI policies, uncertainties and serious challenges persist in jointly promoting STI collaboration among BRICS countries. Additionally, with the increasingly expanded STI collaborative demands among BRICS countries, it is urgent to set up a regular mechanism for technology collaboration and transfer as both the governments and a regularly organized conference on technology transfer could no longer satisfy the needs in this regard.

Therefore, the suggestions are proposed as follows to address challenges and materialize social-economic development through S&T progress achieved by integrating complementary strengths of BRICS countries.

(1) **Strengthening Coordination and Overall Planning**

It is suggested BRICS countries strengthen strategic research, top design and country- specific policy studies on and for STI collaboration, and formulate joint national plan for STI collaboration under BRICS mechanism. It is encouraged to fully play out the comparative advantages of BRICS countries, set up an interactive cooperation mechanism across regions, and build up shared STI community of BRICS countries through coordinated and scientific planning so to have the peoples of BRICS countries share the collaborative STI achievements

(2) **Intensifying Joint STI Efforts**

It is suggested BRICS countries expand and deepen intergovernmental STI dialogues and carry out multilateral cooperation in the areas jointly identified on STI Ministers Meeting of BRICS countries. The number of international STI joint projects among BRICS countries and the related project funding should be increased under BRICS mechanism, and more international bases with BRICS countries are expected strongly. The efforts of jointly building STI parks should been encouraged.

(3) **Expanding People-to-People Exchanges**

It is suggested BRICS countries enhance policy communication and coordination, innovate the STI people-to-people exchanges mechanism and set up information network so to actively create a more efficient exchange platform. STI people-to-people exchanges, the foundation and ushers of substantive collaboration, would offer policy and facilities support in materializing practical STI joint work. It is suggested to set up people-to-people exchange funding of BRICS countries so to encourage more exchanges and visits of young STI talents from BRICS countries.

(4) **Establishing and Sharing Jointly a Platform for Technology Transfer**

It is suggested BRICS countries jointly build up a platform to share STI resources through which we could efficiently integrate STI resources and intensify the sharing of infrastructure, information, S&T results and talents among BRICS countries. S&T authorities of BRICS countries should offer policy incentives to collaborative parties for them to actively engage in building a technology transfer platform; moreover, giving a full play of the market and STI intermediaries, STI authorities could provide systematic and professional information and technology services to enterprises and research entities as well as all other participating entities so to break the bottleneck of free information flow.

(5) **Exercising Influence of BRICS Countries**

It is suggested BRICS countries play out their advantages as countries of important influence in respective regions. The bilateral and multilateral STI cooperation among BRICS countries and with regions where BRICS countries are in could bring forth more active south-south STI collaboration particularly between China and countries in middle Asia, Southern Asia, Southern America and Africa.

(6) **Matching Development Strategies**

It is suggested STI collaboration in "B&R" initiative be more aligned with that in "Eurasian Economic Union" proposed by Russia as well as those proposed by India, Brazil and South Africa in their own development strategies. In respond to the urgent development needs in respective countries, BRICS countries may enhance joint efforts in areas of new energy, transportation, new materials, and support a batch of collaborative STI mega projects with great influence on the platform of joint labs/research centers and through the implementation of key STI measures to breakthrough technology bottleneck.

(7) **Enhancing Collaboration in Basic Research and Mega Scientific and Engineering Projects**

It is encouraged that BRICS countries joint hands to initiate and carry out mega scientific and engineering projects/programs so to help materialize STI breakthroughs. The seminars/symposiums that gather experts of mega scientific and engineering projects and experts in basic research are encouraged to be organized regularly so to promote the exchange of talents and technologies and explore the potential of future collaboration. BRICS countries could continue and deepen their collaboration in ITER[10] and other ongoing mega science projects through sharing experiment equipment and joint research. Making best use of the 5/6 latitude and 60% time zones BRICS countries cover, five countries could together build remote sensing stations to carry the remote sensing and monitoring from the earth to the space.

(8) **Establishing an Enterprises-Oriented STI Collaboration Mechanism with a Well Functioned Industry-Academia-Research Chain**

Enterprises, as the most active market players in the industry-academia-research chain, constitute an important force of international STI cooperation. In addition to collaboration funded by BRICS governments among higher-learning institutions and research entities, BRICS STI cooperation are in need of participation from enterprises. It is suggested a long term mechanism be established to offer in various ways key support to collaborative results featuring technology maturity and market potential.

---

[10]ITER (International Thermonuclear Experimental Reactor) is by far the world's largest fusion experiment.

Reference

Dutta, S., & Calkin, S. (2007). Global innovation index 2007: The world's top innovators: Globalisation has pushed innovation to the top of the agenda, but which countries respond best to the new challenges? *Business World.*

Dutta, S., Lanvin, B., & Wunsch-Vincent, S. (2017). *The global innovation index 2017*. Geneva: Cornell University, INSEAD, WIPO.

Li, J., Li, M., & Zhao, X. (2011). *The report on the group of twenty (G20) national innovation competitiveness development (2001–2016)*. Beijing: Social Sciences Academic Press.

Li, J., Li, M., & Zhao, X. (2017). *The report on the group of twenty (G20) national innovation competitiveness development (2016–2017)*. Beijing: Social Sciences Academic Press.

World Economic Forum. (2002). *The global competitiveness report 2001–2002*. New York: Oxford University Press.

World Economic Forum. (2017). *The global competitiveness report 2001–2002*. Geneva: World Economic Forum.

Chapter 2
S&T Priorities for BRICS Countries: In Search of a Win-Win Strategy

Alexander Sokolov, Sergey Shashnov, Maxim Kotsemir and Anna Grebenyuk

1 Introduction

Most of the developed and developing nations, including BRICS countries, have been devoting considerable attention to S&T priority setting for quite a while now, since such priorities serve as a basis for their science, technology, and innovation (STI) policies (OECD 2010; BILAT-USA 2010; Gassler et al. 2004; Gokhberg et al. 2016; Grebenyuk et al. 2016; Cagnin 2014; Kuwahara et al. 2008; Li 2009; Pouris and Raphasha 2015). Relevant efforts are mainly focused on solving strategic socio-economic problems, and making efficient use of national competitive advantages (OECD 2012; European Forum on Forward Looking Activities 2015; Meissner et al. 2013; Poznyak and Shashnov 2011; Sokolov and Chulok 2016). S&T priorities are currently being set through a comprehensive assessment of their possible contribution to achieving sustainable socio-economic development, and strengthening the country's competitiveness.

Tools facilitating economic and social development include international S&T cooperation, international research and development integration, and establishment of efficient partnerships with foreign R&D centres and organisations, primarily in BRICS countries. Such activities would be more productive when they match identified and agreed priorities for S&T cooperation. Accordingly, identifying S&T priorities which BRICS countries share with each other becomes increasingly relevant for planning their cooperation. This objective is partially accomplished in the

The original version of this chapter was revised: The affiliation of authors "S. Shashnov, M. Kotsemir, and A. Grebenyuk" was changed. The correction to this chapter is available at https://doi.org/10.1007/978-981-10-8078-4_13

A. Sokolov (✉) · S. Shashnov · M. Kotsemir · A. Grebenyuk
Institute for Statistical Studies and Economics of Knowledge, National Research University Higher School of Economics, Moscow, Russia
e-mail: sokolov@hse.ru

X. Zhao et al. (eds.), *BRICS Innovative Competitiveness Report 2017*,
Research Series on the Chinese Dream and China's Development Path,
https://doi.org/10.1007/978-981-10-8078-4_2

scope of various bilateral S&T cooperation programmes implemented by BRICS countries. Developing joint approaches to setting S&T cooperation priorities is becoming particularly important, followed by their successful practical implementation. Especially interesting are cooperation areas where joining forces can potentially produce major synergies. The partner countries' long-term goal is turning BRICS into a full-fledged platform for ongoing and strategic interaction on key issues, including science and technology.

During the last few years Russia and other BRICS countries have significantly stepped up international aspect of their S&T and innovation policies, which made a tangible contribution to implementing their competitive advantages. The Cape Town Declaration of 10 December, 2014 adopted by the first meeting of BRICS education and science ministers (BRICS Science, Technology and Innovation Ministerial Meeting 2014) stresses the need to strengthen the countries' cooperation in the science, technology and innovation sphere, to help meet common global and regional socio-economic challenges on the basis of shared experience, complementary efforts, joint creation of new knowledge and development of innovative products, services, and processes using relevant funding mechanisms and investment promotion tools, and encouraging partnership with other strategic players in emerging countries. The Declaration also outlines the main areas for potential cooperation. In March, 2015, during the second BRICS ministerial meeting in Brasilia (BRICS Science, Technology and Innovation Ministerial Meeting 2015), the Memorandum of Understanding on Cooperation in the Fields of Science, Technology and Innovation was signed, identifying several particularly important areas for international cooperation (such as food security and sustainable agriculture; managing natural disasters; new and renewable energy sources and energy efficiency; nanotechnology; information and computer technologies, etc.).

A number of fundamental documents such as the Moscow Declaration on BRICS Countries' S&T Cooperation approved by BRICS science, technology and innovation ministers in 2015, and the BRICS Science, Technology and Innovation Work Plan for 2015–2018 (BRICS Science, Technology and Innovation Ministerial Meeting 2015; BRICS Science, Technology and Innovation Work Plan 2015) play a major role in promoting international activities. Agreeing priority S&T areas is also necessary for implementing the BRICS Multilateral Research Initiative in the scope of the BRICS Framework Programme. It's important for BRICS countries to participate in joint priority- and objective-setting for international cooperation, since such exercises yield immediate and potential advantages resulting from international STI cooperation. Also important is to have reliable information for drafting agendas understandable and acceptable to all BRICS countries. This, in turn, requires comprehensively analysing BRICS countries' socio-economic requirements, their S&T potential, and priorities for S&T cooperation.

This section presents an analysis of BRICS countries' S&T development potential and their possible interest in stepping up international cooperation, suggests approaches to identifying a system of priorities for BRICS countries' S&T cooperation, and relevant tools for implementing these priorities.

2 BRICS Countries' R&D Resources

All BRICS countries except South Africa are among the world's largest economies, and have a significant potential for meeting current challenges—especially if they pool and efficiently apply their available resources.

Key S&T development indicators include the following:

Input indicators:

- GERD as % of GDP
- GERD in current USD PPP
- R&D personnel

Output indicators:

- Publication activity
- International scientific collaboration
- Citation impact
- Patent activity

Below, BRICS countries are characterised on the basis of these indicators.

We take data on R&D indicators in BRICS from the world's largest database on science, technology and innovation indicators: UNESCO Institute of Statistics (UNESCO UIS) "Research and Experimental Development Database[1], and OECD Main Science and Technology Indicators (MSTI)[2]. Data on publication activity we take from Scopus database. Scopus is one of the largest science citation databases in the world. Scopus is owned by Elsevier publishing company[3]. As the end of June 2017, Scopus indexed more than 68.1 mln documents.

China is a major economic and scientific power. Its internal R&D expenditures (GERD) are three times higher than combined expenditures of all other BRICS countries (Figs. 1 and 2).

In 2015, Chinese GERD exceeded combined GERD of EU28 countries, having reached $408.8 billion (PPP). In the US (the world leader in terms of R&D expenditures), relevant figure was $502.9 billion. In 2015, China's GERD were many times higher than relevant expenditures by other BRICS countries. Russian, Indian, and Brazilian GERD in recent years were comparable, at about $35–$50 billion (PPP). South Africa's R&D investments were much smaller, at $4.5–$5.0 billion (PPP) during the last few years. Main indicators describing BRICS countries' R&D sectors are presented in Table 1.

In China, in the last 15 years GERD have increased 11.2 times; in other BRICS countries the growth was much lower, from 1.85 times in India to 4.23 times in

[1]Free access to this database is available on http://data.uis.unesco.org/Index.aspx?queryid=115.

[2]OECD MSTI database is available here: http://stats.oecd.org/Index.aspx?DataSetCode=MSTI_PUB.

[3]https://www.scopus.com/home.uri.

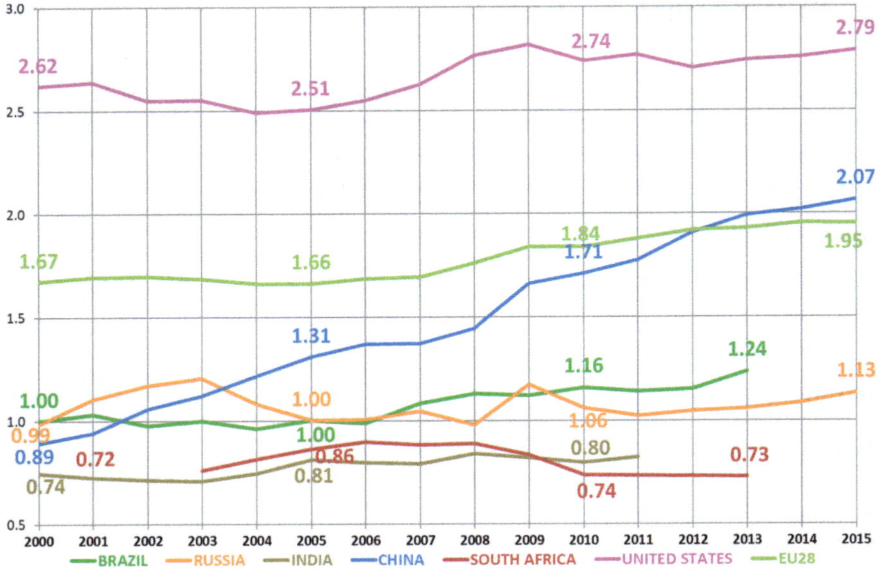

Fig. 1 GERD as % of GDP in BRICS, EU28 countries, and the USA. *Source* USA, EU28, China, Russia, SAR—OECD MSTI (Main Science and technology Indicators database) (last update: April 2017); Brazil, India—UNESCO Institute of Statistic database (section "Science, technology and innovation") (last update: July 2016)

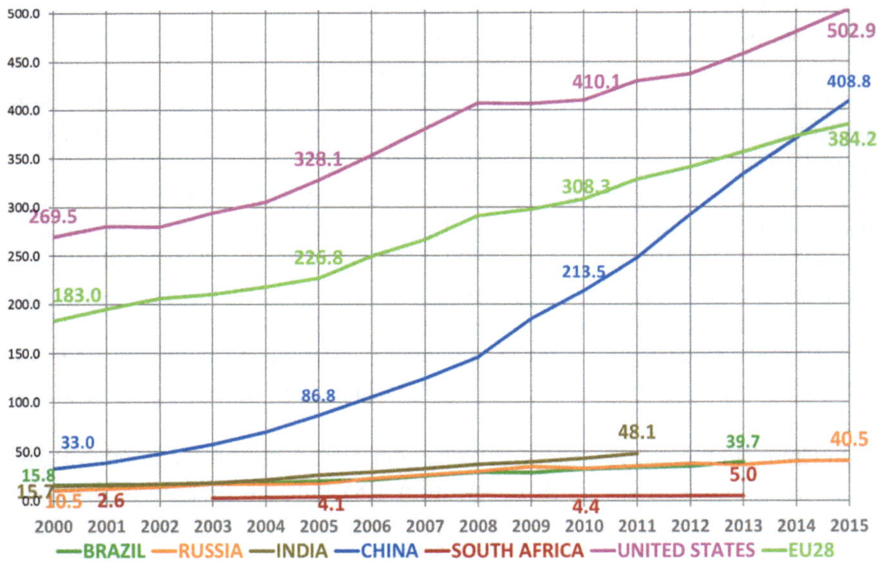

Fig. 2 GERD (USD, PPP) in BRICS, EU28 countries, and the USA. *Source* USA, EU28, China, Russia, SAR—OECD MSTI (Main Science and technology Indicators database) (last update: April 2017); Brazil, India—UNECO Institute of Statistic database (section "Science, technology and in-novation") (last update: July 2016)

Table 1 Key indicators of BRICS countries' R&D potential

Country	2000	2005	2010	2011	2012	2013	2014	2015
Gross expenditures on research and development (GERD), billion USD (PPP), in current prices								
Brazil	15.8	20.5	32.5	33.9	35.5	39.7		
Russia	10.5	18.1	33.1	35.2	37.9	36.6	39.9	40.5
India	15.7	26.5	42.8	48.1				
China	33.0	86.8	213.5	247.8	292.2	334.1	370.1	408.8
South Africa	2.6 (2001)	4.1	4.4	4.7	4.8	5.0		
USA	269.5	328.1	410.1	429.8	437.1	457.6	479.4	502.9
EU28	183.0	226.8	308.3	328.4	341.2	356.0	372.6	384.2
GERD as % of GDP								
Brazil	1.00	1.00	1.16	1.14	1.15	1.24		
Russia	0.99	1.00	1.06	1.02	1.05	1.06	1.09	1.13
India	0.74	0.81	0.80	0.82				
China	0.89	1.31	1.71	1.78	1.91	1.99	2.02	2.07
South Africa	0.72 (2001)	0.86	0.74	0.73	0.73	0.73		
USA	2.62	2.51	2.74	2.77	2.71	2.74	2.76	2.79
EU28	1.67	1.66	1.84	1.88	1.92	1.93	1.95	1.95
Number of researchers (full-time equivalents)								
Brazil	73.9	109.4	138.7					
Russia	506.4	464.6	442.1	447.6	443.3	440.6	444.9	449.2
India	115.9	154.8	192.8					
China	695.1*	1 118.7*	1 210.8	1 318.1	1 404.0	1 484.0	1 524.3	1 619.0
South Africa	14.2 (2001)	17.3	18.7	20.1	21.4	23.3		
USA	983.3	1 101.1	1 198.8	1 253.1	1 264.2	1 305.9	1 351.9	
EU28	1 117.8	1 374.8	1 601.1	1 626.8	1 681.6	1 730.7	1 759.1	1 805.3
GERD per researcher, thousand USD (PPP), in current prices								
Brazil	214.3	187.8	234.5					
Russia	20.7	39.0	74.9	78.6	85.5	83.1	89.6	90.2
India	135.1	171.4	222.0					
China	47.5	77.6	176.3	188.0	208.1	225.2	242.8	252.5
South Africa		234.1	236.8	231.3	225.7	213.1		
USA	274.1	298.0	342.1	343.0	345.7	350.4	354.6	
EU28	163.8	164.9	192.5	201.9	202.9	205.7	211.8	212.8

Note From 2009, researcher data in Chinaare collected according to the Frascati Manual definition of researcher. Beforehand, this was only the case for independent research institutions, while for the other sectors data were collected according to the UNESCO concept of "scientist and engineer"

Source USA, EU28, China, Russia, SAR—OECD MSTI (Main Science and technology Indicators database) (last update: April 2017); Brazil, India—UNECO Institute of Statistic database (section "Science, technology and innovation") (last update: July 2016)

Russia. It should be noted that in China annual GERD growth (at about $30–$40 billion) in recent years was comparable with the total annual GERD in Russia, India, and Brazil.

During the last 15 years R&D intensity steadily grew in China, while in other BRICS countries relevant indicators remained largely unchanged, especially during the last 5 years. E.g. GERD as a percentage of the GDP in China has grown from 0.9% in 2000 to 2.07% in 2015, exceeding the relevant figure for the EU28 countries for 2013. In the EU and the US GERD measured as a share of the GDP during the last 15 years grew insignificantly.

China has the largest number of researchers in the world. In 2015 there were 1.62 million researchers in China (in full-time employment equivalents). In the US the relevant figure (for 2014) is 1.35 million, and the total for the EU28 countries is 1.81 million. Russia has 446.2 thousand researchers (in full-time employment equivalents), which is among the largest figures in the world. By this indicator

Russia is behind only China (1.62 million), the US (1.35 million), and Japan (662.1 thousand). The numbers of researchers in India (192.8 thousand in full-time employment equivalents, 2010) and Brazil (138.7 thousand in full-time employment equivalents, 2010) are comparable. South Africa (as in the case of GERD) has much fewer researchers than other BRICS countries, just 23.3 thousand in full-time employment equivalents (2013).

In terms of R&D expenditures per researcher (in full-time employment equivalent) Russia has the lowest figure among BRICS countries, at \$80–\$90 thousand (PPP) during the last 5 years. In other BRICS nations relevant figures in recent years were between \$200 and \$250 thousand, which is comparable with the average for the EU28 countries (\$200–\$210 thousand) but much lower than in the US (\$340–\$355 thousand).

3 BRICS Countries' Publication Activity

The number of publications authored by BRICS countries' researchers has significantly increased since 2000, along with their share in the global research community (see Fig. 3)[4]. In 2010 the total number of publications by BRICS researchers exceeded the number of publications in the US, and in 2014 came very close to the relevant figure for the EU28 countries. This was largely due to the exceptionally high growth of Chinese publication activity. In 2000–2015, the number of publications by Chinese authors grew 8.5 times, while the overall growth rate of global publication activity during the last five years has declined. Accordingly, between 2000 and 2015 China has moved up from the 6th to the 2nd place in terms of the total number of publications. Due to the relatively high growth of publication activity in recent years China has managed to come much closer to the US, which recently was displaying a rather low growth of the number of publications.

The number of Russian publications indexed in Scopus has grown just 1.86 times in 2001–2015, with the bulk of the growth occurring during the last five years. Despite this fact Russia has moved down in the "Number of publications" rating from the 9th to the 13th place during that period. In 2000–2012 the number of publications by Russian researchers remained at about 30–38 thousand a year, and only in the last years of the period in question Russian publication activity began to increase rapidly.

India, and to lesser extent Brazil, along with China display a high growth rate of publication activity. The number of publications by Brazilian authors indexed in Scopus in 2000–2015 grew from 14.1 thousand to 62.0 thousand. In the global "Number of publications" rating Brazil has moved up from the 17th place in 2000

[4]All calculations are based on the Scopus data. Types of publications included: articles, reviews, and conference papers.

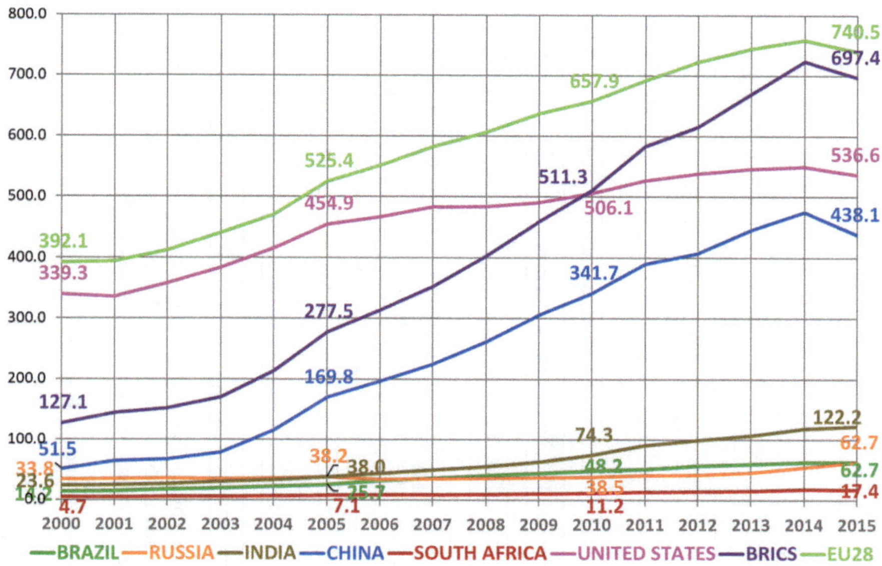

Fig. 3 Growth of the number of publications by BRICS, EU28, and US researchers in 2000–2015 (thousands, indexed in Scopus). *Source* HSE calculations based on Scopus SciVal Benchmarking Toolbox. Types of publications included: articles, reviews and conference papers (last update: March 2017)

to the 14th in 2015. In 2000–2015, the number of Indian Scopus-indexed publications has grown from 23.5 thousand to 122 thousand.

South Africa is also showing a quickly growing publication activity. However, the high growth rate is largely due to the "low start" effect. The number of publications by South African researchers has grown 3.75 times between 2000 and 2015 (from 4.6 thousand to 17.1 thousand). In the overall "Number of publications" rating South Africa falls in the fourth ten of countries.

Generally, in 2015 BRICS countries produced almost 29% of the world's total number of Scopus-indexed publications; out of that, China's share was 18%, India's—5%, Russia's and Brazil's—2.6% each, and South Africa's—0.72% (see Table 2). In terms of the total number of Scopus-indexed publications BRICS countries came very close to the EU28 (30.5% of the world's total in 2015). Currently China is leading in 10 out of 27 top-level Scopus classes subject areas (see Table 2).

For the first time China became the world leader in 2005, in the ENER subject area, and since then steadily improved its positions in other areas. Other BRICS countries' results are more modest. E.g. India holds the third place in terms of the number of publications in 8 subject areas, Brazil—2 s places and 1 third place, and Russia only has 7 places in the first ten countries producing the largest numbers of publications.

Table 2 Positions of BRICS countries in 27 Scopus subject areas (2015)

Field of science	Brazil			Russia			India			China			South Africa			Global leader by number of publ.
	Ranking place	% to global leader	Number of publ.	Ranking place	% to global leader	Number of publ.	Ranking place	% to global leader	Number of publ.	Ranking place	% to global leader	Number of publ.	Ranking place	% to global leader	Number of publ.	
Agricultural and Biological Sciences	3	29.3%	12.3	16	7.9%	3.3	8	20.1%	8.4	2	73.4%	30.8	25	6.2%	2.6	USA
Arts and Humanities	14	4.7%	1.1	9	10.7%	2.6	25	2.8%	0.7	10	8.9%	2.2	13	4.7%	1.1	USA
Biochemistry, Genetics and Molecular Biology	14	9.0%	7.0	16	6.4%	5.0	6	19.3%	15.0	2	73.4%	57.2	36	2.1%	1.6	USA
Business, Management and Accounting	12	10.3%	1.1	32	3.4%	0.4	7	15.3%	1.7	3	26.5%	3.0	20	5.2%	0.6	USA
Chemical Engineering	15	5.9%	2.2	13	7.6%	2.8	3	24.7%	9.0	1	100.0%	36.7	39	1.3%	0.5	China (since 2009)
Chemistry	16	6.8%	4.2	9	12.6%	7.8	3	25.6%	15.9	1	100.0%	62.1	36	1.9%	1.2	China (since 2007)
Computer Science	15	7.7%	3.6	16	7.2%	3.4	3	29.0%	13.4	1	100.0%	46.4	47	1.5%	0.7	China (since 2006)
Decision Sciences	10	17.4%	0.8	27	4.9%	0.2	6	22.6%	1.1	2	53.8%	2.6	38	2.3%	0.1	USA
Dentistry	2	63.5%	1.4	66	0.4%	0.0	3	42.1%	0.9	6	28.8%	0.6	51	0.9%	0.0	USA
Earth and Planetary Sciences	15	7.9%	2.0	8	18.7%	4.7	12	12.9%	3.2	1	100.0%	25.0	24	4.5%	1.1	China (since 2015)
Economics, Econometrics and Finance	20	5.3%	0.5	4	20.6%	1.8	10	13.4%	1.2	7	18.9%	1.6	23	4.6%	0.4	USA
Energy	16	5.5%	1.2	12	9.2%	2.0	3	16.2%	3.5	1	100.0%	21.6	38	1.6%	0.4	China (since 2005)
Engineering	18	5.1%	6.0	11	8.6%	10.1	3	22.4%	26.4	1	100.0%	118.0	45	1.0%	1.2	China (since 2007)
Environmental Science	12	13.9%	3.6	19	7.8%	2.0	5	24.9%	6.4	2	94.7%	24.4	29	4.5%	1.2	USA
Health Professions	10	11.9%	1.3	18	4.7%	0.5	20	4.2%	0.5	8	12.4%	1.3	31	2.1%	0.2	USA
Immunology and Microbiology	12	13.3%	2.4	21	4.6%	0.8	10	14.0%	2.5	2	57.1%	10.4	26	4.0%	0.7	USA
Materials Science	17	4.8%	3.5	8	13.5%	10.0	5	18.5%	13.7	1	100.0%	74.0	40	1.3%	1.0	China (since 2007)
Mathematics	15	9.2%	3.0	9	17.2%	5.6	7	19.6%	6.4	1	100.0%	32.5	43	2.1%	0.7	China (since 2012)
Medicine	14	9.2%	15.4	32	2.3%	3.8	11	11.5%	19.2	2	39.2%	65.5	27	2.5%	4.2	USA
Multidisciplinary	15	6.5%	0.5	16	6.3%	0.5	3	26.9%	2.2	1	100.0%	8.1	34	1.7%	0.1	China (since 2015)
Neuroscience	13	7.7%	1.6	25	2.4%	0.5	15	5.2%	1.1	2	29.8%	6.2	40	0.9%	0.2	USA
Nursing	7	12.4%	1.4	43	1.0%	0.1	20	3.2%	0.4	8	11.5%	1.3	28	1.9%	0.2	USA
Pharmacology, Toxicology and Pharmaceutics	11	13.2%	2.2	19	7.0%	1.2	3	56.9%	9.5	2	81.9%	13.6	33	2.9%	0.5	USA
Physics and Astronomy	15	8.5%	5.6	5	25.4%	16.9	8	21.1%	14.0	1	100.0%	66.3	39	2.3%	1.6	China (since 2013)
Psychology	11	4.7%	1.0	33	1.3%	0.3	29	1.7%	0.4	10	7.1%	1.5	26	2.0%	0.4	USA
Social Sciences	12	7.0%	3.4	11	8.3%	4.1	13	6.7%	3.3	6	13.6%	6.7	16	4.8%	2.3	USA
Veterinary	2	49.4%	1.7	61	0.9%	0.0	5	26.5%	0.9	4	29.7%	1.0	26	5.2%	0.2	USA
Total	13	11.7%	57.0	14	11.4%	55.5	5	23.2%	113.1	2	82.5%	401.9	34	3.2%	15.6	USA

Notes 1. For each BRICS country the first column is the position of this country in a global ranking by number of publications in each of 27 Scopus subject areas ("Subjarea"). 2. The second column is the relation of publication of a given country in a given subject are to a number of publications in the global leader by number of publications in this subject area. 3. The third column is the number of publications (in thousands). 4. Global leaders by number of publications in each of 27 subject areas are shown in the last column

Source HSE calculations based on SCImago Journal and country Rank database, based on Scopus (last update: June 2016). Types of publications included: articles, reviews and conference papers

Table 3 Thematic structure of publications, and values of Relative Comparative Advantages Index for BRICS countries (2011–2015)

Subject area	World Struct.	Brazil		Russia		India		China		South Africa	
		Struct.	RCA	Struct.	RCA	Struct.	RCA	Struct.	RCA	Struct.	RCA
Agricultural and Biological Sciences	7.9%	***20.3%***	***2.57***	5.7%	0.72	8.7%	***1.09***	6.4%	0.81	***16.4%***	***2.07***
Arts and Humanities	3.7%	1.9%	0.51	2.1%	0.57	0.6%	0.16	0.5%	0.14	8.7%	***2.31***
Biochemistry, Genetics and Molecular Biology	***12.0%***	***11.2%***	0.93	9.3%	0.77	12.9%	***1.07***	11.1%	0.93	9.8%	0.82
Business, Management and Accounting	2.2%	1.6%	0.72	0.9%	0.41	1.7%	0.80	1.2%	0.55	3.2%	***1.49***
Chemical Engineering	4.5%	3.5%	0.78	5.1%	***1.12***	6.8%	***1.51***	6.7%	***1.48***	3.0%	0.66
Chemistry	8.9%	7.2%	0.82	***15.0%***	***1.69***	***14.2%***	***1.60***	12.6%	***1.42***	7.6%	0.86
Computer Science	***12.4%***	8.9%	0.72	6.9%	0.56	***15.4%***	***1.24***	***15.5%***	***1.25***	6.8%	0.55
Decision Sciences	1.0%	1.3%	***1.29***	0.5%	0.48	0.9%	0.89	0.9%	0.96	0.8%	0.81
Dentistry	0.5%	2.8%	***5.50***	0.0%	0.02	1.0%	***1.90***	0.1%	0.28	0.2%	0.32
Earth and Planetary Sciences	4.4%	3.7%	0.84	10.0%	***2.25***	3.7%	0.83	5.6%	***1.27***	7.5%	***1.68***
Economics, Econometrics and Finance	1.5%	0.8%	0.53	1.3%	0.88	0.9%	0.62	0.5%	0.31	4.1%	***2.68***
Energy	3.3%	2.3%	0.70	4.2%	***1.29***	3.5%	***1.08***	4.9%	***1.49***	2.6%	0.81
Engineering	***21.3%***	***11.7%***	0.55	***18.5%***	0.87	***21.8%***	***1.03***	***38.4%***	***1.80***	***10.1%***	0.48
Environmental Science	5.0%	5.8%	***1.17***	3.2%	0.64	5.8%	***1.17***	5.1%	***1.03***	7.3%	***1.46***
Health Professions	1.1%	1.7%	***1.49***	0.7%	0.66	0.4%	0.39	0.3%	0.29	1.0%	0.93
Immunology and Microbiology	2.8%	4.3%	***1.54***	1.6%	0.58	2.6%	0.94	2.1%	0.76	4.6%	***1.65***
Materials Science	10.3%	6.3%	0.61	***18.1%***	***1.77***	12.4%	***1.20***	***15.8%***	***1.54***	6.2%	0.60
Mathematics	6.9%	5.3%	0.78	***10.5%***	***1.54***	6.2%	0.90	8.1%	***1.18***	5.1%	0.74
Medicine	***28.1%***	***29.5%***	***1.05***	8.5%	0.30	***19.8%***	0.70	***14.8%***	0.53	***25.6%***	0.91
Multidisciplinary	1.0%	0.7%	0.72	1.2%	***1.26***	1.6%	***1.58***	1.4%	***1.42***	0.5%	0.53
Neuroscience	2.4%	2.7%	***1.15***	0.9%	0.39	0.9%	0.36	1.2%	0.51	1.0%	0.43
Nursing	1.5%	2.5%	***1.64***	0.3%	0.19	0.3%	0.22	0.3%	0.17	1.2%	0.82
Pharmacology, Toxicology and Pharmaceutics	3.3%	3.8%	***1.13***	1.6%	0.47	9.8%	***2.93***	3.1%	0.92	2.8%	0.84
Physics and Astronomy	***12.3%***	***10.0%***	0.82	***33.4%***	***2.72***	***13.8%***	***1.13***	***15.5%***	***1.26***	***10.0%***	0.82
Psychology	2.2%	1.9%	0.86	0.4%	0.20	0.4%	0.16	0.3%	0.12	2.7%	***1.25***
Social Sciences	7.5%	6.0%	0.79	4.1%	0.54	3.3%	0.43	2.3%	0.30	***16.6%***	***2.20***
Veterinary	0.8%	3.7%	***4.42***	0.1%	0.06	1.2%	***1.44***	0.3%	0.34	1.4%	***1.74***
Total number of papers for 2011-2015	**11 668 705**	**285 454**		**2 092 672**		**239 799**		**525 853**		**75 234**	

Source HSE calculations based on Scopus. Types of publications included: articles, reviews and conference papers (last update: September 2016)

4 Structure of BRICS Countries' Publications

Structures of publications by BRICS countries' scientists were assessed using 27 major subject areas of the SCOPUS database, and compared with the global publication structure to calculate the countries Relative Comparative Advantages Index (RCA index) (see Table 3)[5].

Russian research sector has a predominantly "physics and technology" profile whose origins go back to the Soviet period. The subject area with the highest presence of Russian researchers (Scopus-indexed publications in 2011–2015) was Physics and Astronomy—33.4% of all Russian publications. Other major subject areas being researched in Russia include Engineering (18.5% of all Russian publications in 2011–2015), Materials Science (18.1%), and Chemistry (15%). Such fields as Neuroscience, Business, Management, and Accounting, Health, Decision Making, Psychology, Nursing, Veterinary, and Dentistry are represented in the

[5]RCA—Revealed comparative advantage index. RCA of country J in scientific field I, calculated as the relationship between the share of its publications in scientific field I in the total number of publications of country J and the equivalent global figure. Those fields where the RCA value is higher than 1 are classified as areas of scientific specialization of a country.

structure of Russian publications very poorly (less than 1% of the total number of published works). The share of Physics and Astronomy publications in all Scopus-indexed publications by Russian researchers (33.4%) is much higher than the relevant world's average figure (12.3%).

China's status as the "global manufacturer" is supported by its Scopus thematic profile. The main area of Chinese research is Engineering (38.4% of all publications). Other prominent areas in the structure of publications by Chinese authors include Materials Science (15.8%); Computer Science (15.5%), Physics and Astronomy (15.5%); Medicine and Health (14.8); Chemistry (12.6); Biochemistry, Genetics, and Molecular Biology (11.1%). At the same time numerous subject areas are very poorly represented in the structure of Chinese publications (under 1% of the total number in 2011–2015): Decision Making; Humanities; Economics, Econometrics, and Finance; Health; Veterinary; Psychology; Nursing; Dentistry.

India shows a more balanced structure of publications than Russia or China. The largest subject area (Engineering) accounts for 21.8% of all Scopus-indexed publications in 2011–2015. Other major areas of Indian research include Medicine (19.8%), Computer Science (15.4%), Chemistry (14.2%), Physics and Astronomy (13.8%); Biochemistry, Genetics, and Molecular Biology (12.9%); and Materials Science (12.4%). Analysis of the country's Revealed comparative advantages Indices for the 27 top-level subject areas clearly reveals India's profile's shift towards pharmaceutics and chemical sciences.

Brazil's and South Africa's publication structures are quite different from other BRICS countries'. Brazil gravitates towards medical and biological research, with major Scopus-indexed areas being Medicine (29.5% of all publications by Brazilian researchers in 2011–2015) and Agricultural and Biological Sciences (20.3%). Other important fields include Engineering (11.7%), Biochemistry, Genetics, and Molecular Biology (11.2%), and Physics and Astronomy (10.0%).

In South Africa the main research area, like in Brazil, is Medicine (25.6% of all Scopus-indexed publications by South African scientists in 2011–2015). Other important research areas covered in Scopus-indexed publications by South African scientists include Social Sciences (16.6%), Agricultural and Biological Sciences (16.4%), Engineering (10.1%), Physics and Astronomy(10.0%), etc.

5 International Cooperation of BRICS Countries

South Africa shows the most active involvement in international research cooperation among all BRICS countries (see Fig. 4). Since 2005 more than 40% of this nation's Scopus-indexed publications were co-authored with scientists from other countries. Note that the share of internationally co-authored publications in South Africa was growing for the last five years (Fig. 5).

In Russia the share of internationally co-authored publications for the last 15 years remained at 25–35%. Note that in Russia, unlike in South Africa, China, and Brazil, this figure was steadily decreasing in recent years—from 33.6% in 2005

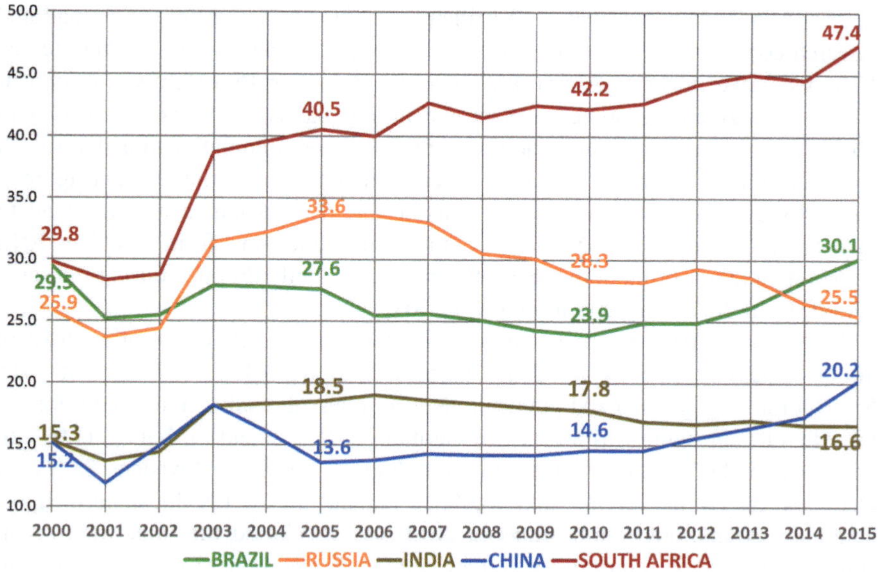

Fig. 4 Share of internationally collaborated publications in the total number of Scopus-indexed publications in BRICS countries (2000–2015). *Source* HSE calculations from Scopus SciVal Benchmarking Toolbox (last update: March 2017). Types of publications included: articles, reviews and conference papers

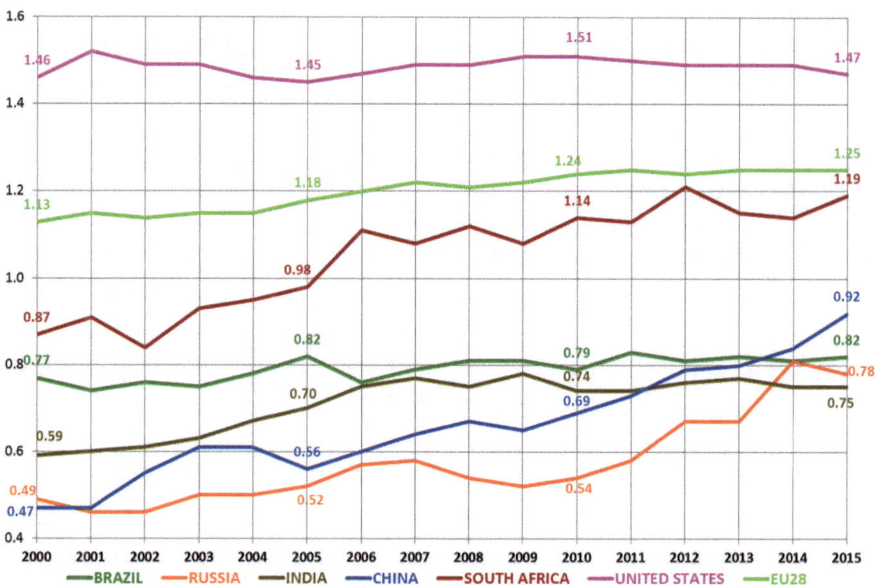

Fig. 5 Field-Weighted Citation Impact of Scopus-indexed publications by BRICS countries' researchers (2000–2015). *Note* Field-Weighted Citation Impact: the ratio of received citations to the expected world average for the subject field, publication type and publication year. *Source* HSE calculations based on Scopus SciVal Benchmarking Toolbox (types of publications included: articles, reviews and conference papers; last update: March 2017)

to 25.5% in 2015. The level of Brazilian scientists' participation in international research cooperation was somewhat lower than in Russia (25–20% during the last 15 years). Like in South Africa, the share of internationally co-authored publications by Brazilian researchers has grown during the last 5 years (from 23.9 to 30.1%). In India and China scientists are integrated into international research cooperation to a lesser extent than in other BRICS countries (the relevant figure is about 15–20% for the last 15 years). In the last 5 years China has managed to increase the share of internationally co-authored publications from 14.6% in 2010 to 30.1% in 2015. In India the relevant figure has slightly dropped during the same period, from 28.5% in 2005 to 16.6% in 2015.

BRICS countries' involvement in international research cooperation (except South Africa) is much lower than European countries'. E.g. in France in 2015 51.8% of all Scopus-indexed publications were internationally co-authored; for the UK the relevant figure was 50.0%, for Germany—48.5%, and for Italy—43.9%. In Scandinavia the relevant values are even higher: 59.1% in Sweden, 58.5% in Denmark, 57.1% in Norway, and 56.0% in Finland. In the US the share of internationally co-authored publications in 2015 was 32.8%. At the same time Asian countries with advanced research systems tend to display rather low participation in international scientific cooperation. E.g. in 2015 only 20.9% of Scopus-indexed publications by Iranian authors were internationally co-authored; for Turkey the relevant figure was 21.1%, for Japan—26.6%, for the Republic of Korea—26.5% (Science and Technology Indicators 2017).

BRICS countries do not yet make key research partners for each other (see Table 4).

The main partner for all BRICS countries in 2015 was the US (just as in all other years). E.g. 44.6% of all internationally co-authored Chinese publications were

Table 4 Key scientific partners of BRICS countries (Scopus-indexed publications, 2015)

BRAZIL		RUSSIA		INDIA		CHINA		SOUTH AFRICA	
Top-5 partners									
1. USA	35.5	1. USA	25.4	1. USA	32.3	1. USA	44.6	1. USA	33.1
2. UK	13.8	2. GER	23.7	2. UK	12.2	2. UK	9.9	2. UK	22.7
3. FRA	12.5	3. FRA	14.1	3. GER	9.8	3. AUS	8.8	3. GER	12.7
4. ESP	11.8	4. UK	13.2	4. KOR	8.5	4. HKG	7.5	4. AUS	12.0
5. GER	11.6	5. ITA	9.6	5. SAU	6.9	5. CAN	7.1	5. FRA	10.8
Partners from BRICS									
12. CHI	4.8	6. CHI	8.4	8. CHI	6.6	18. IND	1.53	8. IND	8.0
17. IND	3.7	19. BRA	3.9	15. SAR	3.5	19. RUS	1.49	13. CHI	5.9
21. RUS	3.2	20. IND	3.8	16. BRA	3.4	23. BRA	1.04	15. BRA	4.7
26. SAR	2.2	35. SAR	2.1	19. RUS	2.9	37. SAR	0.60	20. RUS	3.8

Note Share of publications in collaboration with a given country in the total number of internationally collaborated publications of each of BRICS countries is given
Source HSE calculations based on Scopus (types of publications included: articles, reviews and conference papers; last update: September 2016)

written jointly with American scientists, while the share of China's second biggest partner (the UK) was just 9.9%. No BRICS country was among China's ten biggest research partners.

Russia's structure of research partners is different from China's, Brazil's, and India's. Russia has two key research partners—the US and Germany, with 25.4 and 23.7% internationally co-authored publications in 2015, respectively. Then, in descending order, follow France (14.1%), the UK (13.2%), Italy (9.6%), and China (8.4%). Other BRICS countries play much smaller roles in Russia's international cooperation. The share of internationally co-authored Russian publications written jointly with Brazilian scientists is 3.9%; the relevant figure for India is 3.8%, and South Africa—2.1%.

South Africa has the highest field-weighted citation impact figures—more than the world's average, and in recent years practically on a par with the EU countries. Other BRICS countries' citations figures remain below the global average. China used to be the leader, ahead of all other BRICS countries after 2001. Russia and India also increased there relevant values during the period in question; in Russia this value's growth recently was more pronounced. Citation figures for publications by Brazilian researchers remained practically unchanged during the last 15 years.

The above data allows to make the following conclusions:

- Extremely fast growth of R&D inputs, and consequently outputs, allowed China to increase its presence on the global arena, gaining the status of a new "scientific superpower"; none of the other BRICS countries could match it, though all of them became more "visible" in terms of internationally recognised publications and patents;
- Thematic structures of basic research in BRICS countries are moving closer to the global scientific agenda; however, their influence measured in field-weighted citation impact remains insufficient and grows mostly due to active participation in international scientific collaborations;
- All BRICS countries have unique collaboration profiles based on traditional leaders and historical partnerships; BRICS-to-BRICS collaborations are not very common except with China, due to expansive nature of its S&T development strategy;
- Differences in scientific specialisations of BRICS countries can be seen as possible connection points for establishing stronger partnerships, to deal with growing global challenges and "universalisation" trends.

6 Setting Priorities for BRICS Countries' Cooperation

For the purposes of this paper, the system of priorities for BRICS countries' S&T cooperation is defined as a set of major S&T development areas with a potential to make a radical contribution to ensuring BRICS countries' security, increasing growth rate of their economies, strengthening their competitiveness, and meeting

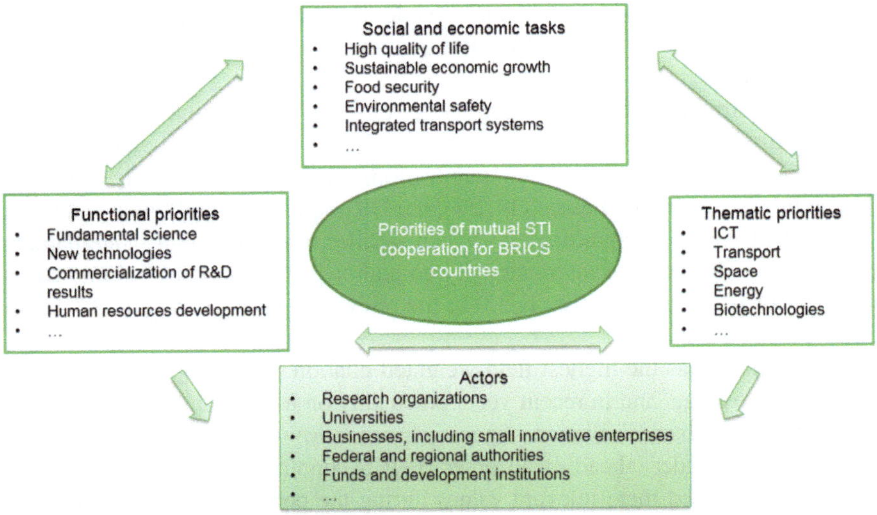

Fig. 6 Priority system for BRICS countries' S&T cooperation

other key socio-economic and S&T challenges. Priority-setting for BRICS countries' S&T cooperation is seen in the context of designing a long-term strategy for their sustainable development.

Priorities for BRICS countries S&T cooperation can be subdivided into thematic and functional ones. Thematic priorities comprise the most important R&D areas (such as ICT, space systems, etc.), investing in which could produce the highest social and/or economic effects in the medium to long term. Functional priorities include objectives aimed at facilitating development and performance of national research and innovation systems, e.g. accelerated development of human potential, commercialisation of R&D results, etc. Joint implementation of such projects would help accomplish major socio-economic objectives (see Fig. 6).

S&T priority setting is aimed at promoting development of BRICS countries' S&T potentials, and focusing them on major socio-economic development areas, taking into account the expected technological breakthroughs. Particular attention should be paid to fully utilising national competitive advantages; a limited number of the most important S&T priorities should be chosen, to provide all necessary support and required resources.

An integrated approach was applied to design the system of priorities, based on goals and objectives reflected in official international and national documents, assessing their S&T potential, and taking into account opinion of the expert community. The following main techniques were used for priority setting: document analysis, bibliometric analysis, and various expert-based procedures (Fig. 7).

The following information sources were used for setting overall priorities for BRICS countries' S&T cooperation:

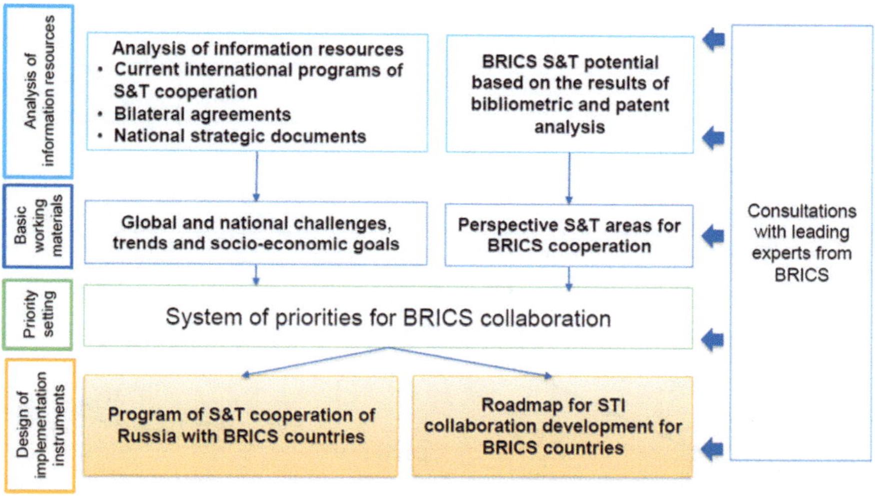

Fig. 7 Priority setting scheme

- Results of bibliometric and patent analysis;
- Official documents of BRICS countries on S&T cooperation (bilateral and multilateral), approved by the countries' governments or government ministries responsible for shaping and implementing S&T and innovation policies;
- Strategic national documents on BRICS countries' S&T and innovation development.

7 BRICS Countries' Cooperation: The Current State and Potential Prospects

In terms of internationally co-authored publications by BRICS countries' scientists (see Fig. 8) China is the leader: the number of such publications by Chinese authors grew from 7.8 thousand in 2000 to 88.7 thousand in 2015. In Russia, Brazil, and India the number of such publications was several times smaller—20.3 thousand, 18.9 thousand, and 16.0 thousand in 2015, respectively. South Africa's figure was smaller still, at just 8.2 thousand publications in 2015.

BRICS countries' international cooperation has a significant development potential (Fig. 9). The share of Chinese publications co-authored with researchers from BRICS countries in the total number of internationally co-authored publications is just 3%; for Brazil the relevant figure is 9%, for Russia—10%, for India—11%, and for South Africa—15%.

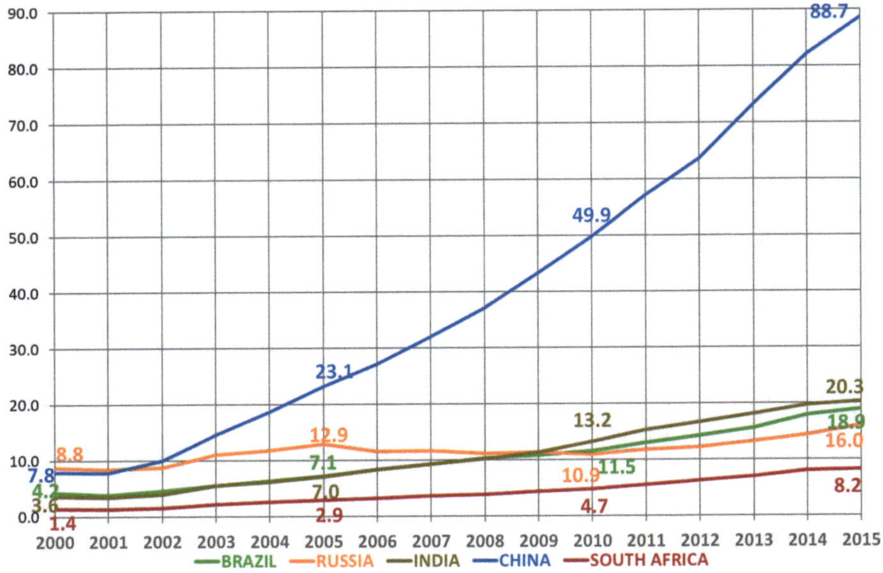

Fig. 8 Dynamics of number of internationally co-authored publications in the total number of BRICS countries' Scopus-indexed publications (2000–2015). *Source* HSE calculations based on Scopus SciVal Benchmarking Toolbox (types of publications included: articles, reviews and conference papers; last update: March 2017)

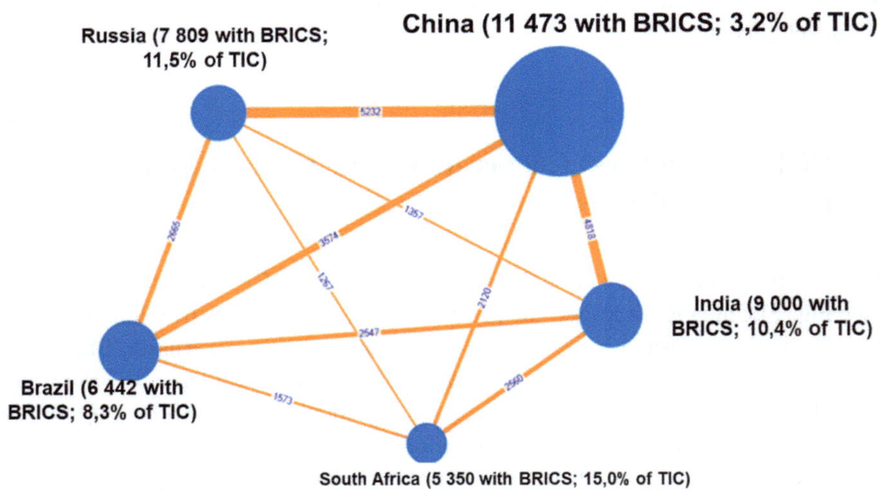

Fig. 9 Map of intra-BRICS collaboration (Scopus-indexed publications, 2011–2015). *Note* TIC means "Total number of Internationally Collaborated publications". In parenthesis we show—total number of publications with BRICS countries and the share of these publications in total number of internationally collaborated publications of a country. *Source* authors' calculations based on Scopus (types of publications included: articles, reviews and conference papers; last update: September 2016). *Source* HSE calculations based on Scopus (types of publications included: articles, reviews and conference papers; last update: September 2016)

Thematic structure of intra-BRICS publications strongly gravitates towards the Physics and Astronomy area (see Table 5). This area's share in the total number of intra-BRICS publications in 2011–2015 amounted to 35.8%. It remains the biggest area in all possible pairs of BRICS countries, and in many cases dominates their S&T cooperation. The importance of Physics and Astronomy subject area is particularly evident in the structure of Russia's cooperation with BRICS countries—the overall share of relevant publications is 55.9%, while in the total number of joint Russian-Brazilian publications the share of this subject area is 75.6%; for joint Russian-Indian publications it's 72.3%.

Table 5 Thematic structure of intra-BRICS Scopus-indexed collaborative publications (2011–2015) (total numbers of joint publications (for two countries, and for a specific country jointly with all BRICS countries), and shares of publications devoted to specific subject areas)

Subject areas	Intra-BRICS collab	BRA-BRICS	CHI-BRICS	IND-BRICS	RUS-BRICS	SAR-BRICS	BRA-CHI	BRA-IND	BRA-RUS	BRA-SAR	CHI-IND	CHI-RUS	CHI-SAR	IND-RUS	IND-SAR	RUS-SAR
Agricultural and Biological Sciences	10.6%	9.2%	10.9%	9.0%	7.6%	23.0%	7.2%	6.8%	3.1%	13.2%	10.3%	7.6%	11.9%	3.3%	7.0%	11.3%
Arts and Humanities	0.6%	0.5%	0.7%	0.6%	0.3%	1.2%	0.6%	0.6%	0.1%	0.5%	0.6%	0.4%	0.8%	0.1%	0.7%	0.4%
Biochemistry, Genetics and Molecular Biology	11.3%	10.5%	11.3%	12.0%	8.0%	17.2%	8.0%	9.8%	5.0%	9.7%	12.1%	7.5%	10.0%	6.0%	10.7%	6.5%
Business, Management and Accounting	1.0%	0.9%	1.0%	1.0%	0.3%	1.7%	1.0%	0.6%	0.2%	0.7%	1.2%	0.2%	1.1%	0.1%	0.7%	0.7%
Chemical Engineering	4.1%	2.0%	3.7%	5.2%	2.5%	7.6%	1.1%	2.7%	0.8%	0.8%	4.2%	2.5%	3.9%	1.9%	7.2%	1.4%
Chemistry	11.2%	5.0%	10.6%	12.8%	9.0%	17.0%	3.9%	4.5%	2.3%	1.9%	9.6%	9.9%	7.9%	5.4%	20.2%	3.2%
Computer Science	5.9%	4.7%	6.2%	6.3%	3.3%	7.1%	3.9%	3.7%	1.9%	2.5%	7.1%	3.0%	5.2%	2.2%	4.1%	1.8%
Decision Sciences	0.6%	0.4%	0.7%	0.6%	0.3%	1.2%	0.5%	0.2%	0.1%	0.3%	0.9%	0.2%	0.8%	0.1%	0.4%	0.4%
Dentistry	0.5%	1.1%	0.6%	0.2%	0.0%	0.6%	1.7%	0.4%	0.0%	0.3%	0.2%	0.0%	0.0%	0.0%	0.1%	0.1%
Earth and Planetary Sciences	10.6%	8.0%	9.2%	9.1%	12.3%	21.2%	5.5%	5.8%	4.9%	8.0%	7.0%	9.3%	8.2%	10.0%	9.6%	19.6%
Economics, Econometrics and Finance	0.6%	0.5%	0.7%	0.7%	0.2%	1.4%	0.5%	0.4%	0.1%	0.5%	0.9%	0.2%	0.7%	0.0%	0.6%	0.4%
Energy	2.7%	1.7%	2.9%	2.4%	2.2%	4.3%	1.7%	1.2%	0.4%	0.8%	2.3%	2.3%	3.4%	1.8%	2.0%	1.4%
Engineering	13.5%	10.4%	14.5%	12.8%	12.8%	20.2%	11.3%	9.1%	10.2%	8.2%	12.1%	14.5%	16.0%	9.9%	13.7%	10.3%
Environmental Science	5.0%	3.9%	5.2%	5.6%	2.6%	9.5%	3.4%	4.2%	0.9%	4.6%	6.0%	2.7%	5.8%	2.0%	5.2%	2.8%
Health Professions	0.4%	0.7%	0.4%	0.4%	0.2%	1.0%	0.7%	0.4%	0.3%	1.3%	0.5%	0.1%	0.6%	0.2%	0.5%	0.3%
Immunology and Microbiology	3.2%	3.7%	3.0%	3.2%	1.7%	8.1%	2.6%	3.2%	1.2%	5.7%	3.2%	1.5%	3.6%	1.0%	3.2%	3.3%
Materials Science	11.8%	7.3%	10.9%	12.8%	11.5%	13.7%	4.8%	7.6%	3.9%	1.8%	8.9%	11.8%	4.3%	8.1%	15.7%	3.5%
Mathematics	7.7%	6.9%	8.4%	7.2%	6.6%	10.6%	5.8%	3.3%	7.5%	2.5%	8.8%	5.1%	7.4%	3.3%	5.1%	5.0%
Medicine	20.8%	24.1%	17.1%	21.5%	10.3%	35.6%	19.9%	26.1%	10.2%	33.1%	21.1%	8.3%	18.3%	9.6%	25.0%	12.7%
Multidisciplinary	1.6%	1.5%	1.8%	1.3%	1.3%	2.4%	1.7%	1.2%	0.6%	1.3%	1.5%	1.5%	1.6%	0.8%	1.3%	1.2%
Neuroscience	1.1%	1.6%	1.3%	0.9%	0.6%	1.7%	1.7%	0.9%	0.3%	1.6%	1.2%	0.6%	1.0%	0.2%	0.6%	0.5%
Nursing	0.5%	0.7%	0.5%	0.8%	0.1%	1.4%	0.6%	1.0%	0.2%	1.4%	0.8%	0.1%	0.6%	0.2%	1.3%	0.4%
Pharmacology, Toxicology and Pharmaceutics	2.7%	2.5%	2.3%	3.3%	1.0%	4.8%	2.4%	2.3%	0.5%	1.9%	2.5%	0.9%	1.5%	0.8%	4.9%	0.7%
Physics and Astronomy	35.7%	42.6%	36.2%	33.8%	55.9%	42.7%	52.4%	47.8%	75.6%	41.1%	34.3%	58.0%	39.0%	72.3%	30.9%	54.2%
Psychology	0.7%	1.1%	0.7%	0.6%	0.5%	1.6%	1.1%	1.0%	0.7%	2.0%	0.7%	0.5%	1.0%	0.3%	0.7%	0.7%
Social Sciences	2.0%	2.1%	1.9%	2.2%	1.0%	4.9%	1.7%	2.1%	1.0%	2.6%	2.1%	0.9%	2.8%	0.9%	2.6%	1.3%
Veterinary	0.5%	0.7%	0.5%	0.5%	0.1%	1.5%	0.5%	0.7%	0.2%	1.0%	0.6%	0.0%	1.0%	0.1%	0.7%	0.3%
Total	**18 181**	**6 442**	**11 473**	**9 000**	**7 809**	**5 350**	**3 505**	**2 519**	**2 617**	**1 536**	**4 741**	**5 137**	**2 061**	**2 400**	**2 502**	**1 948**

Source HSE calculations based on Scopus (types of publications included: articles, reviews and conference papers; last update: September 2016)

Table 6 Prospective areas for further cooperation

Index of scientific specialization

Thematic areas	Brazil	China	Russia	India	South Africa
Agricultural and Biological Sciences	2,57	0,81	0,72	1,09	2,07
Arts and Humanities	0,51	0,14	0,57	0,16	2,31
Biochemistry, Genetics and Molecular Biology	0,93	0,93	0,77	1,07	0,82
Business, Management and Accounting	0,72	0,55	0,41	0,8	1,49
Chemical Engineering	0,78	1,48	1,12	1,51	0,66
Chemistry	0,82	1,42	1,69	1,6	0,86
Computer Science	0,72	1,25	0,56	1,24	0,55
Decision Sciences	1,29	0,96	0,48	0,89	0,81
Dentistry	5,5	0,28	0,02	1,9	0,32
Earth and Planetary Sciences	0,84	1,27	2,25	0,83	1,68
Economics, Econometrics and Finance	0,53	0,31	0,88	0,62	2,68
Energy	0,7	1,49	1,29	1,08	0,81
Engineering	0,55	1,8	0,87	1,03	0,48
Environmental Science	1,17	1,03	0,64	1,17	1,46
Health Professions	1,49	0,29	0,66	0,39	0,93
Immunology and Microbiology	1,54	0,76	0,58	0,94	1,65
Materials Science	0,61	1,54	1,77	1,2	0,6
Mathematics	0,78	1,18	1,54	0,9	0,74
Medicine	1,05	0,53	0,3	0,7	0,91
Multidisciplinary	0,72	1,42	1,26	1,58	0,53
Neuroscience	1,15	0,51	0,39	0,36	0,43
Nursing	1,64	0,17	0,19	0,22	0,82
Pharmacology, Toxicology and Pharmaceutics	1,13	0,92	0,47	2,93	0,84
Physics and Astronomy	0,82	1,26	2,72	1,13	0,82
Psychology	0,86	0,12	0,2	0,16	1,25
Social Sciences	0,79	0,3	0,54	0,43	2,2
Veterinary	4,42	0,34	0,06	1,44	1,74

Index of technological specialization

Thematic areas	Brazil	China	Russia	India	South Africa
1 - Electrical machinery, apparatus, energy	0,60	1,00	0,52	0,28	0,63
2 - Audio-visual technology	0,40	0,59	0,18	0,53	0,39
3 - Telecommunications	0,40	0,84	0,46	0,69	0,34
4 - Digital communication	0,21	1,44	0,11	1,06	0,28
5 - Basic communication processes	0,31	0,63	0,98	1,59	0,70
6 - Computer technology	0,33	0,83	0,28	2,54	0,45
7 - IT methods for management	0,97	0,53	0,26	2,85	3,07
8 - Semiconductors	0,05	0,47	0,21	0,14	0,26
9 - Optics	0,15	0,50	0,24	0,13	0,12
10 - Measurement	0,63	1,36	1,41	0,44	0,59
11 - Analysis of biological materials	0,74	0,69	2,91	1,10	0,60
12 - Control	1,25	1,25	0,99	0,82	0,97
13 - Medical technology	1,32	0,46	1,52	0,66	1,22
14 - Organic fine chemistry	1,15	0,95	0,69	5,10	1,23
15 - Biotechnology	1,07	0,93	0,80	1,91	1,07
16 - Pharmaceuticals	1,25	1,16	1,26	4,35	1,00
17 - Macromolecular chemistry, polymers	0,87	1,21	0,50	0,73	0,41
18 - Food chemistry	1,53	2,08	7,34	0,40	0,95
19 - Basic materials chemistry	1,56	1,51	1,15	1,25	1,88
20 - Materials, metallurgy	1,48	1,88	2,47	0,57	2,55
21 - Surface technology, coating	0,64	1,02	0,91	0,54	0,76
22 - Micro-structural and nano-technology	1,29	1,29	3,94	0,63	0,82
23 - Chemical engineering	1,59	1,32	1,49	1,01	2,50
24 - Environmental technology	1,42	1,46	1,31	0,52	1,50
25 - Handling	1,88	1,00	0,45	0,32	1,66
26 - Machine tools	0,75	1,65	1,10	0,28	0,88
27 - Engines, pumps, turbines	1,68	0,63	1,66	0,86	0,86
28 - Textile and paper machines	0,95	1,16	0,27	0,32	0,38
29 - Other special machines	1,99	1,29	1,63	0,32	1,79
30 - Thermal processes and apparatus	1,46	1,24	1,06	0,43	1,32
31 - Mechanical elements	1,35	0,91	1,06	0,33	1,02
32 - Transport	1,41	0,61	1,04	0,31	0,80
33 - Furniture, games	1,82	0,77	0,44	0,14	1,64
34 - Other consumer goods	2,33	0,99	1,23	0,22	1,78
35 - Civil engineering	1,76	1,18	1,93	0,20	2,05

Another major area of BRICS countries' research cooperation is Medicine: it accounts for 18.9% intra-BRICS publications in 2011–2015. Medicine is particularly important for joint Brazilian–South African publications (33.1%), and least important for joint publications by Russian and Chinese researchers (8.3%). The share of medical publications co-authored by Russian and BRICS countries' scientists (10.3%) is much lower than relevant figures for other BRICS nations: 17.1% for China, 21.5% for India, and 24.1% for Brazil.

Thematic structure of Russia's research cooperation with BRICS countries matches both the overall structure of Russian Scopus-indexed publications, and the structure of internationally co-authored publications by Russian scientists. As to other BRICS countries (especially Brazil and China), there is a certain mismatch between thematic structures of intra-BRICS collaboration and the overall structure of internationally co-authored publications by these countries' researchers.

Calculated scientific and technological specialisation indices[6] allowed to identify R&D areas with a significant potential for cooperation between BRICS countries (Table 6).

Russia's Relative comparative advantages index (RCA) for the Physics and Astronomy subject area is 33.4%:12.3% = 2.72. It's the highest specialisation level in this area among all BRICS countries. To compare, China's RCA for this area is 1.26, India's—1.13, and in South Africa and Brazil it's less than 1 (at 0.82 in each country). A very high RCA in the structure of Russian Scopus-indexed scientific publications was noted for Earth and Planetary Sciences—2.25 in 2011–2015. Again, it's the highest value among all BRICS countries. RCA ranging between 1.5 and 2.0 were noted in such subject areas as Material Sciences (1.77), Chemistry (1.69), and Mathematics (1.54). At the same time several subject areas have very low RCA values in the structure of Russian publications (under 0.20), specifically Psychology (0.20), Nursing (0.19), Veterinary (0.06), and Dentistry (0.02).

In the structure of Scopus-indexed publications by Chinese scientists, the main specialisation areas include Engineering (RCA of 1.80 in 2011–2015), Material Sciences (1.54), Chemical Technologies (1.48), and Chemistry (1.42). Less important subject areas include Earth and Planetary Sciences (1.27), Physics and Astronomy (1.26), and Computer Sciences (1.25). China has the highest RCA index value in Engineering among all BRICS countries.

India's main specialisation area (in terms of Scopus-indexed publications by the country's researchers) is Pharmacology and Pharmaceutics. India's RCA value in this area in 2011–2015 was 2.93. It is the highest among all BRICS countries; to compare, the relevant figure for Brazil was 1.13, and in other BRICS nations it is below 1. Other areas Indian scientists specialise in include Dentistry (1.90); Chemistry (1.60); Interdisciplinary Studies (1.58); Chemical Technologies (1.51); and Veterinary (1.44).

[6]Index of Technological Specialisation is calculated as the relationship between the share of patent applications in a specific subject area in the total number of national patent applications, and the share of all patent applications in the same areas filed globally in the world's total number of patent applications.

Brazilian publications stand out with extremely high RCA index values in Dentistry (5.50 in 2011–2015) and Veterinary (4.42). These are the highest figures among BRICS countries, and among the highest in the world (for countries with a significant number of publications). Other Brazilian specialisation areas include Agricultural and Biological Sciences (2.57), Nursing (1.64—the highest RCA in this area among BRICS countries), Microbiology and Immunology (1.54), and Health (1.49).

South Africa, unlike other BRICS countries, specialises in social sciences and humanities. Its RCA index values in these areas exceeded 2.00 in 2011–2015: Economics, Econometrics, and Finance (2.68), Humanities (2.31), and Social Sciences (2.20). These are the highest RCA values among all BRICS countries: their relevant figures in the above areas remain under 1. South Africa also has relatively high RCA in the following areas: Veterinary (1.74), Microbiology and Immunology (1.65), Management (1.49), and Environmental Sciences (1.46).

Brazil and South Africa stand out with high RCA values in Immunology and Microbiology, compared with other BRICS countries.

All BRICS countries have Technological Specialisation indices in excess of 1 in two subject areas: Pharmaceutics and Chemical Engineering. In four other subject areas four BRICS countries have relevant index values higher than 2, and three countries—in eight more areas.

8 Analysis of BRICS Countries' National, Bilateral, and Multilateral Strategic and Forecasting Documents

More than 90 national, and 20 bilateral and multilateral strategic and forecasting documents adopted by BRICS countries were analysed, including the following:

Agreements on BRICS countries' cooperation:

- Memorandum of Understanding on Cooperation in Science, Technology and Innovation between the Governments of The Federative Republic of Brazil, The Russia Federation, The republic of India. The People's Republic of China and The Republic of South Africa/Brasilia. 18 March 2015;
- First BRICS Science, Technology and Innovation Ministerial Meeting (2014) Cape Town Declaration. 10 February 2014;
- Moscow Declaration of BRICS countries' Science, Technology, and Innovation Ministers of 26 October, 2015;
- BRICS Science, Technology and Innovation Work Plan for 2015–2018, etc.

Brazilian strategic documents:

- National Strategy for ST&I 2016–2019;
- Growth Acceleration Program;
- The Greater Brazil Plan;
- National Program for Space Activities 2012–2021;

- Science Without Borders;
- Nuclear Program 2016–2019;
- Ten-Year Energy Expansion Plan 2024;
- Antarctica Science for Brazil Action Plan 2013–2022;
- National Plan on Climate Change;
- National Health Plan, etc.

Russian strategic documents:

- Russian S&T Development Strategy;
- Priority S&T Development Areas for the Russian Federation;
- National Technology Initiative;
- Russian S&T Foresight 2030;
- Priority S&T Development Areas of the Russian Science Foundation;
- RF National Programme "Development of Science and Technology foe 2013–2020", etc.

Indian strategic documents:

- Science, Technology and Innovation Policy 2013;
- Twelfth Five Year Plan;
- Vision 2030;
- National Action Plan on Climate Change;
- Atal Innovation Mission;
- National Water Mission;
- National Mission for sustainable Agriculture;
- Made in India;
- National Biotechnology Development Strategy 2015–2020, etc.

Chinese strategic documents:

- National Medium and Long-term Plan for the Development of Science and Technology;
- 13th Five-Year Plan;
- Innovation Driven Development Strategy;
- Strategy 2050;
- 20 Strategic Emerging Industries 2010–2020;
- Energy Development Strategy Action Plan;
- National Key Technologies R&D Program;
- Healthy China 2030;
- Made in China 2025;
- New Silk Road Economic Belt, etc.

South African strategic documents:

- Our future—make it work. National Development Plan 2030;
- Innovation Towards A Knowledge-based Economy. The Ten-Year Innovation Plan for South Africa 2008–2018;
- National Energy Efficiency Strategy of the Republic of South Africa;

- National Water Resource Strategy;
- The New Growth Path;
- Strategic Plan for the Department of Agriculture, Forestry and Fisheries;
- South African Research Infrastructure Roadmap;
- Information and Communication Technology. Research & Development and Innovation Strategy;
- Strategic Plan 2016–2021, etc.

The relevant documents (see References) were analysed in terms of thematic or functional priorities they reflect. Major S&T fields and areas specified there were identified. E.g. the first thematic priorities for international cooperation of BRICS countries were set in documents adopted following the first and second meetings of BRICS education and science ministers (BRICS Science, Technology and Innovation Ministerial Meeting, 2014; BRICS Science, Technology and Innovation Ministerial Meeting 2015).

In 2015 in Moscow, BRICS education and science ministers signed the Moscow Declaration on BRICS Countries' S&T Cooperation (BRICS Science, Technology and Innovation Ministerial Meeting 2015) which outlined the main areas for future cooperation and various tools for supporting it, including setting up work groups on major research infrastructures; funding multilateral research projects; technology commercialisation; and innovation. The document paid a lot of attention to setting up a joint research and innovation platform to coordinate BRICS countries' national research communities' approaches in each of the five agreed (and assigned to specific countries) S&T cooperation areas:

- Prevention and management of natural disasters (coordinated by Brazil);
- Water resources, and prevention of water pollution (coordinated by Russia);
- Geospatial technologies and their application (coordinated by India);
- New and renewable energy; energy efficiency (coordinated by China);
- Astronomy (coordinated by South Africa).

Other international and national documents were analysed in a similar way. The analysis allowed identifying prospective areas for future cooperation. The results were summarised and presented in a table, reflecting priority cooperation areas and BRICS countries' national S&T priorities (see Table 7).

The table provided a basis for drafting lists of S&T areas whose development would make most significant contributions to accomplishing socio-economic, S&T, and innovation development objectives shared by all BRICS countries.

Drafting the consolidated list of national and international S&T development priorities for the BRICS countries, it was assumed that it should meet the following requirements:

- Include all major S&T development areas pursued by several BRICS countries;
- Selected areas (fields) must to the maximum possible extent match S&T and innovation development priorities reflected in the national strategic documents;
- The wording of the list should be more or less uniform, and consistent;
- The identified priorities should match BRICS countries' publication and patenting trends;

Table 7 BRICS countries' national S&T and innovation development priorities, and priority cooperation areas

Area	Brazil	Russia	India	China	South Africa	International documents
Information and telecommunication systems	Economics, and digital society Information and communication technologies Cybersecurity	Information and communication technology Big Data systems, machine learning, artificial intelligence Quantum communications Control and management systems	Information and communication technologies Telecommunication technologies	Information technologies Cyberspace, including cybersecurity Advanced electronics Telecommunications	Information and communication technology Digital economy	Information and communication technology High-performance computing Photonics
Life sciences	Health Pharmaceutics Biomes and bioeconomics Biotechnology	Personalised medicine, high-tech healthcare, health-improving technologies Medicine and health Genomics and synthetic biology Neurotechnology Biotechnology	Health Pharmaceutics Medical equipment Biotechnology	Health, healthcare Medicine Neuroscience Pharmaceutics, biopharmaceutics Biotechnology	Health Biotechnology Pharmaceutics Bioeconomy	Medicine and biotechnology Biomedicine and life sciences (biomedical engineering, bioinformatics, biomaterials) Biotechnology and biomedicine, including health care and neuroscience
Agriculture	Food supply Agriculture Biodiversity Biotechnology	Highly productive green agriculture and aquaculture; efficient chemical and biological crops and farm animals protection systems; efficient storage and processing of agricultural products; production of safe, high-quality foods Biotechnology Personal food and water production and delivery systems	Sustainable agriculture Animal farming Biotechnology	Agriculture Agrifood products Food industry Biotechnology	Agriculture Fisheries Food supply Biodiversity Biotechnology	Food security, and sustainable agriculture Biotechnology

(continued)

Table 7 (continued)

Area	Brazil	Russia	India	China	South Africa	International documents
New materials, nanotechnology	Nanotechnology	New materials and design techniques New materials and nanotechnology	Materials	New materials Nanotechnology	Nanosystems and materials Nanotechnology	Nanotechnology Materials science
Efficient environment management	Environment protection Climate change Water resources Ocean, and coastal areas Hydrocarbon production Green economy Preserving biodiversity	Efficient environment management More efficient production and deep processing of hydrocarbons Reducing risks, and managing consequences of natural and anthropogenic disasters Countering anthropogenic and biogenic threats	Climate change Predicting climate change impact Environment protection Water resources Marine studies Geosciences, seismology Himalayan ecosystem Green technologies Non-fuel mineral resources Waste management	Water and mineral resources Ecology Environment Deep prospecting and drilling Deep-water exploration Water resources Mineral production Oceanography Marine technologies	Climate change Production of mineral resources Green economy Water resources Environment Waste recycling	Water resources, managing water pollution; marine and polar areas studies Geospatial technologies and their application Prevention and management of natural disasters Marine and polar studies, and relevant technologies Geospatial technology and its application
Energy	Energy Nuclear energy Renewable energy sources Biofuel	Environmentally safe, resource-saving energy generation; new energy sources; new energy transmission and storage technologies Energy efficiency and energy saving Nuclear and thermonuclear energy New energy sources	Energy Energy efficiency Nuclear energy Solar energy Renewable energy sources	Energy Hydroenergy Energy saving Next-generation nuclear energy, renewable and non-renewable energy sources	Energy	New and renewable energy sources Energy efficiency Clean coal technologies Natural gas and unconventional gas sources

(continued)

Table 7 (continued)

Area	Brazil	Russia	India	China	South Africa	International documents
Transport and space systems	Aerospace technologies Space Transport, including high-speed systems	Smart transport and telecommunication systems; transportation and logistics systems; development of airspace and outer space, oceans, Arctic and Antarctic areas Transport and space systems Distributed unmanned aerial vehicles systems Unmanned transportation systems	Space exploration and technologies Urban transport	Space exploration Aerospace equipment Space technologies Navigation Transport High-speed railways Automobile industry Aircraft engine production High-tech vessels Railway equipment	Aerospace technologies Astronomy	Space exploration and development, aviation sciences Astronomy Earth observation Geospatial technology and its application
Production	Industry	Advanced digital and smart production technologies Additive technologies Smart technologies for robotic and mechatronic systems Instruments and devices based on nano- and microsystem-technologies Sensory systems Bionics	Industrial production Manufacturing	Advanced production technologies Smart production systems Robotics Additive production systems	Advanced production technologies	
Security	Cybersecurity	Countering anthropogenic, biogenic, socio-cultural threats, and cyberthreats		Cyberspace, including cybersecurity	Information security	

Results of bibliometric and patent analysis of BRICS countries' S&T potential were also taken into account in the course of the priority setting exercise, which allowed to identify more promising fields and areas for future cooperation.

9 Common Priorities as the Basis for BRICS Countries' Future S&T Cooperation

Analysis of BRICS countries' national strategic documents, and assessment of their S&T potentials suggest that 14 major S&T development areas can be included in the list of common S&T cooperation priorities of interest to several BRICS countries:

- Information and communication technology
- Nanotechnology and new materials
- Advanced manufacturing and robotics
- Space systems and astronomical observations
- Transport systems
- Energy efficiency and energy saving
- Nuclear energy
- Renewable energy sources
- Search, exploration, development and mining of minerals
- Climate change, environmental protection and disaster management
- Water resources
- Food security and sustainable agriculture
- Healthcare and medicine
- Biotechnology.

The above subject areas are considered priority ones by all (or almost all) BRICS countries, which is confirmed by their national strategic documents (development strategies, strategic plans, five-year plans, initiatives, mission statements, etc.). These areas are also included in most of the bilateral agreements signed by the BRICS countries.

These areas have a wide scope for practical application, and open opportunities for making use of national comparative advantages (such as territory, available resources, S&T potential, etc.). In the framework of the overall priority systems we can also consider the issue of wide complementarity, which would help to tackle existing S&T problems and limitations through increased cooperation and exchanges between participating countries, and sharing their best practices.

Revealed comparative advantages index(RCA) were calculated for the above subject areas on the basis of the Scopus database[7], and citation impact indices (Table 8).

All serial sources (i.e. journals, book series and conference series) in Scopus are classified on 313 specific subject categories ("SC") that are integrated into 27 major

[7]Calculations were made on the basis of the first- and second-level Scopus classifications.

Table 8 Number of publications, relative comparative advantages index values, and field-weighted citation impact values for priority areas in Scopus for 2011–2015

Priority areas	BRA	RUS	IND	CHI	SAR	BRA	RUS	IND	CHI	SAR	BRA	RUS	IND	CHI	SAR
	Number of publications in Scopus					Relative comparative advantages index					Field-weighted citation impact				
1. Information and communication technologies	26,091	17,366	83,235	331,226	5399	0.72	0.56	1.24	1.25	0.56	0.82	0.93	0.75	0.77	0.81
2. Nanotechnology and new materials	18,163	44,339	65,799	335,146	4711	0.61	1.77	1.20	1.55	0.60	0.84	0.65	0.97	1.05	0.91
3. Advanced manufacturing and robotics	24,937	33,356	76,608	417,316	5520	0.69	1.00	1.05	1.69	0.55	0.89	0.68	0.97	0.86	1.27
4. Space systems and astronomical observations	3075	7334	5771	38,976	1795	0.64	1.91	0.66	1.29	1.35	0.98	0.72	0.99	0.71	1.35
5. Transport systems (including aerospace)	1022	395	3004	18,763	214	0.49	0.25	0.76	1.43	0.44	0.86	1.03	0.87	0.77	0.92
6. Energy efficiency and energy saving	4243	7721	9155	66,191	1254	0.74	2.11	0.95	1.66	1.08	1.22	0.34	1.05	1.02	1.30
7. Nuclear energy	858	2678	2604	15,375	230	0.53	1.95	0.87	1.30	0.54	0.99	0.65	1.15	0.99	1.42
8. Renewable energy resources	2007	836	7405	26,125	894	0.70	0.34	1.39	1.24	1.17	1.22	1.10	0.97	1.86	1.19
9. Search, exploration, development and mining of minerals	4189	11,137	7496	70,183	2411	1.03	2.33	0.61	1.60	1.45	0.70	0.69	0.90	0.92	1.26
10. Climate change, environmental protection and disaster management	18,462	10,524	33,668	120,599	6191	1.16	0.73	0.73	0.88	1.73	1.09	0.87	0.87	1.19	1.24
11. Water resources	7222	5708	9856	36,244	2432	1.27	1.34	0.94	0.87	1.48	0.75	0.59	0.71	0.92	1.30
12. Food security and sustainable agriculture	31,091	3337	31,104	55,194	4970	3.34	0.32	1.65	0.73	1.78	0.73	0.80	0.62	0.99	1.07
13. Healthcare and medicine	87,882	22,636	108,790	321,297	20,375	1.27	0.49	0.54	0.41	0.92	0.86	0.66	0.82	0.87	1.45
14. Biotechnology	18,634	12,313	42,775	138,343	3930	0.88	0.83	1.39	0.98	0.80	0.87	0.71	0.76	1.04	1.10

Source HSE calculations based on Scopus (types of publications included: articles, reviews and conference papers; last update: September 2016

subject areas ("SA") according to Scopus® Classification and All Science Journal Classification Codes (ASJC)[8]. Scheme of merging of the studied 14 priority areas and Scopus subject areas and Subject categories is show below.

Scopus subject areas and subject categories used for the bibliometric analysis for priority areas:

Priority areas	Scopus subject areas (SAs) and subject categories (SCs)
1. Information and communication technologies	All subject categories for Subject area "Computer Science"
2. Nanotechnology and new materials	All subject categories for Subject area "Material Science"
3. Advanced manufacturing and robotics	"Control and Systems Engineering"; "Electrical and Electronic Engineering"; "Industrial and Manufacturing Engineering"; "Mechanical Engineering"; "Mechanics of Materials" SCs
4. Space systems and astronomical observations	"Space and Planetary Science"; "Aerospace Engineering" SCs
5. Transport systems	"Automotive Engineering"; "Transportation" SCs
6. Energy efficiency and energy saving	"Energy Engineering and Power Technology"; "Fuel Technology" SCs
7. Nuclear energy	"Nuclear Energy and Engineering" SC
8. Renewable energy resources	"Renewable Energy, Sustainability and the Environment" SC
9. Search, exploration, development and mining of minerals	"Economic Geology"; "Geochemistry and Petrology"; "Geology" "Geophysics"; "Geotechnical Engineering and Engineering Geology" SCs
10. Climate change, environmental protection and disaster management	"Ecological Modelling"; "Ecology"; "Environmental Engineering"; "Global and Planetary Change"; "Management, Monitoring, Policy and Law"; "Nature and Landscape Conservation"; "Pollution" "Atmospheric Science"; "Earth-Surface Processes" SCs
11. Water resources	"Aquatic Science"; "Oceanography"; "Ocean Engineering"; "Water Science and Technology" SCs
12. Food security and sustainable agriculture	"Agronomy and Crop Science"; "Food Science"; "Plant Science"; "Veterinary" SCs
13. Healthcare and medicine	"Medicine"; "Health Professions" SAs
14. Biotechnology	"Biochemistry"; "Biophysics"; "Biotechnology"; "Cell Biology"; "Molecular Biology"; "Molecular Medicine"; "Structural Biology"; "Applied Microbiology and Biotechnology" SCs

[8]See more in https://service.elsevier.com/app/answers/detail/a_id/15181/kw/subject%20categories%20and%20subject%20areas/supporthub/scopus/related/1/.

In one of the above subject areas (Search, Exploration, Development and Mining of Minerals) four BRICS countries have RCA in excess of 1; in seven other areas three countries have RCA values higher than 1; and only in four subject areas just two or one countries have RCA above 1. Citation impact figures in the selected subject areas in most cases are below the global averages.

Only in two areas (Energy Efficiency and Energy Saving, and Renewable Energy Sources) four BRICS countries have citation impact figures higher than the world average values; in two other areas three or two countries have relevant values higher than 1; in the remaining areas either a single country has citation impact in excess of 1, or all of them are below the global averages.

Though BRICS countries show a significant publication activity in most of the selected subject areas (which is evidenced by their RCA index values), these areas do not command particularly high interest of the global research community. For individual BRICS countries the number of areas in which their RCA exceeds 1 varies from 5 (Brazil) to 9 (China), which on the whole indicates that the proposed list has a balanced nature.

BRICS countries' RCA and citation impact profiles are shown in Figs. 10 and 11. As we can see in Fig. 10, specialisation profiles of Brazil and Russia particularly stand out. Other countries' profiles are more uniform, though different from each other (e.g. China's specialisation profile gravitates toward engineering areas, while South Africa's—towards water resources, environment protection, and

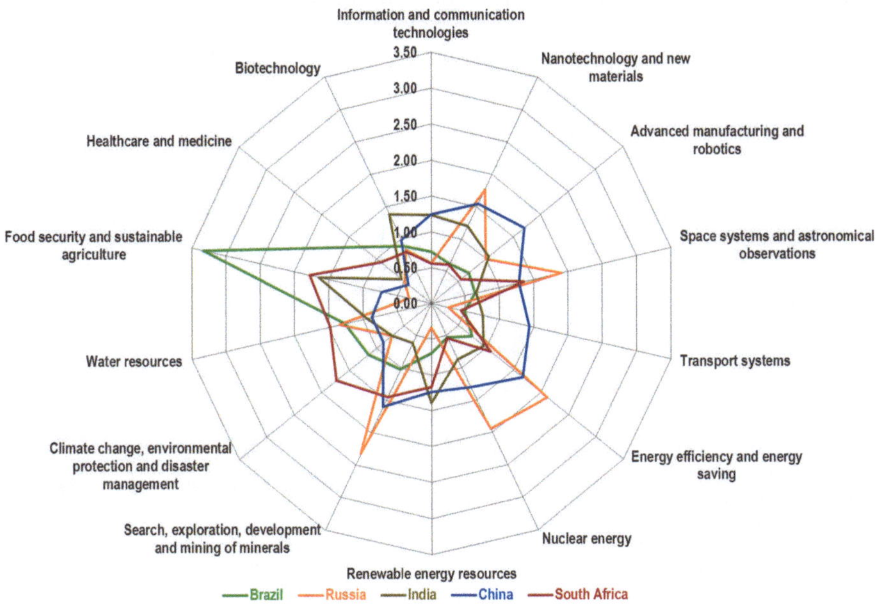

Fig. 10 Relative comparative advantages index values for priority areas in BRICS countries (Scopus-indexed publications, 2011–2015). *Source* HSE calculations based on Scopus (types of publications included: articles, reviews and conference papers; last update: September 2016)

Priority areas	Brazil	Russia	India	China	South Africa
Information and communication technologies	0.82	0.93	0.75	0.77	0.81
Nanotechnology and new materials	0.84	0.65	0.97	1.05	0.91
Advanced manufacturing and robotics	0.89	0.68	0.97	0.86	1.27
Space systems and astronomical observations	0.98	0.72	0.99	0.71	1.35
Transport systems	0.86	1.03	0.87	0.77	0.92
Energy efficiency and energy saving	1.22	0.34	1.05	1.02	1.30
Nuclear energy	0.99	0.65	1.15	0.99	1.42
Renewable energy resources	1.22	1.10	0.97	1.86	1.19
Search, exploration, development and mining of minerals	0.70	0.69	0.90	0.92	1.26
Climate change, environmental protection and disaster	1.09	0.87	0.87	1.19	1.24
Water resources	0.75	0.59	0.71	0.92	1.30
Food security and sustainable agriculture	0.73	0.80	0.62	0.99	1.07
Healthcare and medicine	0.86	0.66	0.82	0.87	1.45
Biotechnology	0.87	0.71	0.76	1.04	1.10

Fig. 11 Field-weighted citation impact values for priority areas in BRICS countries (Scopus-indexed publications, 2011–2015). *Source* HSE calculations based on Scopus (types of publications included: articles, reviews and conference papers; last update: September 2016)

agriculture). In terms of citation impact the picture is more consistent than for RCA. Note that SAR and China display the highest citation figures.

Analysis of RCA index and citation impact figures allowed to make preliminary estimates of the scope for BRICS countries' cooperation on implementing their S&T priorities. On the basis of RCA and citation impact values (taken in combination), subject areas mutually interesting to BRICS countries can be identified. Certain priorities can be important to several partners at the same time, i.e. constitute priorities for multilateral cooperation.

Relevant values for Russia are presented in Fig. 12.

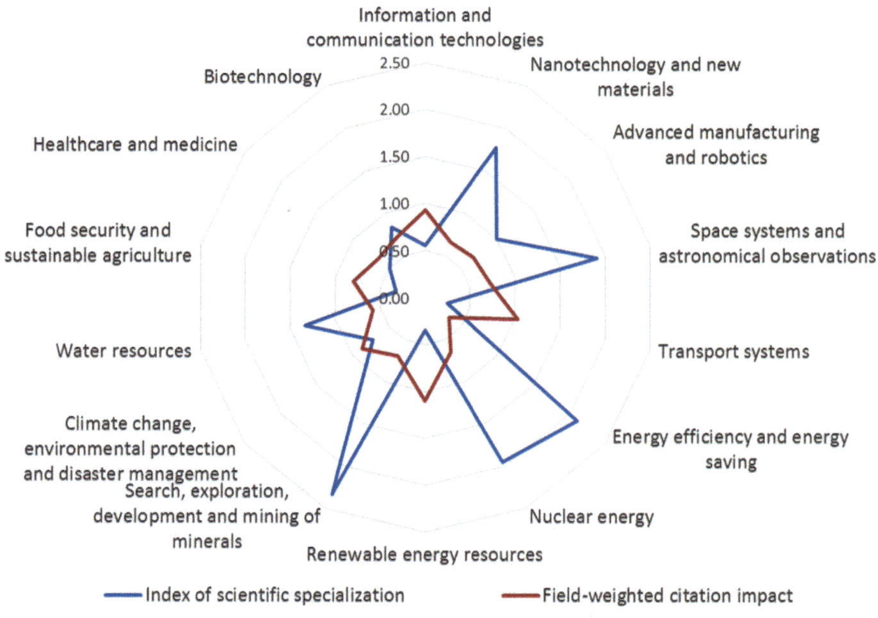

Fig. 12 Potential for BRICS countries' cooperation on implementing S&T priorities

E.g. Russia, China, and India may be interested in parity cooperation in the "Nanotechnology and Next-Generation Materials" area. Russia actively conducts research in energy efficiency and energy saving, but demand for its results remains less significant compared with publications by four other BRICS countries. Russia may be able to considerably improve the quality of, and the demand for its publications on the relevant subjects by stepping up its cooperation with BRICS countries.

China demonstrates a high level of research, and impressive results in a wide range of priority areas. Cooperating with China as principal partner would help to improve productivity of Russian R&D. Cooperation with India, Brazil, and South Africa could also produce positive effects.

Priorities can also be structured on the basis of potentially interested participants and technology readiness level: e.g. cooperation between R&D organisations and universities to develop technologies, which requires public support; public-private partnerships at pre-competitive stages; participation of businesses, including small innovation companies, in developing prototypes and applying advance technologies, etc.

As to more general S&T development priorities for BRICS countries, formulating them in a more formal, standardised way seems to be in order, e.g. in the following format: main goals and objectives, and a brief summary of the subject area; its composition; practical application areas; relevant leading R&D centres in BRICS countries; BRICS countries' potential interest in cooperation; and organisations—potential participants of S&T cooperation.

Shared S&T development priorities create a basis for mutually beneficial cooperation, in the framework of which scientists from different countries would be able to extend the scope of their research, step up collaboration, share experience, and ultimately strengthen Russia's S&T cooperation with other countries.

The list of priorities for BRICS countries' S&T cooperation can be useful for drafting inter-agency agreements with BRICS countries on conducting R&D, preparing work plans (roadmaps) for stepping up S&T cooperation, and applying other relevant tools and mechanisms.

The results of our comparative analysis largely suggest that BRICS countries have reached advanced positions in the S&T and innovation spheres, and still have a potential for improving them further by applying available resources. This raises an issue of identifying practical steps, which on the one hand would allow to make full use of national comparative advantages, while on the other avoid lagging further behind, especially in such respects as researchers' publication activity and creative cooperation with international colleagues.

Available tools for supporting and promoting S&T cooperation of BRICS countries include the following:

- Joint bilateral and multilateral S&T programmes and projects (co-funding mechanism), including:
  - BRICS STI Framework Programme
  - Russian R&D Federal Targeted Program Action 2.1 "Support for research in the framework of international multilateral and bilateral cooperation"
  - Scientific research conducted by international research teams (RSF)
  - National High-tech R&D Programme of China
- Researchers mobility, including:
  - BRICS Young Scientist Forum
  - Leading/Young Researchers International Fellowships
  - Research infrastructures
    - Development of BRICS Global Research Advanced Infrastructure Network (BRICS GRAIN).

Application of the following tools would also contribute to increasing efficiency of cooperation:

- International scientific seminars (establishing new contacts)
- Network for information collection, analysis and exchange
- Research starting grants [for scientific teams headed by early-career researchers (2–7 years of experience after completion of Ph.D.)]
- Industry-academia partnerships
- International Innovation Programme (joint industrial R&D with high potential for commercialization)
- Innovation and technology entrepreneurship networks
- Exchange Programme for Innovative Talents and Entrepreneurs
- International Advanced Research Centres (including virtual ones) in priority research areas
- Joint S&T Foresight programmes
- Strategic collaborative research (addressing global challenges)
- Access to research infrastructures—travel & training grants.

10 Conclusions

As Russia‘s and other BRICS countries' experience shows, S&T priorities are usually set in the context of designing long-term sustainable development strategies, to support accomplishing key national and global socio-economic objectives.

The results of our analysis allowed to identify a number of prospective S&T areas where BRICS countries may be interested in stepping up cooperation with Russia, to more efficiently implement their national priorities. Similarity of S&T

and innovation development priorities in BRICS countries and in Russia is a major factor promoting establishment of sustainable long-term partnership between them.

Furthermore, recent cooperation practice shows that such partnerships tend to strengthen specifically in the scope of projects implemented in priority subject areas, with a potential to produce significant economic and social effects. BRICS countries' cooperation would be more efficient and productive if it covers all stages of innovation cycle—from creating new basic knowledge to its practical application to develop and market new technologies, products, and services. This implies that such stages may be "distributed" between BRICS countries, in line not only with their respective S&T priorities but also their production potential.

Subsequently, an information database may be created on the basis of the obtained results to support various participants of national innovation systems, so they'd be able to quickly identify suitable subject areas for S&T cooperation with BRICS countries, find partners (including R&D organisations, universities, industrial enterprises specialising in various sectors of the economy, etc.), and identify the best formats and mechanism for cooperation.

Despite significant differences between the BRICS countries in terms of their S&T development level, interaction and cooperation forms which would ensure their equal rights at all cooperation stages and in all relevant processes must be actively supported and implemented; this would also help to increase individual countries' potentials. Acting as a single group on the international arena, BRICS countries could become a major global S&T and innovation development centre.

References

BILAT-USA. (2010). Analysis of S&T Priorities in Public Research in Europe and the USA.

Brazilian Development Bank. (2007). Growth acceleration program.

BRICS. (2016). BRICS STI Framework Programme Coordinated call for BRICS multilateral projects—Pilot call 2016.

BRICS. (2017). BRICS STI Framework Programme Coordinated call for BRICS multilateral projects.

BRICS Science, Technology and Innovation Ministerial Meeting. (2015, March 18). Memorandum of understanding on cooperation in science, technology and innovation between the Governments of The Federative Republic of Brazil, The Russia Federation, The republic of India. The People's Republic of China and The Republic of South Africa/Brasilia.

BRICS Science, Technology and Innovation Work Plan 2015–2018 (2015).

Cagnin, C. (2014). STI foresight in Brazil. *Foresight-Russia, 8*(2), 46–55.

Chan, L., & Daim, T. (2012). Exploring the impact of technology foresight studies on innovation: Case of BRIC countries. *Futures, 44*(6), 618–630.

Department of Agriculture, Forestry and Fisheries. (2015). Strategic plan for the Department of Agriculture, Forestry and Fisheries 2015/16 to 2019/20.

Department of Industrial Research and Promotion. (2015). Make in India, India.

Department of Minerals and Energy. (2009). National energy efficiency strategy of the Republic of South Africa.

Department of Science and Technology. (2007). Information and communication technology. Research & development and innovation strategy.

Department of Science and Technology. (2008). The Ten-Year innovation plan for South Africa 2008–2018, South Africa.
Department of Science and Technology. (2015). National biotechnology development strategy 2015–2020.
Department of Science and Technology. (2016). South African research infrastructure roadmap: First Edition.
Department of Water Affairs. (2013). National Water Resource Strategy.
European Forum on Forward Looking Activities. (2015). How to design a European foresight process that contributes to a European challenge driven R&I strategy process, European Forum on Forward Looking Activities, Policy Brief No. 2.
First BRICS Science, Technology and Innovation Ministerial Meeting. (2014, February 10). Cape Town declaration.
Gassler, H., Polt, W., Schindler, J., Weber, M., Mahroum, S., Kubeczko, K., et al. (2004). Priorities in science & technology policy—an international comparison. Project report commissioned by the Austrian Council for Research and Technology Development.
Gokhberg, L., Meissner, D., & Sokolov, A. (2016). *Deploying foresight for policy and strategy makers: Creating opportunities through public policies and corporate strategies in science, technology and innovation*. Netherlands: Springer International Publishing Switzerland.
Government of Brazil. (2008). National Plan on Climate Change.
Government of India. (2012). Twelfth Five Year Plan (2012–17).
Government of India. (2013). Science, Technology and Innovation Policy. New Delhi.
Government of India, Ministry of Water Resources, River Development and Ganga Rejuvenation. (2010). National water mission.
Government of India, National Institution for Transforming India. (2016). Atal Innovation Mission, India.
Grebenyuk, A., Pikalova, A., Sokolov, A., Shashnov, S., & Kaivo-oja, J. (2016). Priority setting in the EU countries and the Russian Federation: The best practices, M.: National Research University Higher School of Economics.
III BRICS Science, Technology and Innovation Ministerial Meeting. (2015). Moscow declaration signed at the 3rd Meeting of BRICS Ministers for Science, Technology and Innovation held at Moscow on 28 October, 2015.
Kuwahara, T., Cuhls, K., & Georghiou, L. (2008). *Foresight in Japan. The Handbook of Technology Foresight Concepts and Practice* (pp. 170–183).
Li, L. (2009). *Research priorities and priority-setting in China*. Vinnova: Vinnova Analysis.
MCTI. (2016). "Estratégia Nacional de Ciência, Tecnologia e Inovação 2016–2019", Brasilia.
Meissner, D., Gokhberg, L., & Sokolov, A. (Eds.). (2013). *Science, technology and innovation policy for the future. Potentials and limits of foresight studies*. New York, Dordrecht, London, Heidelberg: Springer.
Ministry of Agriculture and Farmers Welfare. (2010). "National Mission for Sustainable Agriculture", India.
Ministry of Earth Sciences. (2016). Vision for 2030, India.
Ministry of Education and Science. (2014). Federal Targeted Programme "R&D in Priority Development Areas for the Russian S&T Complex in 2014–2020". Available at: http://rosmu.ru/activity/attach/events/1079/fcp_ir_2014-2020.pdf. Accessed 25 November 2015 (in Russian).
Ministry of Environment, Forest and Climate Change. (2008). National Action Plan on Climate Change.
MINISTÉRIO DA SAÚDE. (2011). Plano Nacional de Saúde—2012–2015.
National Planning Commission. Republic of South Africa. (2011). Our Future-make it work. National Development Plan 2030.
OECD. (2010). *Priority setting for public research: Challenges and opportunities*. Paris: OECD Publishing.
OECD. (2012). *Meeting global challenges through better governance*. Paris: International Co-operation in Science, Technology and Innovation, OECD Publishing.

OECD. (2014). Brazil. OECD Science, Technology and Industry Outlook 2014.
Pouris, A., & Raphasha, P. (2015). Priorities setting with foresight in South Africa. *Foresight and STI Governance, 9*(3), 66–79.
Poznyak, A., & Shashnov, S. (2011). S&T priorities for modernization of the Russian Economy. *Foresight-Russia, 5*(2), 48–56. (in Russian).
President of the Russian Federation. (2011). Order of the President of the Russian Federation No. 899, dated 7 July 2011, On approval of the policy priorities for the development of science, technology and engineering in the Russian Federation and the list of critical technologies of the Russian Federation.
President of the Russian Federation. (2016). Scientific and technological development strategy of the Russian Federation (order of the President of the Russian Federation No. 642, dated 1 December 2016).
Science and Technology Indicators. (2017). Data book. In Y. Voynilov, N. Gorodnikova, & L. Gokhberg, et al. (Eds.), *National Research University Higher School of Economics*. Moscow: HSE.
Sokolov, A., & Chulok, A. (2016, June). Priorities for future innovation: Russian S&T foresight 2030. *Futures* 80, 17–32.
South Africa Government. (2010). The new growth path.
The Russian Government. (2014a). Russia 2030: Science and technology foresight (resolution of Prime Minister of the Russian Federation No. DM-P8—5 of January 3, 2014).
The Russian Government. (2014b). The State Programme of the Russian Federation «Development of Science and Technology» 2013–2020 (resolution of the Russian Government No. 301 of 15 April, 2014).
中华人民共和国科学技术部. (2006). 实施《国家中长期科学和技术发展规划纲要 (2006–2020年)》 的若干配套政策, 中国.
国务院. (2016). 国务院关于印发"十三五"国家科技创新规划的通知, 中国.
中华人民共和国科学技术部. (2016). 《国家创新驱动发展战略纲要》 "三步走"到2050年建成世界科技创新强国, 中国.
中国科学院. (2009).《创新2050:科学技术与中国的未来》中国科学院战略研究系列报告发布, 中国.
国务院. (2010). 《国务院关于加快培育和发展战略性新兴产业的决定》, 中国.
国家发展和改革委员会. (2015). 《丝绸之路经济带》, 中国, 哈萨克斯坦.
国务院办公厅. (2014). 《国务院办公厅关于印发能源发展战略行动计划 (2014–2020年) 的通知》, 中国.
国务院,中共中央. (2016). 《"健康中国2030"规划纲要》, 中国.

Part II
Country Reports

Chapter 3
Brazil Report on Science, Technology and Innovation

Lei Wang, Changlin Gao, Quanchao Dong and Tao Shi

China and Brazil are both major countries in the developing world, and the two countries, though far away from each other, have maintained strong partnership. China and Brazil established diplomatic ties in 1974, built strategic partnership in 1993 and elevated it to comprehensive strategic partnership in 2012. The two nations have maintained close communication and coordination on major international and regional affairs, and worked side by side in international organizations and multilateral mechanisms like the United Nations, the World Trade Organization, G20 and BRICS. Both countries are committed to advancing the international system towards a more equal and reasonable direction, defending developing countries' interest and playing an increasingly important role to promote world peace and development. The China-Brazil comprehensive partnership has developed rapidly in recent years. Political mutual trust has deepened and pragmatic bilateral cooperation has produced fruitful results. China has been Brazil's largest trading partner for seven consecutive years, while Brazil is now China's tenth largest trading partner. Bilateral cooperation has been intensified in areas of science & technology, education, culture, tourism and sports. The cooperation has been productive and become an increasingly important part of the bilateral ties.

In the report we gave an account of the basic situation of science andtechnology innovation in Brazil, the status quo of Brazil and China's bilateral cooperation, and finally offered suggestions on advancing bilateral cooperation on science andtechnology innovation.

L. Wang (✉) · C. Gao · Q. Dong · T. Shi
China Science and Technology Exchange Center, Beijing, China

X. Zhao et al. (eds.), *BRICS Innovative Competitiveness Report 2017*,
Research Series on the Chinese Dream and China's Development Path,
https://doi.org/10.1007/978-981-10-8078-4_3

1 Basic Situation of Science and Technology Innovation in Brazil

1.1 Brazil National Innovative Competitiveness: Appraisal and Analysis

For the 20 years from 2001 to 2020, the analysis of changes in scores of Brazilian national innovations competitiveness and the rankings of Brazil among the BRICS are as follows:

Among the BRICS, the rankings of Brazil in terms of national innovative competitiveness with changes in scores (see Fig. 1).

The forecast of Brazilian innovation indexes, see Table 1.

(1) From the changes in general rankings, Brazil ranked the 3rd among the BRICS for national innovative competitiveness. Compared with 2001, the ranking went one place up. In 2007, the ranking rose to the 3rd place and only dropped to the 4th in 2013. Generally speaking, the trend is upward with some fluctuation within the period of appraisal.

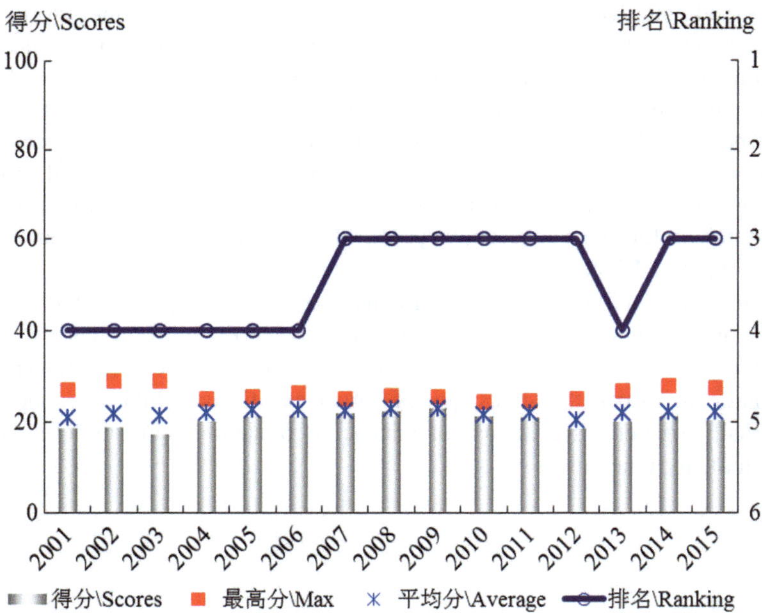

Fig. 1 The changes in scores and rankings of Brazilian national innovative competitiveness from 2001 to 2015

Table 1 Forecast of Brazilian innovation indexes in next five tears

Year	2016	2017	2018	2019	2020
Innovation index	26.59	27.01	27.40	27.80	28.22

(2) Regarding the scores, Brazil got 20.35 points for national innovative competitiveness in 2015, 6.96 points lower than the highest among the BRICS and 1.74 points lower than the average. Compared to 2001, Brazil scored 1.75 points more, narrowing both the gap with the highest score in 2001 by 1.44 points and the gap with the average among the BRICS by 0.46 point.

(3) According to the forecast, the general innovation index of Brazil is expected to grow from 26.59 to 28.22 in 2016–2020.

1.2 National Strategy on Science and Technology Innovation

Since the beginning of the 21st century, the Brazilian government has rolled out several four-year strategic development plans on science & technology. *The National Strategy on Science & Technology Innovation (2016–2019)* (hereinafter referred to as the "Strategy") was published in May 2016.

The Strategy is a mid-term guidance to deliver public policies on S&T innovation, a comprehensive blueprint to establish, strengthen and expand the framework on science, technology and innovation (STI), and also a guideline to pursue coordinated development of the five core elements of a well-developed innovation system (research, infrastructure, funding, human resources and innovation). With the objectives of breaking innovation barriers and encouraging public-private partnership to pursue innovation, the strategy focuses on enhancing Brazil's S&T strength as well as its ability to foster innovation activities and overcome social challenges. Based on the strategy, Brazil selected 11 fields (see Table 2) for preferential development and rolled out concrete measures to make sure these fields will be supported by relevant policies.

Table 2 Key fields of preferential development in Brazil (2016–2019)

No.	Priority fields of development
1	Aviation & aerospace and national defense
2	Drinking water
3	Food
4	Biocoenosis and biological economy
5	Social science & technology
6	Climate
7	Digital economy & digital society
8	Energy
9	Nuclear energy
10	Healthcare
11	Integration & advanced technologies

Table 3 Primary objectives of S&T development in Brazil (as of 2019)

No.	Indicator	Latest statistics	As of 2019
1	National R&D spending as a percentage of GDP	1.24% (2013)	2.00%
2	Business R&D spending as a percentage of GDP	0.52% (2013)	0.90%
3	Government R&D spending as a percentage of GDP	0.71% (2013)	1.10%
4	Federal government R&D spending as a percentage of GDP	0.50% (2013)	0.80%
5	Corporate innovation rate	35.7% (2011)	48.6%
6	Number of companies with persistent R&D activities	5600 (2011)	10,000
7	Proportion of innovative enterprises using at least one government supported innovation tool	34.2% (2011)	40.0%
8	Number of corporate R&D staff and researchers	103,290 (2011)	120,000
9	Proportion of engineering-related college graduates to total graduates	7.2% (2013)	12.0%
10	Number of researchers per one million residents	709 (2010)	2100

The Strategy aims to expedite fundamental research and R&D, upgrade or add innovation infrastructure, increase innovation technology investment on training, attracting and retaining human resources, and motivate enterprises to pursue technological innovations. The strategy hopes to increase the benefits of innovation activities by improving research conditions, reduce social disparity and regional imbalance through S&T innovation, and promote sustainable development by encouraging the development of popular end-user products.

In order to monitor the implementation of strategic goals, the federal government worked out quantitative indicators to measure the level of S&T development (see Table 3).

1.3 Policy Incentives for Innovation and Entrepreneurship

1.3.1 Improving Laws and Regulations to Encourage Entrepreneurship Among Researchers

For a long period of time, Brazil's S&T resources have not been distributed in a balanced manner, with most researchers and developers working at universities, research institutes and foundations, while the large majority of enterprises do not have their own research team. Scientific research has been disconnected with corporate demand, which has restrained the process to convert scientific results into actual productivity and hindered the improvement of corporate entities' innovation capability.

In order to encourage innovation and entrepreneurship, especially scientists and researchers to quit their jobs and start a new business, the federal government of Brazil enacted the Law of Science and Technology Innovation (Detailed Rules) in October 2015. According to the 14th article of the law, researchers at public institutions can leave their jobs and engage in R&D activities at a S&T organization, where they would hold the same position as they were in the public institution; soldiers and public institution staff should hold the same position as before when they leave their jobs to conduct joint research at other institutions; researchers who leave for a specific period of time would still get paid by their former employer with additional subsidies and social insurance benefits. According to the 15th article of the law, as long as researchers of the public institution can prove they are not working under the probationary period, they can set up their own innovation R&D, but their former employers would not provide any remuneration.

1.3.2 Capitalizing Financial Policies to Support Entrepreneurship Among SMEs and Micro Businesses

Small and micro businesses as well as individual operations have thrived in Brazil in recent years. The rapid increase has been driven not only by Brazilians' strong entrepreneurship, but also by a remarkable improvement in the country's overall business environment. In order to encourage wide-spread entrepreneurship among small businesses, the government has adopted a series of measures such as simplifying administrative approval processes and introducing preferential tax policies.

In 2014 the federal government of Brazil established the "secretariat of small and micro enterprises and economic cooperation", announcing that the annual interest rate on government loans to small and micro enterprises to be reduced from 8 to 5%, in order to give bigger support to small business owners.

According to the National Confederation of Commerce, 4000 small and micro firms are created every day in Brazil. These small businesses, as they increase in number and grow in size, are playing a prominent role in driving the economy, creating employment opportunity, spurring domestic demand and increasing export. They are becoming a pivotal motivating factor of Brazil's economic and social development.

SMEs can also gain support from private institutions. Founded 43 years ago, SEBRAE (Brazilian Center of Assistance to Micro and Small Enterprises) is a privately-held non-profit organization supporting small and micro companies to join in the competition and achieve sustainable business development. SEBRAE provides long-term financial support, trade services, as well as training courses related to entrepreneurship and consulting services to small businesses. Its service offerings primarily cover seven key areas including creative design, productivity development, intellectual property protection, product quality, knowledge innovation, sustainable development and communication technologies. As an intermediary platform, SEBRAE also helps small business owners find business partners and financing institutions and get access to information, capital, technology and market to improve their product competitiveness.

According to a survey conducted by SEBRAE, the percentage of entrepreneurial population in Brazil had hugely increased from 23% in 2004 to 34.5% in 2014. That means among the age group of 18–64, every three in ten people have set up or are in the process of setting up their own businesses. Most of the entrepreneurs started as small business owners. The country is now home to roughly six million micro and small enterprises, or 97% of the total enterprises; those micro and small enterprises employed 52% of the total urban workforce.

1.3.3 Supporting SMEs and Micro Firms' Startup & Innovation Initiatives with Research Funds

The Brazil Ministry of Science, Technology and Innovation (MCTI) launched in November 2012 the "Startup Brazil Initiative" to provide funding support to startups and aimed to transform the country into a powerful innovation-based economy. As of June 2014, the initiative had offered totally US$18 million funds to 150 small and micro firms. In addition to direct investment and infrastructure support, the initiative has introduced one-on-one business mentorship and built a network to facilitate partnership between large and small companies. The Brazilian Trade and Investment Promotion Agency also joined the initiative to help startups take their products to the world.

SMEs and micro firms face lower threshold and more flexible procedures when applying for assistance from Brazil's research funds. For instance, FAPESP announced a R&D assistance program in May 2015, which was open to small firms (with a workforce of less than 250) doing research work in the state of Sao Paulo. Even if a small company has yet to be registered, it can still apply for R&D assistance under the condition that "the company is preparing to be established". Researchers have to provide evidence of them possessing the relevant knowledge and technology, but education diploma (bachelor degree or internship period) is not required. The corporate applicants spare no efforts to commercialize their products during the R&D process. The process is divided into two stages. The first stage lasts nine months where the foundation will provide 200,000 reals in financial subsidy; the second stage is extended to 24 months and the foundation will provide one million reals. If researchers can prove that their technologies are practically feasible (meeting the R&D objectives in the first stage), they will be able to apply to move on to the second stage. Evidently, FAPESP encourages entrepreneurship and innovation, especially valuing the applicant's practical research capability.

According to a report published by the News of Economic Value on February 13, 2016, the Brazilian Development Bank planned to make a total investment of 200 million reals into 36 projects in four years. As the third phase of the CRIATEC fund, the projects' priority areas involve information communication, agricultural trade, nanometer, biotechnology and new materials. The CRIATEC fund was set up in January 2007 to provide financing to innovation-oriented SMEs. The fund provided 68 million reals to 36 firms in the first stage (2008–2014), and 20 million reals in the second stage (2014–2015). Enterprises bankrolled by the fund have obtained 37 domestic patents, 9 foreign patents and generated an average annual return of 30%.

1.4 R&D Spending, Output and S&T Personnel

1.4.1 Statistics on R&D Spending

According to statistics released by the Ministry of Science, Technology and Innovation, Brazil's R&D spending has increased at an annual rate of 16% since 2007 and reached 54.93 billion reals in 2012, but the corresponding proportion of R&D spending to GDP has only edged up from 1.11 to 1.24%. In the 2013–2014 fiscal year the federal government provided 32.9 billion reals as funds to innovation activities carried out by the corporate community. The federal and local governments accounted for 56.2% of the country's total R&D spending, and the enterprises contributed 43.8%. R&D activities jointly conducted by research institutes and enterprises are funded by the government's tax revenue. 77% of enterprises can get the federal and state governments' financial support when they participate in joint R&D initiatives, although such support is much less than that received by their counterparts in the developed world.

We can conclude from Table 4 that from 2008 to 2014, Brazil's R&D spending has grown rapidly. It more than doubled since 2008, hitting 73.4 billion reals in 2014. However the R&D spending as a percentage of GDP only increased by 0.14 percentage during the six years to 1.27% in 2014. Generally speaking, it has grown rather slowly and even shown a downward trend in some years. The country has set an objective to raise the percentage to 2% by 2019, but that goal is almost unachievable in light of the historical trend and the country's current difficulties in increasing fiscal revenue and expanding the stagnant economy.

Brazil's public-sector spending on R&D activities accounts for 52% of the total R&D expenditure, and the federal government's contribution is 2.2 time more than that of state governments. In Brazil, the public sector has always been spending more on R&D than the business sector, and the country has a long way to go as it seeks to encourage more private investments in R&D and improve its overall innovation capability.

1.4.2 Output from R&D Activities

At the 67th Annual Meeting of the Brazilian Society of Scientific Progress in mid-July 2016, Hernan Chaimovich, Chairman of the National Council for Science & Technology Development, made remarks on the output of the country's S&T innovation activities: during 2007–2009, every US$1 million investment in Brazil

Table 4 Brazil's R&D spending as a proportion of GDP during 2008–2014 based on current prices

Year		2008	2009	2010	2011	2012	2013	2014[a]
R&D spending		35,110.8	37,285.3	45,072.9	49,875.9	54,254.6	63,748.6	73,461.2
Proportion of R&D spending		100.00	100.00	100.00	100.00	100.00	100.00	100.00
Proportion of GDP		1.13	1.12	1.16	1.14	1.15	1.24	1.27
Public spending	Total	17,680.7	19,498.1	23,039.2	26,382.6	29,802.9	36,783.7	38,784.0
	Proportion of R&D spending	50.36	52.29	51.12	52.90	54.93	57.70	52.80
	Proportion of GDP	0.57	0.59	0.59	0.60	0.63	0.71	0.67
	Federal government	12,069.1	13,461.9	16,039.5	17,784.2	20,020.7	25,802.5	26,100.1
	Proportion of R&D spending	34.37	36.11	35.59	35.66	36.90	40.48	35.53
	Proportion of GDP	0.39	0.40	0.41	0.41	0.42	0.50	0.45
	State government	5611.7	6036.2	6999.7	8598.4	9782.2	10,981.3	12,684.0
	Proportion of R&D spending	15.98	16.19	15.53	17.24	18.03	17.23	17.27
	Proportion of GDP	0.18	0.18	0.18	0.20	0.21	0.21	0.22
Corporate spending		17,430.1	17,787.2	22,033.6	23,493.2	24,451.7	26,964.9	34,677.2
Proportion of R&D spending		49.64	47.71	48.88	47.10	45.07	42.90	47.20
Proportion of GDP		0.56	0.53	0.57	0.54	0.52	0.52	0.60

Note R&D spending is measured by million reals, proportion is measured by percentage point
[a]*Source* Brazilian Ministry of Science, Technology and Innovation website (http://www.mct.gov.br/index.php/content/view/2068.html), statistics updated on April 5, 2017, and 2014 figures were estimates

could bring 26 new products to the market, publish 23 academic papers, produce 19 innovative processes, provide 10 services, obtain eight patents as well as a number of technology application standards. Based on statistics from the Thomson/ISI and Scopus databases, Table 5 presented a comparison of academic papers published by Brazil and by Latin America and the whole world during 2000–2014. We can see that Brazil is the largest academic paper producer in Latin America, but its influence on the international academic world remained small. Table 6 presented Brazil's patent data during 2007–2014. We can see that Brazil has been a laggard in terms of both patent application and patent grant. Brazil's domestic invention patent grants were only one fifth or one fourth of foreign invention patent grants, which is an indirect indication that the country somewhat lacks the capability of original innovation.

1.4.3 S&T Personnel

The federal government has established an information system on science, technology and innovation, Plataforma Lattes, which is named after Cesare Mansueto Giulio Lattes, a well-known Brazilian physicist in remembrance of his contribution to the discovery of π Mesons. As of November 30, 2016 the platform had registered 3.1 million resumes, including 218,562 from doctorate degree holders (132,631 people hold doctorate degree in research and teaching) and 364,740 from master degree holders (82,818 hold master degree in research and teaching).

The federal government launched the "Science without Border" program in 2011, under which the government will send talented personnel to further their studies abroad. The government planned to spend US$1.9 billion to help 100,000 students (a per capita scholarship of US$19,000) to pursue their studies in science, technology, engineering, mathematics, etc. at foreign universities to train excellent S&T personnel, strengthen Brazil's S&T development and enhance the internationalization of its higher education system. As of January 2017, the "Science without Border" program had provided scholarships to 95,843 students specialized in subjects that are strategically important to Brazil, such as earth science, healthcare, and computer and information technology. The scholarship recipients represent nearly 30 countries. Through the exchange study of international students, the program has stabilized and expanded Brazil's S&T innovation system and enhanced the level of quality of high-end S&T personnel with international vision.

The National Council for Science & Technology Development has introduced and implemented in 1987 the "human resource training program in strategic fields", encouraging master and doctoral degree holders to do research at relevant enterprises. The program has granted 240 million reals of entrepreneurial scholarships to fund the researchers' R&D activities. In 2007, these researchers were involved in 700 R&D projects, and the figure soared to nearly 2000 in 2013. From 2007 to 2013, the council provided entrepreneurial scholarships to 4274 people, of which

Table 5 Academic papers published by scientists from Brazil, Latin America and worldwide

Year	Thomson/ISI					Scopus				
	Brazil	Latin America	Worldwide	Brazil/Latin America (%)	Brazil/Worldwide (%)	Brazil	Latin America	Worldwide	Brazil/Latin America (%)	Brazil/Worldwide (%)
2000	10,521	24,529	777,827	42.89	1.35	14,042	31,966	1185,354	43.9	1.18
2001	11,581	26,478	796,862	43.74	1.45	14,668	33,104	1235,962	44.3	1.19
2002	12,929	28,620	797,668	45.17	1.62	17,071	37,268	1293,820	45.8	1.32
2003	14,288	31,591	875,756	45.23	1.63	19,211	41,671	1369,012	46.1	1.40
2004	14,995	31,655	854,703	47.37	1.75	22,024	46,187	1489,170	47.7	1.48
2005	17,714	37,250	982,533	47.55	1.80	24,920	52,063	1674,645	47.9	1.49
2006	19,294	38,743	983,424	49.8	1.96	32,187	63,125	1763,602	51.0	1.83
2007	19,510	39,367	981,932	49.56	1.99	34,902	67,439	1862,243	51.8	1.87
2008	30,422	55,757	1158,057	54.56	2.63	40,197	76,759	1938,656	52.4	2.07
2009	32,100	58,985	1191,707	54.42	2.69	44,018	83,842	2040,665	52.5	2.16
2010	–	–	–	–	–	47,362	88,934	2152,206	53.3	2.20
2011	–	–	–	–	–	51,342	96,004	2288,154	53.5	2.24
2012	–	–	–	–	–	56,436	104,161	2355,174	54.2	2.40
2013	–	–	–	–	–	59,012	108,347	2438,393	54.5	2.42
2014	–	–	–	–	–	61,418	114,224	2490,293	53.8	2.47

Thomson/ISI & Scopus databases, 2000–2014

Table 6 Brazil's R&D patent data during 2007–2014

Patent category	Unit	2007	2008	2009	2010	2011	2012	2013	2014
Domestic invention patent applications	Piece	24,840	26,641	25,885	28,099	31,881	33,569	34,050	33,182
Domestic invention patent grants	Piece	393	527	690	666	725	654		
Overseas invention patent grants	Piece	1470	2302	2472	2956	3088	2484		
PCT patents	Piece	15,270	16,837	16,171	18,718	21,371	22,688	22,572	

70% were working at small and micro enterprises, 14% at medium-sized enterprises and 6% at large enterprises. Some experts said that Brazil has already established a foundation of its human resource base made up of talented technology innovation and R&D personnel.

1.5 Operating Model of Technology Transfer Institutions

In order to promote integration of science and technology with socioeconomic development and further build the national technology transfer system, the Brazilian government has released or amended relevant laws over the past 10 years, initiating diversified financial services, building and perfecting technology transfer platforms, and encouraging independent construction of R&D and technology transfer centers to perfect the nation's industry-university-research cooperative mechanisms as well as science and technology innovation system.

1.5.1 Improving Laws and Regulations to Remove Institutional Barriers of Technology Transfer

To promote effective integration of policies concerning innovation, industry and foreign trade, Brazil released in 2003 an industrial policy named the "Guiding Principle for Industry, Technology and Foreign Trade Policies (PITCE)". PITCE proposed that corporate technology innovation should be motivated through promulgation of favorable policies regarding taxes, subsidies and loans, etc. so is to achieve integration of technology with industry. Also, the government appropriated BRL160 million to implement the "Plan of Strengthening Corporate Exporting Capabilities", aiming to solve the problem of insufficient technology support and management of Small and Mid-Size Enterprises (SMEs).

In December 2004, the Brazilian government promulgated the Innovation Law (Lei № 10.973, de 2 de dezembro de 2004) and in October 2005, the Interpretation of the Innovation Law (Decreto № 5.563, de 11 de outubro de 2005). These moves were meant to motivate research institutions and enterprises to cooperate and participate in whole process of innovation to shorten the cycle of transforming scientific and technological research findings to actual outcomes, change the passive situation of delayed or even postponed transformation of such sci-tech achievements.

In 2005, the federal government promulgated the Friendliness Law (Lei do Bem) to implement tax exemption for R&D facilities purchase of Brazilian research institutions and enterprises, motivating private sectors to invest in the field of innovation and cooperate closely with public research institutions as well as colleges and universities. In the same year, Brazil also released new incentive tax policies to promote technology innovation, mainly including plans to lower corporate tax burdens and optimize distribution of intellectual property revenues among enterprises, universities and research institutions.

Prior to this, there was the previously mentioned the 85th Amendment to Constitution, promulgated on Feb. 26, 2015 and the Amended Decree of the Innovation Law, promulgated in January 2016. These laws clarified the preemptive input of national financial budget into scientific and technological research that contributes to public interests and social progress. These are of great significance to regulating relationship between government and private sectors in scientific research, improving legal security and transparency, and increasing efficiency when reviewing scientific and technology innovation while contributing to the promotion of economic development and employment improvement.

1.5.2 Initiating Diversified Financial Services to Support Technology Transfer

Brazil mainly relies on institutions like Scientific Programs Loan Bureau and Brazilian National Socioeconomic Development Bank to provide financial support to science and technology innovation and technology transfer. Meanwhile, state governments and some industry funds have also played an important role in this regard.

The Brazil Scientific Programs Loan Bureau supports cooperation in technology innovation between Brazilian enterprises, universities, research institutions and other public and private entities, concerning fundamental research and applies research, innovation, technology transfer, new technology development (products, process and services), capacity building of HR capital, international exchange of scientific personnel, and purchase, maintenance and repairs of infrastructure for research in science and technology innovation. FINEP can provide totally free capital, payable capital, and enterprise special capital, of which the totally free funds offer 16 types of funds that play a great role in improving corporate competitiveness and strengthening relationship between public and private entities. These types of funds also include a "Green-Yellow Fund" that is committed to promoting cooperation between industry, universities and research institutions, while actively promoting implementation of technology transfer. Also, FINEP directly participates in investment fund and the "Startup Investment and Loan Common Fund" (FMIEE), realizing sustainable innovation through helping innovative enterprises build R&D centers, promoting cooperative R&D between enterprises and research institutions, and making long-term investments in R&D programs.

To encourage technology transfer and transformation and stimulate economic development, the Brazilian government also formed two production-intensive investment funds of R&D innovative economy (FIP-PD&I). One was jointly contributed by BNDES, FINEP, Desenvolve-SP and Brazil Air Industry Corporation (EMBRAERO) to push forward significant cooperation of strategic enterprises in space industry. With a total amount of BRL5.3 million, the fund aimed to promote the nation's satellite technology transformation to improve corporate capabilities of innovation and strengthen the system integration of air and space sectors and national security. The other was contributed by innovative enterprises (FIP Inova Empresa). The government made use of capital formative means such as corporate stocks and convertible bonds to motivate innovative enterprises to participate in critical technology-intensive industries that are expected to grow rapidly.

1.5.3 Building Multi-tiered Platforms to Facilitate Technology Transfer

(1) **Strengthening Construction of S&T Zones and Business Incubators to Vigorously Promote Industry-University-Research Cooperation**

To fix the loose connection between scientific research and market, the Brazilian government built S&T parks and zones and business incubators in universities and research institutions. Such movement promotes industry-university-research cooperation from the R&D end and also creates an important channel of promoting transformation, commercialization, and industrialization of scientific and technological research findings. According to statistics of Anprotec, Brazil now has 369 incubators with 2310 under incubation and 2815 incubated, generating 53,280 employment opportunities. In 2013, Brazil has operated 28 S&T parks and zones, with 28 under construction and 24 to be constructed; 84% of them are located in the south and southeast of Brazil.

(2) **Setting Up Platform to Support Small to Micro Enterprises in Technology Innovation and Technology Transfer**

As mentioned above, the "Brazil Small to Micro Enterprises Support and Service Center" established in Brazil provides small enterprises with generally-beneficial, all-round free or low-cost services to promote technology innovation. SEBRAE is committed to motivating independent entrepreneurship, competition participation, and realization of sustainable development of Brazilian small to micro enterprises; in addition to technical training concerning business operation skills, entrepreneurship planning, corporate internal management, and market opening-up, it provides enterprises with technology transfer services, helps extend industry chain, offers loans and capital services, and provides information consultation concerning entrepreneurship and legal matters. Besides, it has built a giant data base of enterprises and products, creating entrepreneurial culture in whole society.

(3) **Establishing the "Technical System" and "Brazilian Technical System Shop"**

The Brazilian Ministry of Science and Technology Innovation and Communication (MCTIC) has established a technical system that supports enterprises in improving innovation capabilities. The system consists of 400 research institutions, which are located in 54 localities (13 innovation centers, 19 technical service stations, and 22 technical stations), and provides services to 175,000 enterprises on average each year. The federal government and state governments respectively contributed BRL105.9 million and BRL32.9 million to the system. Besides, Sibratec, SEBRAE and Brazilian Industrial Training Service Center (Senai) have jointly established the Brazilian Technical System Shop (SibratecShop), co-sharing software and research facilities with enterprises to facilitate the R&D, realization and transfer of new production.

(4) **Utilizing S&T Innovation Institutions' Intellectual Property Policy Information Form (Formict) to Keep Track of Technology Transfer Dynamics of Research Institutions**

MCTIC has designed an annual statistical form of S&T innovation institutions regarding intellectual property policy information, requiring all research institutions at a certain point at the beginning of each year fill in relevant information of the previous year online, including the status of intellectual property, technical development, signing information of permission agreements and technology transfer contracts, and construction of "technology innovation offices" etc. According to Formict 2015, a total of 264 institutions completed the form in 2014, including 194 public research institutions and 70 private institutions, of which 68.2% have set up "technology innovation offices", 20.5% are in the process of setting up, and 11.4% have not initiated such project; 216 institutions have signed a total of 2171 technical contracts while 48 institutions have not signed anything. Compared to the data in the past few years, the numbers of "technology innovation offices" as well as technical contracts signed have grown. However, growth rate is quite low and it still takes some time for the Brazilian society to form the culture of promoting technology transfer and transformation.

1.5.4 Encouraging Construction of R&D and Technology Transfer Centers to Promote Implementation of Technology Transfer

The Brazilian government motivates research institutions to transfer new technologies to production sectors, requiring that only joint development projects led by research institutions with at least two enterprises involved are eligible for the state funding of 50% of the project expenditure. The All-Brazil Industrial Federation will have all the new technologies developed by research institutions registered in the three areas of biology, new materials and information technology, recommending the technologies to production sectors so that enterprises can better know and utilize them. In recent years, Brazil has formulated the "Innovative Enterprises Program" to accord economic compensations to corporate R&D from the ends of products, production and market. It was meant to guide enterprises in strengthening technical cooperation with research institutions, encourage independent construction of R&D and technology transfer centers, and promote innovation activities in scientific research.

1.6 Developments of S&T Innovation in Key Areas

Since the beginning of the 21st Century, Brazil has achieved notable progress from the policies of S&T innovation, consolidating the foundation of human resources in various fields while expanding infrastructure of R&D innovation to enhance capabilities of S&T development in key areas.

In terms of infrastructure of scientific research, national-level large infrastructure of scientific research has been constructed to promote decentralization and redistribution of research institutions to all states of Brazil. The large such infrastructures under construction include the third-generation synchrotron SIRIUS, multi-purpose reactor, UVX synchrotron, and Brazil Navy oceanographic research vessel Vital de Oliveira. Besides, the Proinfra project was approved, inputting BRL1.2 billion into funding and expanding scientific research facilities in the past four years. In the field of climate change, the National Earth System Science Center (CCST) was established to collect monitoring data and forecast the trend of climate change.

In the area of health, the National Online Institute of Substitution Method (Rename) was established to allow partial use of animals as substitutes in scientific experiments. Three research centers have been built with 23 affiliated labs. By the end of 2017, the existing "Health Innovation Program" will have a registered input of BRL 3.6 billion into public and private sectors for funding medical and hygienic R&D innovations.

In the field of information technology, remarkable progress has been made in electronically improved economic system, national software technology accreditation, support of start-ups in fundamental science, electronic education, and attracting the world's R&D innovation centers.

In the field of nanotechnology, the nation has combined multiple action plans, integrated and strengthened government roles in the activities within the area of nanotechnology, founded the National Nanotechnology Laboratory System (SisNANO), and made nanotechnology formative suggestions to provide research institutions and corporate users the opportunity to access nanotechnology.

The nation has established the National Natural Disaster Warning and Monitoring Center; a nationwide warning system covering sensitive areas is basically completed. From November 2011 to January 2016, the Center has provided over 4200 pieces of warning information.

In terms of aerospace, the No. 4 China-Brazil earth resources satellite, CBERS-4, was successfully launched into orbit. Besides, Brazil has formulated the "technological program of geostationary satellite for national defense and strategic communication".

In terms of petroleum and natural gas, Brazil has implemented human resource program and national bio-diesel production and utilization program. BRL 1 billion has already been invested in ethanol energy.

In the field of nuclear medical science, National Nuclear Commission (Cnen) has provided over 400 local clinics and medical centers with radioactive medicines for treatment of cancer, heart diseases and mental diseases, with over 1.5 million clinical experiments of various kinds conducted nationwide each year. Although the key raw material, Mo-99, still needs to be imported at this point, with the successful construction of multi-purpose reactor, Brazil will become self-sufficient soon.

2 Cooperation with China in S&T Innovation

2.1 History of Cooperation Between China and Brazil in S&T Innovation

China and Brazil established diplomatic ties in 1974. Since the signing of the Inter-Governmental Agreement on S&T Cooperation in March 1982, the two sides have seen steady progress in S&T cooperation, with increasingly expanded areas of cooperation and ever remarkable achievements. In 1993, the two sides established strategic partnership and with frequent exchanges of high-level visits, fruitful results have been achieved in S&T cooperation, with the goal of mutual benefits fulfilled. With successful launch of the first jointly developed earth resource satellite CBERS-01 in Taiyuan in October 1999, the two countries have headed for deeper, higher and more practical S&T cooperation on the basis of the results of the high-tech cooperation.

2.2 Status Quo of STI Cooperation Between the Two Countries

In 2004, the two countries released the MOU on establishing the high-level coordination and cooperation commission. As a mechanism of the highest-level political dialogue between China and Brazil, the commission has a science and technology committee, which called the first meeting in 2006. As the ties between two countries were upgraded into comprehensive strategic partnership in 2012, the bilateral STI cooperation has registered vigorous progress. At the moment, the main mechanisms of cooperation between China and Brazil in S&T innovation are as follows:

(1) The Science and Technology Committee, under China-Brazil High-Level Coordination and Cooperation Commission, with three sessions already held;
(2) China-Brazil High-level Dialogue in S&T Innovation, with two sessions already held;
(3) BRICS STI Ministers Meeting, with four sessions already held;
(4) BRICS Senior Officials Meeting for STI Cooperation, with six sessions already held;
(5) BRICS STI Working Group Meeting;

(6) G20 STI Ministers Meeting, with one session already called;
(7) The Space Cooperation Committee, under China-Brazil High-Level Coordination and Cooperation Commission, with four sessions already held;
(8) The Industry and Information Technology Committee, under China-Brazil High-Level Coordination and Cooperation Commission, with four sessions already held.

For main agreements and MOUs concerning China-Brazil cooperation in S&T innovation, see Table 7.

The bilateral R&D cooperation like China-Brazil Joint Laboratory (Research Center) has been going smoothly. At the moment, China and Brazil have established the research center for innovation in climate change and energy technology, space weather joint laboratory, nanotechnology joint research center, bio-mass combustion 3-D monitoring joint research center, joint laboratory of agricultural technology, and deep-sea technology joint research institute. Meanwhile, the two sides have maintained close contacts in bamboo industry. Besides, there are joint centers under construction, including meteorological satellite joint center and China-Brazil biotechnology center.

In 2015, when the two sides signed the MOU on bilateral cooperation in area of science parks, the substantial cooperation between two countries formally began. Since then, the delegation of Association of Science Parks and Business Incubators and the China's Torch Center have paid frequent mutual visits. In 2015, the Brazilian Delegation of Science Parks and Business Incubators arrived in Beijing to attend the 32nd annual meeting of IASP-International Association of Science Parks and Areas of Innovation; in 2016, the Shijingshan District of Beijing Municipality and Sao Jose dos Campos signed agreement on cooperation in area of science parks and business incubators while representatives from Tsinghua University S&T Park arrived in Brazil to attend the annual meeting of Brazilian Science Parks and Business Incubators. In the same year, the Brazilian Association of Science Parks and Business Incubators also held in Sao Paulo the "China-Brazil Science Parks and Business Incubators Forum" and representatives from the Torch Center, Zhangjiang Science Park, and Tianjin Science Park attended the event.

Technology-oriented enterprises play an increasingly greater role in bilateral S&T cooperation. For example, (1) Huawei and Brazilian Ministry of Science & Technology and Innovation signed the Agreement on Cooperation in Cloud Computing and Big Data Technology, signed with Rio Grande do Sul Agreement on Strategic Cooperation, partnered with The Catholic University of Rio Grande do Sul (PUCRS) in building smart city joint research lab, and built joint labs including cloud computing lab together with several Brazilian universities. Also, (2) Baidu and Brazilian Ministry of Science & Technology and Innovation signed the Agreement on S&T Cooperation in Internet Technology, launching search engine in Portuguese language. Besides, (3) BYD Auto made investment in Campinas City of Brazil to build the electric bus plant and solar battery plant while discussing with Campinas University to build solar joint lab. In addition, (4) for the all-new security solutions to online banking launched by Brazil's largest Internet company PSafe,

Table 7 The main agreements and MOUs concerning China-Brazil cooperation in S&T innovation

No.	Year	Name of Agreements
1	1982	Agreement on Intergovernmental S&T Cooperation
2	1984	Additional Articles to the Agreement on Intergovernmental S&T Cooperation
3		Agreement on S&T Cooperation between Chinese State Science and Technology Commission and Brazilian Science & Technology Development Commission
4		Chinese Academy of Sciences and Brazilian Science & Technology Development Commission Agreement on Cooperation in Pure Science and Applied Science
5		Agreement on Cooperation in Peaceful Use of Nuclear Energy
6	1985	Agreement between the Chinese Ministry Geology and Mineral Resources and the Brazilian Ministry of Mines and Energy on Technical Cooperation in Offshore Petroleum Development
7	1986	Protocol on Cooperation in Geological Science
8	1988	Protocol on Joint Development of Earth Resources Satellite
9		Protocol on Cooperation in Industrial Technologies
10		Protocol on S&T Cooperation of Electric Power (including hydropower)
11		Protocol on S&T Cooperation in Transportation
12		Protocol on S&T Cooperation in Prevention of Serious Epidemics
13		Protocol on Cooperation in Traditional Medicine
14	1990	Intergovernmental Agreement on Economic and Technical Cooperation
15	1994	Intergovernmental Agreement on S&T Cooperation in Peaceful Use of Outer Space
16	1995	Agreement on S&T Cooperation
17		Protocol on Cooperation in Small Hydro
18		MOU on Strengthening and Expanding China-Brazil Space Technology
19	1996	Joint Statement on Peaceful Use of Outer Space Science and Technology
20	2000	Intergovernmental Protocol on Cooperation in Space Technology
21	2001	MOU on S&T Cooperation between Ministries of Science and Technology of the Two Countries
22	2004	MOU on Establishing China-Brazil High-level Coordination and Cooperation Commission
23	2009	Intergovernmental Protocol on Cooperation in Energy and Mining Industry
24	2010	Intergovernmental Joint Action Plan 2010–2014
25		MOU on Establishing Joint Laboratory and Promoting Cooperation in Agricultural S&T Innovation
26	2011	MOU on Implementing Bilateral S&T Cooperation in Bamboo Industry
27		MOU on Establishing China-Brazil Nanotechnology Joint Research Center
28		MOU on Cooperation in Water Resources
29	2012	Ten-year Cooperation Plan between the People's Republic of China and the Federal Republic of Brazil
30		MOU on Establishing Meteorological Satellite Joint Center
31		MOU on Establishing China-Brazil Biotechnology Center

(continued)

Table 7 (continued)

No.	Year	Name of Agreements
32	2014	MOU on Cooperation in Remote Sensing Satellite Data and Application
33	2015	Intergovernmental Joint Action Plan 2015–2021
34		MOU on Bilateral Cooperation in Science Parks
35		Agreement on Joint Development of Earth Resources Satellite—04A
36		MOU on Establishing Joint Work Mechanism in Remote Sensing Long-distance Communication and Information Technology

the core technology was provided by China's Qihu 360 Corporation. Also, (5) China's State Grid and Brazilian State Electric Power Corporation (Electrobras) formed a consortium to win the bid of Brazilian Hydroelectric Power Ultrahigh Voltage DC Output Project, signifying the significant breakthrough of the outreach of China's ultrahigh voltage technology.

In recent years, the cooperative mechanisms of BRICS have been enhanced and under framework of S&T and innovation of BRICS, the mechanism of funding research and development has been initially formed through joint funding of cooperative projects, with an aim to support and promote cooperation between at least three countries. In 2016, State Natural Science Foundation will fund a maximum of 10 research projects involving cooperation with BRICS, concerning basically the five areas of water resources and sewage treatment, astronomy, biotechnology and bio-medicine (including human health and neurological science), marine and polar science and technology and material science (including nanotechnology). The fund (direct fund) for Chinese scientists is maximum RMB2 million (including RMB2 million) per project for purposes of research, international exchange and cooperation, and small-scale symposiums; the cycle of implementation is three years (January of 2017–Dec. 31, 2019).

In 2016, the five BRICS countries formed a sponsor work group for funding S&T innovation, signed the BRICS Framework Plan of Science & Technology Innovation and the Implementation Scheme, having decided to jointly collect multi-lateral R&D projects under the framework. At the moment, the BRICS cooperative project has been initiated in the key category of the Intergovernmental International Cooperation in S&T Innovation, with five areas selected as prioritized and China is considering to provide RMB13 million as research funds.

2.3 The Role Model of China-Brazil Cooperation in S&T Innovation—China-Brazil Earth Resources Satellite (CBERS)

In 1988, Chinese and Brazilian governments signed The People's Republic of China and The Federal Republic of Brazil Agreement on Reviewing the R&D of

Earth Resources Satellite, starting the road of cooperation between two countries in space industry. This was the first-time China was engaged in all-round international cooperation with foreign countries in the field of space technology. The agreement between China and Brazil on cooperation in space technology was signed by China's State Space Agency and Brazilian Space Bureau; the specific institutions in charge of implementation are China's Academy of Space Technology and Brazilian Space Technology Institute.

According to articles of the Agreement, China and Brazil would jointly develop two earth resources probe satellites with a total investment of USD 300 million, of which the Chinese side would contribute 70% of the capital and the Brazilian side 30%; both sides would be entitled to use of the satellite after it was put into operation. In 1994, the two governments signed "The People's Republic of China and The Federal Republic of Brazil Framework Agreement on Cooperation in Peaceful Use of Outer Space Science and Technology". On October 14, 1999, the first earth resources satellite (CBERS-01) jointly developed by the two countries was successfully launched into orbit in Taiyuan, China. Its delivery and usage came into effected in March 2000.

On September 21, 2000, the two governments signed "The People's Republic of China and The Federal Republic of Brazil Protocol on Cooperation in Area of Space Technology"; in 2004 the Additional Articles to Agreement on Cooperation in R&D of Earth Resources Satellite, initiating the R&D of China-Brazil earth resources satellite—02B was signed; in 2010 The China-Brazil Joint Action Plan to specify the two sides' continuous and expanded cooperation in space sector was signed. In 2013, although an unexpected problem occurred during the launch of China-Brazil earth resources satellite—03, the two governments immediately expressed their great confidence in continuing cooperation in the space sector; on December 7, 2014, the China-Brazil earth resources satellite -04 was successfully launched into orbit; on Dec. 9, the space authorities on both sides signed a letter of intent on continuing satellite cooperation projects.

For capital contribution and proportion in this project, see Table 8.

The China-Brazil Earth Resources Satellite (CBERS) is mainly used for research in areas like disaster monitoring, environmental protection, agriculture and forestry, water conservancy, national land resources, and urban planning; it has made great contribution to the economic growth and social development of both countries. For China, the satellite data is used in all sectors of the national economy, including agriculture, forestry, water conservancy, geological and mineral resources, energy, land, marine, environmental protection, surveying and mapping, urban & rural planning, and disaster monitoring. The users cover over 20 ministries and commissions or provinces, municipalities and autonomous regions including the Ministry of Land Resources, with many of the nation's major projects utilizing enormous amount of data from the resources satellite. On the other hand, Brazil also fully utilized the data obtained from resources satellite to map out its unique path of resources satellite application designated to best fits the country. For example, since 2000, Brazil has been using data from the resources satellite to accomplish monitoring and management of the Amazon rain forest. This is the most successful

Table 8 Capital contribution, proportion and launch of China-Brazil earth resources satellite

Type of satellite	Cbers-01	Cbers-02	Cbers-02B	Cbers-03	Cbers-04
Date	Oct. 14, 1999	Oct. 21, 2003	Sept. 19, 2007	Dec. 9, 2013	Dec. 7, 2014
Place	Taiyuan launch center	Taiyuan launch center	Taiyuan launch center	Taiyuan launch center	Taiyuan launch center
Launch vehicle	Long March-4 (LM-4)	Long March-4 (LM-4)	Long March-4B (LM-4B)	Long March-4B (LM-4B)	Long March-4B (LM-4B)
Launch description	Success	Success	Success	Failure	Success
China's capital contribution (%)	70	70	70	50	50
Brazil's capital contribution (%)	30	30	30	50	50

application of the resources satellite in Brazil and has effectively controlled illegal tree cutting to ensure the regrowth and utilization of forest could get back to the virtuous circle of development. Brazil's emissions of greenhouse gas mainly come from tree cutting and wild grass and plants burning; ever since Brazil started using CBSRS to conduct real-time monitoring of tree cutting in Amazon area, the annual tree cutting rate of forest has been dropping year on year. From 2009 to 2015, the average annual forest cutting area was 6080 km^2; in particular, the tree cutting area in 2012 was 4571 km^2, ranking the lowest in the past 20 years.

Starting from CBERS-02, China and Brazil jointly announced that the satellite's 20-meter definition data will be provided to all countries in the world free of charge, allowing the data to be more widely used in more countries. Since China's State Space Agency joined the International Charter Organization of Space and Big Disasters, CBERS has, on behalf of Chinese satellites, provided global disaster reduction and monitoring services numerous times, supplying enormous amount of satellite remote sensing data to the Australian forest fire, Pakistan flood, Japan earthquake and tsunami. This has shown a responsible image of developing country in utilizing space technology to provide international humanitarian aid.

After the successful launch of CBERS-01, President Jiang Zemin and the Brazilian President, Fernando Cardoso made a positive comment that the R&D and the successful launch of the satellite was an epitome of the South-South S&T cooperation. Such high praise continuously inspired scientists and researchers in both countries to devote into the research, focus on cooperation, seek the potential, and finally make great achievements.

2.4 Main Difficulties and Barriers in the Course of Cooperation

(1) China and Brazil share considerable differences in their political system as well as culture on top of the long distance in between; consequently, the cooperation in S&T innovation is substantially subjected to objective conditions. Brazil has high tax rates and one of the most complicated tax laws in the world; the legal system is highly developed but complex and the interest rate has remained above 11% in recent years. Meanwhile, the trade unions are powerful, the social security level is beyond the nation's stage of development, and there are few researchers (only about 140,000 Ph.D. holders for science and technology and teaching nationwide). Besides, the official language is Portuguese and the people know little about the development of science and technology in China. All these factors constitute big barriers for Chinese technology-oriented enterprises, research institutions and scientific personnel to start cooperation with Brazilian counterparts. Without the establishment of effective cooperative mechanisms to provide incentives, the cooperation will be very limited in scope and depth.

(2) **The political instability in Brazil in recent years and the high financial deficit combined to hinder the deepening of cooperation in S&T innovation.** In recent years, there has been political instability in Brazil. The former president was impeached and abdicated while the ministers of the federal government were changing frequently. Meanwhile, the Ministry of S&T Innovation and Communication was no exception; the implementation of government polices of S&T development and foreign cooperation failed to be continued effectively. In 2016, the GDP of Brazil was BRL 6.2669 trillion, 3.6% lower than that of 2015 while the cumulative economic recession in two years has exceeded 7.2%. Meanwhile, the preliminary financial deficit of both the Federal and the local governments reached BRL155.7 billion, accounting for 2.47% of the GDP as the annual average unemployment rate reached 11.5%. As a result, research funds in Brazil has dropped accordingly and a good many research projects and scientific personnel training programs have been suspended, e.g. the Science and Technology Knows No Boundary Project.

(3) There has not been a stable, bilateral intergovernmental joint funding mechanism between China and Brazil. Although China-Brazil joint laboratories are getting large in number and covering quite a few new high-tech areas, these labs are still functioning in isolated spots. Every lab has tried its best to apply for national funds and only when the application was passed can the cooperation be continued or the regular exchange and cooperation may not survive at all. The BRICS countries have initially set up multilaterally funded research mechanism, but it requires the participation of three countries in the research, which raises the bar of application for the participating parties. In May 2014,

China's State Natural Science Foundation and the Brazilian National S&T Development Council signed the Agreement on S&T Cooperation, hoping to collectively fund joint research projects and symposiums, giving priority to areas of biodiversity, green energy, air and space, and marine researches. However, at this point the relevant financial support has not yet been practically initiated.

Chapter 4
Russia Report on Science, Technology and Innovation

Qiang Chen, Shimin Zheng, Yi Xiao and Yefeng Yang

In recent years, the development of science and technology in Russia has undergone a series of significant changes. Basic scientific research, which is traditionally an area of strength for Russia, has not only preserved its position as the foundation of national science and technology development but also become more competitive globally. According to the international rankings of innovation countries released by Bloomberg in late January 2016, Russia jumped to the 12th position, and this upward trend was further consolidated during the year.

- The number of published papers has been increasing year by year. Since 2013, Russia has reversed the decline in the number of papers published in international journals, and issued more papers each year for four years straight. Between 2013 and 2014, the number of papers published by research personnel from Russia in international scientific journals accounted for 2.11% of the total; it further increases to 2.28% in the beginning of 2016. In the past three years, 29,010, 30,044 and 31,542 papers respectively went through peer review. In 2015, 39.8% of the scientific papers in Russia were published by universities, 32.8% were by scientific research institutes under the jurisdiction of the Federal Scientific Research Institution, and 14.4% were joint papers. The number of papers published by Russian universities has exceeded that of scientific research institutes.
- Arresting the decline in the number of research personnel. From 2014 onwards, the number of Russian researchers saw the first increase in the past 25 years, adding 4500 research personnel in that year, and up by another 6588 in 2015. Among them, young research personnel under 39 years of age increased by 3.6%, accounting for 44.9% of the total.

Q. Chen (✉) · S. Zheng · Y. Xiao · Y. Yang
China Science and Technology Exchange Center, Beijing, China

X. Zhao et al. (eds.), *BRICS Innovative Competitiveness Report 2017*,
Research Series on the Chinese Dream and China's Development Path,
https://doi.org/10.1007/978-981-10-8078-4_4

- Scientific research funding has been growing steadily. Total R&D spending in Russia stood at 847.5 billion rubles in 2014, accounting for 1.19% of GDP, up by 13% year-on-year; in 2012–2015, the amount of funding provided by state-level foundations has almost quadrupled; in 2015, funding support for basic and exploratory research increased from 85 billion rubles in 2012 to 100 billion rubles. According to the Russian Ministry of Education and Science, Russia now ranks the fifth in the world in terms of state budgetary spending on science.

In 2016, although Russia continued to experience weak growth due to economic sanctions imposed by the West, its understanding of the strategic significance of the development of science and technology did not change, and the country remained committed to building a modern system for innovation development. The Russian government has set up professional institutions and formulated science and technology policies to the benefit of the whole society in order to put in place a modern system for innovation development. In 2016, Russia set up a new technology development agency, promulgated the relevant regulations to support legal institutions to develop education and science, and introduced the funding program for young scientist and other measures.

1 Overview of STI Development

1.1 Evaluation and Analysis of Russia's National Innovative Competitiveness

Detailed analysis of the changes in Russia's score of national innovative competitiveness and ranking in BRICS during the 20 years between 2001 and 2020.

The change of Russia's ranking in BRICS countries and score of national innovative competitiveness are displayed in Fig. 1.

Forecast of Russia's innovation index is displayed in Table 1.

(1) In terms of ranking changes, Russia ranked the second in national innovative competitiveness among BRICS countries in 2015, down by one place compared to 2001. On the whole, it was on a downward trajectory during the evaluation period.
(2) Score-wise, Russia got 24.21 points in national innovative competitiveness in 2015, which is 3.00 points lower than the best-performing BRICS country and 2.22 points higher than average; compared with 2001, Russia's score of national innovative competitiveness fell by 2.69 points, widening its gap with the highest score of the year by 3.00 points, and narrowing its gap with the "BRICS" average by 3.98 points.
(3) In terms of forecast, excluding the impact of economic factors, Russia's performance in comprehensive innovative competitiveness is expected to see a certain amount of growth during 2016–2020.

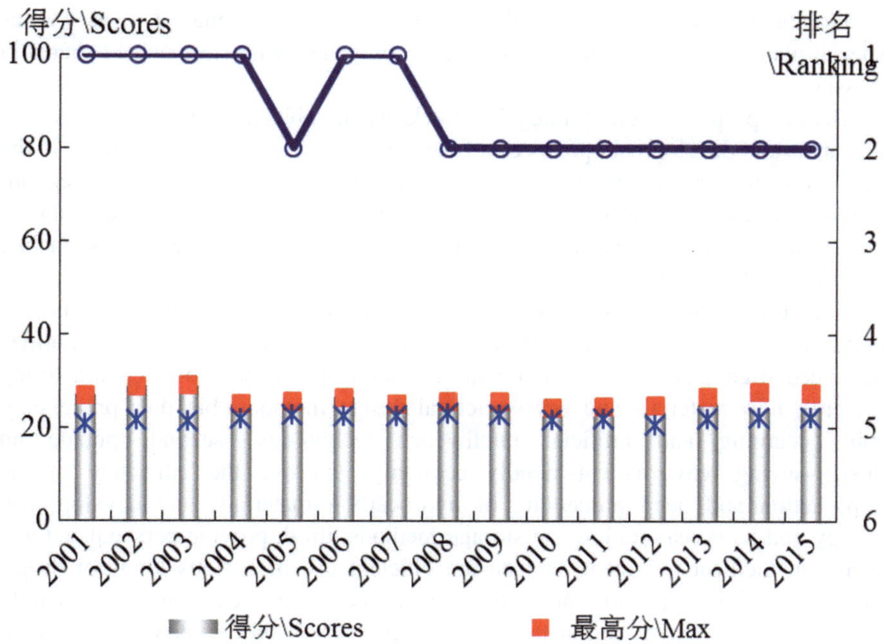

Fig. 1 Change of Russia's position in BRICS countries and score of national innovative competitiveness in 2001–2015

Table 1 Forecast of Russia's innovation index

Year	2016	2017	2018	2019	2020
Index	52.66	56.06	59.66	63.42	67.37

1.2 STI Policies, Strategies and Development Plans

1.2.1 Strategy for Science and Technology Development of the Russian Federation

On December 1 2016, Russian President Vladimir Putin signed the Presidential Decree No. 642, formally approving the implementation of the *Strategy for Science and Technology Development of the Russian Federatio* (hereinafter referred to as the "Strategy"). As early as the initial stage of the strategy's formulation, Russia had made "setting a goal that could enable Russia to resolve global issues with its own science and technology capability" as a basic task. President Putin pointed out that the "strategy" is legally equivalent to the national security strategy. The "strategy" is the foundational document for the formulation of medium-to-long-term (to 2035) industrial strategies and plans for science and technology development, federal-level national plans, federal-based national plans as well as special plans of state-owned enterprises and companies with state-owned shares in Russia.

The key purpose of the "strategy" is to clarify the objectives, strategic direction, key areas and development prospects of the country's science and technology in the medium-to-long-term, set out the basic principles, main contents and safeguard measures of the national science and technology policy, lay out the implementation steps and assess the expected results to ensure Russia's long-term, sustainable, rapid and balanced development.

According to the strategy, Russia plans to achieve a series of changes in the key areas of science and technology development in the next 10–15 years. First, utilize advanced digital and intelligent manufacturing technology, robot manufacturing systems, new materials and new structural design methods, big data processing, robot learning and artificial intelligence technology; second, pursue an energy-saving, environment-friendly economy, improve the efficiency of the exploitation and deep processing of hydrocarbon materials, and develop new energy and its transportation and storage methods; third, provide personalized and precision medical treatment, and ensure efficient use of various kinds of drugs, including antibiotics; fourth, promote efficient processing of environmental friendly agricultural and aquatic products, and conduct research on the methods of animal and plant protection that promotes balanced use of chemical and biological agents; fifth, guard against and tackle various sources of danger that may pose a threat to the society, economy and the country as a whole; sixth, build an intelligent transport and communication system, and occupy leading position in the development of airspace, space, ocean, the Antarctic, the Arctic and others; seventh, promote sound interactions between man and nature, between man and technologies and between social organizations with methods used in both cultural and social sciences. The implementation of the strategy is divided into two stages, with the first stage being 2017–2019, and the second stage 2020–2025 and beyond.

1.2.2 Long-Term Strategy for Social and Economic Development 2020 (November 2008)

The part of the development strategy on the "national innovation system and technology" has set out the objective of building a national innovation system, which is to pursue and promote innovation in all economic sectors, conduct large-scale transformation of production technologies based on the advance of science and technology, and build competitive R&D department in Russia. Russia must address two issues in pursuing innovative development, one is filling the gap in technology development, the other is creating conditions for breakthroughs in areas that could determine its unique position in the global economy.

The basic elements of the national innovation system proposed by the strategy include: integrating scientific research and development with higher education according to the needs of economic development, promoting engineering technology services, and strengthening the building of innovation infrastructure, intellectual property market and innovation incentive mechanism, building competitive scientific and technological complex, seeking and maintaining Russia's leading position in frontier scientific research and technological field, cultivating a globally competitive processing industry center, boosting the competitiveness of company's cooperative business department, expanding long-term financing channels for companies, cultivating high-quality management personnel, engineering technicians and general workers for various economic sectors, supporting the export of high-tech products and high value-added products, and providing legitimate protection to the domestic market according to the universally recognized international rules in the relevant fields.

1.2.3 Innovative Development Strategy 2020 (October 21, 2011)

The strategy has set out the following innovation targets for 2020: the proportion of enterprises engaged in technological innovation increases to 40–50% (it was 10.4% in 2009); Russia's high-tech products and services (including nuclear, aeronautical technology, space technology and services, special shipbuilding, etc.) account for at least 5–10% of the world market share; Russia's high-tech export increases to 2% of the world's total; total output of innovative products accounts for 17–20% of GDP; the share of innovative products in total industrial output rises to 25–35%; domestic R&D expenditure increases to 2.5–3% of the total GDP, and more than half of which will come from private businesses; the papers published by Russian researchers in international scientific journals rise to 5% of the total; average citation of academic papers published by Russian researchers increases to 5 times; total number of patent registrations made by Russian legal and natural persons in the patent agencies of the EU, the United States and Japan reach 2500–3000 pieces per year; Russian universities will have 25% of their funds obtained through R&D and experimental design; the proportion of university research funding will increase to 30%. The strategy also envisages that by 2020, 20% of the Russian officials will be able to speak a foreign language, and the state functionaries will be younger in age, with the elderly accounting for no more than 30%, and about 10% of the civil servants will receive education in foreign countries.

1.2.4 Law of the Russian Federation on Science and State Policies on Science and Technology

Since its promulgation in August 1996, the Law of the Russian Federation on Science and the State Policies on Science and Technology has been amended and expanded several times. In the amendment in 2009, the main objectives of Russia's science and technology policies were established as the following: to develop, efficiently allocate and effectively utilize scientific and technological capabilities, to improve the contribution of science and technology to the development of national economy, to solve important social problems, to provide guarantee for sustained structural adjustment in the field of material production, to boost the efficiency of material production and the competitiveness of products, to improve environmental conditions and protect the country's information resources, to strengthen national defense and national, social and public security, to integrate science and education.

In accordance with the law, the implementation of the state science and technology policies is mainly based on the following principles: recognize that science is important to the society, and determines the productivity of the country; promote and use various forms of public debate to choose the priority direction of science and technology; review the progress of competitive research programs and projects; give priority to the development of basic science research; integrate science and education in the R&D process through building platforms underpinned by teaching-research complex and laboratories in higher-learning institutions as well as scientific research institutions of the state academy of sciences and federal administrative agencies and on the basis of the various forms of participation of teaching staff, graduate students and students in higher learning institutions; support competition and business activities in the field of science and technology; concentrate resources on science and technology priorities; set up national science centers and other institutions; promote the development of science and technology and innovation activities; promote Russia's participation in international scientific and technological cooperation.

In 2016, Russia amended the law again. The proposer of the amendment suggested to improve the state policy for academic degree management by transferring some of the power for awarding, cancelling and restoring academic degrees to agencies that have affiliated thesis defense committee. According to the interpretation of the draft law, if the proposal is adopted, these agencies will be able to independently adjust the mechanics of the defense committee, including changing its composition, establishing its authority, as well as monitoring, stopping and resuming its activities. The procedures for awarding, cancelling and restoring degrees, the evaluation criteria of the thesis and the requirements for the production and issuance of the diploma will also be completed at the grassroots level. At the same time, the rules for awarding degrees, which will be made independently, shall include the criteria for thesis quality and requirements for the academic expertise of the members of the defense committee, which shall not be lower than the current level.

1.3 Policies for Innovation and Entrepreneurship

The separation between scientific research and industrial development, which has been formed under a deep-rooted planned economy, has made it difficult for Russia to translate its strong scientific research capability into real productivity. This is a typical example of disconnect between scientific research education and production. The 2008 global financial crisis has highlighted the importance and urgency of changing the model of economic growth and building a business-driven, market-oriented technological innovation system featuring cooperation among producers, universities and research institutions to Russia. The Russian authorities have vigorously advanced the building of the national innovation system, which involves the legal guarantee for integrating production, education and research, innovation incentives, distribution of innovation risks, innovation ownership and sharing of benefits.

(1) **Building the Operation System**

To promote the building of the national technological innovation system and strengthening cooperation between industry, academia and research community, the Russian government has implemented a mechanism called "development agency", which can be seen as a combination of institutions with the ability to invest. It consists of eight institutions, including the strategic planning agency for advancing Russia's autonomous non-commercial institutional projects, Russian foreign economic bank, Russian venture capital firm, Russian nanotechnology company, Moscow Stock Exchange innovation and investment market, Russian technology development fund, the new technology development and commercial development fund (Skolkovo) and the fund for small businesses in science and technology. It can also be seen as Russia's national policy instrument for addressing "market failure", developing investment cooperation between state and private capital, encouraging cooperation between industry, academia and research community, cultivating growth points for high-tech innovation companies, providing solutions that cannot be realized under market economy, and ensuring steady progress of the national innovation system. Take Russian Venture Company for example, it is a wholly state-owned company, with a statutory capital of 30 billion rubles, and part of the asset is managed by the Russian State-owned Asset Management Agency. It has 15 funds with the total value 25.2 billion rubles, in which the company accounts for 15.7 billion rubles, and has invested 14.4 billion rubles in 156 innovative enterprises.

The main operation model of the "development agency" is to combine state-owned capital with private capital through policy leverage to set up funds that provide investment and financing services and infrastructure services in order to cultivate and support all kinds of creative designs, technology research and development, R&D outcome conversion, production application, marketing and other activities, enhance the technological innovation capability of SMEs, address the problem of backward technology, ensure the smooth implementation of the national innovation strategy, and promote the transition toward economic diversification and modernization. The agency provides continued, uninterrupted services for innovation projects at all stages, from "conception, creation, nurturing to growing and maturing". The main functions of the "innovation elevator" include: (1) to provide financial support for each stage of the innovation projects; (2) to find potential innovation projects to be reviewed by the "development agency"; (3) to attract private investment in projects initiated by the "development agency"; (4) to formulate unified plan for the screening, assessment, launching and implementation of innovation projects.

(2) **Building Platforms for Cooperation**

To leverage the comparative advantages of industry, academia and research community and promote the development of industry innovation strategy, Russia has established technology development cooperation platforms in 13 areas. The participating organizations include national science centers such as the Russian Academy of Sciences, the Moscow State University and the Kurchatov Institute, scientific and educational institutions like federal research universities, business representatives such as Russian nanotechnology company, Russian Venture Company, Russian natural gas company and other public and private companies. Through these public cooperation platforms, representatives of Russian industries, academic institutions and research agencies meet to discuss and determine development strategies and objectives in their respective areas of cooperation.

The platforms have identified the following 13 areas for R&D: medical and biotechnology, information and communication technology, photonics, aerospace technology, nuclear and radiation technology, energy, transportation technology, metallurgical technology and new materials, oil and gas and natural resource extraction, electronics and machinery, ecological protection, agriculture and food industry, and industrial technology. A total of 34 research priorities have been identified, with two to four priorities in each field.

The Russian Federal government gives each platform project funding support with varying amount based on the results of evaluation conducted with reference to indicators such as research goals and potential. The funds allocated will come from the budget of the Russian Federal Special Plan for R&D Priorities 2007–2012, the Federal Plan for Basic Research before 2020, as well as the Russian Academy of Sciences, Russian Basic Research Fund, Russian nanotechnology company, Russian venture capital firm, the development fund for small businesses in science and technology and the Skolkovo Fund.

The 13 technology development cooperation platforms have horizontally linked the industry, academia and research community based on their areas of studies, covering all the priority areas of science and technology development, and leveraged the advantages of national academic disciplines. In the same token, the 25 regional innovation clusters approved by the Russian government have served as cooperation platforms that give scope to the existing advantages of local industries and vertically consolidate the strengths of industry, academia and research community. They share the same feature of the clustering of companies, R&D and service agencies of the same industrial chain in a designated region, where industrial organizations with pulling effect across industries and regions and global competitiveness can be formed through division of labor and collaborative innovation.

Regional innovation clusters have been built according to the order issued by the Russian government on August 28, 2012. Out of the 94 initial proposals submitted nationwide, 25 clusters were approved through feasibility studies, competitive evaluation and other procedures. The federal and local governments provide preferential support in policies, capital, personnel and infrastructure to the participating organizations of the cluster.

(3) Lifting Restrictions on Research Institutes and Higher Learning Institutions to Run Business

Higher learning institutions and scientific research institutes are the foundation of cooperation between industry, academia and research community; it is the source of innovation and the talent pool of the country. In 2009, the State Duma adopted the Decree on Revising the Rules Concerning the Establishment of Companies for the Industrialization of Scientific Research Outcomes by Research and Education Institutions Funded by the State in the third reading. In August, the Russian president approved the Amendment to the Law of the Russian Federation on the Establishment of Economic Entities for the Conversion of Scientific and Technological Outcomes by Research and Education Institutions Funded by State Budget, permitting state-owned scientific research institutes and universities to run independent companies with R&D outcomes obtained by the funding support of the state and conduct R&D outcome conversion and innovation activities. The specific provisions include: scientific research institutes and universities that receive state funding can now set up companies and engage in the industrialization of their own R&D outcomes without state approval; the scientific and technological outcomes will be counted as fixed-asset investment after the assessment of their right to use, but non-transferrable to the third party; scientific and technological outcomes that are deemed as fixed-asset investment may be in the forms of computer software, database, invention patent, applicable model, industrial sample, new seed variety, mini integrated circuit and critical technologies. Scientific research institutes and universities are also allowed to run joint ventures with other companies, and hold no less than 25% of the shares. The partner company's fixed-asset investment in the joint venture shall have at least half in the form of capital. Scientific research institutes and universities that receive state funding can only transfer their shares in the company after getting the approval of the state. The joint venture may spend the business revenue at their own disposal, but the revenue must be accounted independently and only used within the approved business scope. In the meantime, Russia has canceled the relevant restrictions on the establishment of companies by state-funded scientific research institutions and universities in the Law on Limited Liability Companies and the Law on SME Development.

The promulgation of the new law has removed the shackles on state scientific research institutes and universities in conducting technology innovation activities, provided the legal ground for the industrialization of scientific and technological outcomes and the participation of research personnel in commercial activities, and strengthened the links between higher learning and research institutions and production activities. According to statistics, between 2011 and 2013, Russian universities invested approximately 220 million US dollars on the building of innovation infrastructures, launched 5223 R&D institutions and 120 technology outcome transfer centers, and designated 29 universities as "national research-intensive universities" with the aim of building world-class education and research complex.

(4) **Cultivating an Innovation-Friendly Culture**

Starting from 2011, the Russian government has invested in the organization of competitions of creative TV and film products, requesting all participating films, TV programs and cartoons to feature scientific inventions, commercial innovations and the stories of successful innovators. The Russian government requires government-funded TV stations to launch science education channels, buy and broadcast foreign science documentaries, and produce TV programs that promote Russia's scientific achievements. All channels on the national TV station, including the children's channel, shall broadcast programs that can help boost the public awareness of the importance of innovation. Special programs have been launched to support the promotion of scientific knowledge through literature, journals, radio and the internet and the report of scientific inventions, research outcomes and creative designs. Science, technology and innovation museums have been built across the country to boost public interest and awareness, especially among the young people. The management model of the Moscow Darwin Museum and the Industry Museum (building joint laboratories with universities and research institutes, where the young people can learn practical skills by experiment) has been promoted, and museums in big cities and regional centers have been renovated and expanded to allow better and more space for the demonstration of technology and natural science outcomes. Subsidies have been provided for the publication of reading materials for youth in order to introduce the latest advances in science and technology and the achievements of modern science and technology, stimulate their interest in science, and encourage them to embrace the vision of working on science and technology in the future. In 2012, a national innovation award was initiated by an authoritative expert panel and chamber of business, which includes individual awards for innovation consumer product, technology breakthrough, improvement in life quality and expansion in overseas market. Every year, the award ceremony is widely covered on the media to promote the innovation achievements of the entrepreneurs.

(5) **Fostering a Policy Environment Friendly to Business Innovation**

The Russian government has put forward the vision of building an "innovation-friendly" environment, and set out the following goals: removing the obstacles for unleashing the innovation dynamism of companies and promoting advanced applicable technologies, encouraging companies to raise their competitiveness through the R&D and use of advanced technologies, creating favorable conditions for the building of high-tech companies and market for new products (services). The key measures introduced include the following:

First, leveling the playing field and motivating the innovation dynamism of companies. To restrict the intervention in innovation activities by some enterprise groups and improve the efficiency of anti-monopoly agencies in countering acts of monopoly; to determine the principles and standards of ex-gratia enterprises and prevent negative impact on the environment of competition and incentives for innovation; to conduct regular evaluations on policies that restrict innovation activities and compare the level of obstacles between different regions; the support for inefficient enterprises will be gradually pulled back, and for those companies with high social significance, business adjustment will be conducted on the condition of protecting the interests of private investors to strengthen the guiding effect of innovation policies; the official representatives of the board of directors and the board of supervisors of state-owned enterprise shall be responsible for formulating their own innovation plans and supervising the implementation by the enterprise; the investment projects and development plans of state-owned natural monopoly enterprises should go through qualitative examination.

Second, improving market regulation of products (services) and ensuring the promotion and application of advanced technologies.

To conduct regular assessment of the barriers in the application standards of key technologies in the economic field and formulate plans of follow-up actions for management improvement; to assess management effectiveness and improve management practices in collaboration with industry associations, national and foreign investors; to strengthen cooperation between government, scientific research institutions and enterprises in innovation management.

Third, accelerating the update of obsolete regulations and standards that impede business innovation. To harmonize the legal standards of Russia and the European Union and enable mutual recognition of the certification results of laboratories and certification centers; to streamline the procedures for market access of new products with reference to similar basic European standards for technology management and simplify the procedures of export certification for manufacturers within the scope of management authority to create maximum opportunities for product export; in the import of technology products, it is no longer required to provide the certified list of imported equipment previously required by the federal government; to build the policy "technology corridor" that guides the operation of enterprises, regulate the use of natural resources by companies, and ensure the safety of products and services to the people and the environment; to reduce energy and material consumption; to build a reward and punishment system; to unify Russia's domestic and international standards, especially the standards for the industries that can potentially boost the export of innovation products. In addition to restrictive measures, the "technology corridor" also provides safeguards for enterprises, including: coordinating the relationship between producers, encouraging the establishment of production enterprise associations, rewarding enterprises that purchase and utilize new technologies, reducing or canceling tariffs on the import of advanced equipment, supporting R&D activities, organizing personnel training, providing preferential policies and national procurement opportunities to products or enterprises that use specific technology solutions.

Fourth, building an intellectual property trading platform with state investment, streamlining the mechanism for the transfer of state-owned (obtained through state investment) intellectual property rights, and better protecting the right in the transfer of invention patents held by institutions to the inventors as natural persons prior to commercial transformation.

Fifth, improving the taxation environment in support of the technology progress of enterprises, and reducing the tax burden on innovation SMEs and newly-established high-tech companies.

Sixth, improving the investment environment, reducing the influence of administrative power on the economy, and expediting the efforts to attract the participation of strategic investors with technology advantages in the privatization process, enhancing the transparency of foreign investment management, and giving special preference to the establishment of high-tech joint ventures between Russia and foreign partners.

Seventh, revising legal provisions, simplifying the immigration system, and attracting people with professional skills to Russia.

(6) **Setting up Motivation Schemes to Promote Innovation in Scientific Research Institutions**

On November 1 2013, Russia issued the government decree No. 979, which approved the implementation of the revised Methods for the Performance Evaluation and Supervision of Scientific Research Institutions Engaged in Civilian Scientific Research, Experiment Design and Technology Development. Compared to the old evaluation methods, the new methods have borrowed the performance evaluation experience of developed countries by introducing a unified evaluation system for all national civilian scientific research institutions based on their areas of focus. The old practice in which the higher authorities make the final verdict has been terminated. Under the new system, the final results will be reviewed and approved by an inter-agency committee. The new methods have also moved beyond the limitations of the affiliations between different scientific research institutions. Various reference groups have been designated based on the similarity of institutions in organizational structure, legal status, research direction and professional task, and minimum indicators have been set for comparative evaluations of institutions in the same category. The evaluation is conducted every 5 years, and a new annual supervision system has been introduced to track the performance of the institutions in a dynamic process. The results of evaluation and supervision are published on the website of the Supervision Agency of the Federal Ministry of Education and Science for pubic supervision.

The evaluation system consists of four parts: the first part is the effect of scientific research and demand for R&D outcomes, which has 8 indicators including the results of intellectual property and their application, participation in the creation of innovative small businesses and paper citation; the second part is the cultivation of talent, which has 4 indicators including occupational skill training, the number of Candidates of Sciences and Ph.D.s, and the number of scientific research personnel working in internationally renowned research institutions; the third part is about internationalization, popular science education and scientific prestige, which has 6 indicators including attracting foreign experts to participate in research, holding international academic conferences, the number of popular science activities, the number of positive comments received on state media and the number of website visits; the fourth part is protection of resources, which has 7 indicators including the number of employees on the payroll, the number of people engaged in research and development, tangible and intangible asset, internal and external R&D expenditure, and the salaries of scientific research personnel. The federal government uses the evaluation results of the subject institutions as the basis for the adjustment of project, funding and personnel policies.

1.3.1 Aerospace

(1) Completed Ion Rocket Engine Test

The engine, which was jointly developed by the Chemical Automatics Design Bureau of Russia and the Moscow Aviation Institute, will be mounted on a rocket developed for space development. The Chemical Automatics Design Bureau of Russia began ion engine development in 2012. As the second-stage thrust device for the Angara rocket, the "RD-0124A engine" was developed and manufactured by the Chemical Automatics Design Bureau.

The test was completed on a vacuum test table, and all test data were in accordance with the technical specifications. The working principle of the engine is to produce a reaction force by ionizing the gaseous working substance and accelerating the flow of the ion under the action of a strong electric field. The engine has the advantages of operating at lower temperature, lighter weight, and lower fuel consumption.

(2) First Atmospheric Satellite Successfully Made its Maiden Test Flight

Russia's first atmospheric satellite, "Sova (owl)", completed its first test flight. Relying on solar panels and batteries, it successfully made uninterrupted cruising for 2 days and nights (50 h) at a 9-km altitude.

The satellite can fly long hours over all latitudes in Russia, including the Arctic Circle. Its cruising time will depend on specific tasks. In other words, "it can fly as long as the task requires". The Sova atmospheric satellite is expected to replace some functions of the existing spacecraft (such as land monitoring, communications, etc.) at lower cost and higher efficiency.

(3) Developed Several Applications of the GLONASS

The Roscosmos State Corporation plans to invest about 1.84 billion rubles to study the possibility of applying the Global Navigation Satellite System (GLONASS) to underground, underwater environment and lunar environments. Cutting-edge technologies are expected to be adopted to expand the application fields of navigation systems, including applications to urban buildings and enclosed interiors, mountainous regions and canyons, as well as to underground, underwater, and outer space.

Underwater navigation is to be realized through the synergy between the GLONASS system and ground radio navigation equipment, underwater data transmission equipment and laser data transmission equipment. To ensure the smooth navigation information communication between lunar spacecraft and interplanetary spacecraft, scientists are working on the possibility of expanding the active regions of near earth satellite systems.

(4) **Heavy Communications Satellite Under Construction**

Academician M. F. Reshetnev Information Satellite Systems of Russia is building the Blagovest, Russia's new heavy communications satellite. The new spacecraft will be assembled in the workshop of Academician Reshetnev Information Satellite Systems before it is tested to check its radio performance. It is expected to be in service in the geosynchronous satellite orbit for more than 15 years. It will be used to make high speed data transmission and provide users with telephone and video conferencing and Internet broadband access services.

The new spacecraft will be built on the "Express-2000" satellite platform. Building on its own strong power supply system, the spacecraft will have a greatly increased effective load capacity. For the first time, it is also equipped with an antenna with contour maps as well as Bands K and Q waveguides. Compared with its previous generation, it has markedly increased the information transmission volume.

(5) **Completed Pulse Detonation Rocket Engine Test**

Russia was the first in the world to successfully test a pulse detonation rocket engine on new-generation eco-friendly liquid fuel. News reports suggest that in 2014, the Russian Advanced Research Foundation (FPI) carried out the test in the Specialized Laboratory on Liquid Rocket Engines (which was established on the basis of the NPO Energomash, a well-known Russian aviation company. The fuel was a mixture of liquid oxygen and kerosene. The research was jointly undertaken by the agency with the New Siberian Lavrentyev Institute of Hydrodynamics of the Siberian Branch of the Russian Academy of Sciences and the Moscow Aviation Institute.

(6) **Increased Investment in the Space Development Program**

The Russian Federal Space Agency said that Russia would invest more than 100 billion rubles a year in its Future National Space Program, which includes lunar, solar and Mars exploration programs. For the 2016–2025 period, the Russian Space Program will cost 1406 billion rubles, with a cut of 845.5 billion rubles in the actual budget of the Program. Among them, the Lunar Exploration Program will have a budget of 38.5 billion rubles and the Mars Exploration Program 28.1 billion rubles, while the Solar Exploration Program be carried out within the framework of the Resonance magnetic storm research project. During the 2019–2024 period, Russia will develop 5 spacecraft to carry out lunar research. Russia will also collaborate with the European Space Agency to conduct research on the Mars. In 2024, a project to get soil samples back from the "Phobos" will be launched. If all goes according to the plan, Russia will achieve Mars landing at the next step.

(7) Preparing for Arctic Exploration Satellite Launch

The Russian NPO Lavochkin plans to launch its first Arctic exploration satellite in 2017. Domestic components will replace imported ones. In 2017, Russia's first Arctic satellite will be blasted into space. The Arctic probe satellite system will include two orbiting satellites and an alternate spacecraft.

1.3.2 Optics

(1) Planning to Build the World's Most Powerful Laser Device by the end of 2017

The world's most powerful laser device UFL-2 M will be built in Sarov City in Nizhny Novgorod, Russia. The Russian Federal Nuclear Center—All-Russian Scientific Research Institute of Experimental Physics is responsible for the R&D, manufacturing and infrastructure construction of the related instruments and equipment. The laser device is expected to be put into operation before the end of 2017, and the localization rate of all equipment will be no less than 95%.

1.3.3 Materials

Launched First Nano Satellite with 3D Printed Shell

Russia's first 3D printed nano satellite Tomsk-TPU-120 was successfully lifted into space on the Progress MS-2 Cargo Spacecraft at the Baikonur Cosmodrome on March 31, 2016. The structure of the Tomsk-TPU-120 nano satellite was designed by Tomsk Polytechnic University, while the materials of the satellite were jointly made by the University, the Institute of Strength Physics and Material Science of the Siberian Branch of the Russian Academy of Sciences, and the RSC Energia.

The satellite was made of several new materials adopting new manufacturing technology, and will be used as an experimental research model for the Institute of Physics and Materials Science at Tomsk Polytechnic University. The shell of the nano satellite was made by 3D printing. The satellite is equipped with a number of batteries and sensors to record temperature changes and accurately track the operation of batteries, parts and electronic components.

1.3.4 Advanced Manufacturing

(1) Aiming for a Maglev Transport System

Scientists at the Russian Railways ("RZD") are studying plans to develop a maglev transport system. Targeting at transportation of more than 1000 km, the system will become a strong competitor against air transport. It is reported that at a Russian railway science committee meeting, experts discussed how to create the best vacuum environment for train operations, reduce aerodynamic drag and other details. The train is expected to reach a speed of 1200 km/h. After full verification, experts at the Siberia Branch of the Russian Academy of Sciences believe that it is technically feasible to manufacture such passenger trains and container cargo and mail trains to travel over a 3000–4000 km distance from central Siberia to central Europe.

(2) **Significantly Improved the Test of Electrification Equipment**

The Russian Federal Experimental Center has increased the tower testing capacity to 110 m in the power tower mechanical test field with independently developed experimental equipment and methods, making it possible to test power transmission lines and tower structures with a height of less than 110 m.

1.3.5 Information Communications

Launched Quantum Computing Technology Development Program

The Ministry of Education and Science of the Russian Federation, Russian State Atomic Energy Corporation and FPI have launched a joint laboratory project to develop and apply quantum computing technologies based on superconducting components and structures. The task of the project participants is to develop superconducting qubit manufacturing technology, as well as qubit initiation, control and reading technologies.

The All-Russia Research Institute of Automation and the Bauman Moscow State Technical University are responsible for the manufacture of multiple Qubit system. The Moscow Institute of Physics and Technology, the Russian Quantum Center, the Institute of Solid State Physics of the Russian Academy of Sciences, the National University of Science and Technology MISIS and the Novosibirsk State Technical University will be responsible for the production of superconducting qubit, for the measurement of open quantum system parameters and the development of open quantum computing algorithms. The Project has a total investment of over 750 million rubles.

1.3.6 Life Sciences

(1) **Molybdenum-Based New Anti-cancer Drugs**

Experts at the Institute of Inorganic Chemistry of the Siberia Branch of the Russian Academy of Sciences and the Novosibirsk State University are developing a new anticancer drug based on molybdenum clusters. The first test has been successful.

The principle of the new method of treatment is to leverage molybdenum clusters to transport singlet linear state oxygen to cancer cells and kill them to play a therapeutic role. Currently, the molybdenum cluster-based drug has been successfully tested on animals in preclinical trials.

(2) **Revolutionary Anti-cancer Medicine**

Russia has developed an anti-cancer drug that can stimulate the body's own ability to resist cancer cells, thereby treating cancer. Russian media said the discovery of the drug is a breakthrough in the field of human cancer treatment. It is reported that animal trials, including monkeys, were completed for the drug in early 2016. The results showed that the treatment effect is better than that of existing anti-cancer drugs.

(3) **New Artificial Heart**

Novosibirsk scientists have successfully developed an artificial heart using original technology, which can greatly reduce the risk of blood clots forming after an artificial heart transplant. Its selling price is only a fraction of the sales price of imported artificial hearts in Russia.

1.3.7 Polar Research

Developed Arctic-Use Ship

The Krylov State Science Center of Russia has launched a new ship suitable for use in the shallow seas of the Arctic. The ship is equipped with an innovative four-propeller integrated power system capable of navigating 4 m deep in the Arctic sea. The ship is 90 m long and is equipped with an 8-seat hovercraft and dual fuel engines.

2 STI Cooperation with China

2.1 History of China-Russia Cooperation in Science, Technology and Innovation

There has been a long history of China-Russia cooperation in science and technology cooperation. Upon the founding of the People's Republic of China in 1949, China and the then Soviet Union signed an intergovernmental agreement on science and technology cooperation. The subsequent China-USSR cooperation in science and technology made contributions to the establishment and development of PRC's modern industrial foundation. In 1992, *the Agreement Between the Government of the People's Republic of China and the Government of the Russian Federation on Science and Technology Cooperation* was signed. Since then, China-Russia cooperation in science, technology and innovation has maintained a steady momentum of development.

With the continuous deepening of political mutual trust between China and Russia and the sustained development of China-Russia strategic partnership, entities of China-Russia cooperation in science, technology and innovation have been expanding, while the fields of cooperation are widening and the forms of cooperation are being continuously innovated, gradually forming a pattern of multi-channel cooperation. Science, technology and innovation has become a field of cooperation with apparent complementarity and huge development potential. Cooperation in science, technology and innovation has become one of the most important, enduring and active links in maintaining the strategic partnership of cooperation between the two countries. "To be science and technology partners for common innovation" has become a consensus of the governments, and science and technology and business communities of the two countries.

China-Russia cooperation in science, technology and innovation has gone through the following three stages:

(1) **Stage of Establishment**

Shortly after the collapse of the former Soviet Union in April 1992, a science and technology delegation of the Chinese government visited Russia and promptly established intergovernmental science and technology cooperation ties with Russia on the historical basis of China-Soviet cooperation in science and technology. On December 18, 1992, China and Russia signed an *Agreement Between the Government of the People's Republic of China and the Government of the Russian Federation on Science and Technology Cooperation*. Since then, the two sides have set up a standing sub-committee on science and technology cooperation under the deputy prime minister-level China-Russia Intergovernmental Commission on Economic and Trade Cooperation. Between 1993 and 1996, the two sides held four regular sessions of Standing Sub-Committee on Science and Technology Cooperation of the China-Russia Joint Commission on Economic, Trade, and Science and Technology Cooperation. During the period, the two sides agreed on a total of 245 intergovernmental science and technology cooperation projects involving machinery, electronics, new materials, agriculture, biotechnology, instrument manufacturing, medicine and other fields.

(2) **Stage of Standardized Cooperation**

The establishment of the China-Russia strategic partnership oriented to the 21st century has opened a new page in China-Russia cooperation in science, technology and innovation. To adapt to the new development situation and the new development needs, at the first meeting of the China-Russia Committee on Prime Ministers' Regular Meetings held in Beijing in June 1997, China and Russia formally decided to set up a sub-committee on science and technology cooperation within the framework of the China-Russia Committee on Prime Ministers' Regular Meetings to take responsibility for unified coordination and management of bilateral science and technology cooperation. The Ministry of Science and Technology of the People's Republic of China and the Ministry of Science and Technology of the Russian Federation were respectively designated as the specific implementing departments of the sub-committee.

To better standardize the cooperation between the two sides and make China-Russia cooperation in science, technology and innovation develop better in line with the principles and on the basis of equality and mutual benefit, results sharing and intellectual property rights protection, China and Russia agreed upon and signed in 1999 a *Protocol on the Protection of Intellectual Property Rights and the Distribution of Rights Within the Framework of the Agreement on Science and Technology Cooperation of the Government of the People's Republic of China and the Government of the Russian Federation.*

(3) **Stage of Cooperation in High Technology Industrialization and Innovation**

Today's world sees the fast advancement of science and technology and the close integration of science and technology and the economy. As fields of bilateral cooperation in science and technology innovation continue to expand, the governments of China and Russia have gradually shifted the focus of their cooperation in science, technology and innovation to the fields of high technology and innovation, and have committed themselves to pushing forward the R&D of high and new technologies and the industrialization of science and technology results. To this end, the two sides have made many valuable explorations in jointly establishing centers (bases) for cooperation in science, technology and innovation and industrialization, in promoting cooperation between China's high-tech development zones and Russia's science, technology and innovation parks.

2.2 *Overview of China-Russia Science and Technology Cooperation Projects*

The total number of projects between 2007 and 2015 stood at 600, with a total investment of RMB2.6 billion. From the time point of view, the number of projects and funding levels were on a growing trend year by year. In terms of the geographical distribution of project implementation organizations, the organizations were from 30 provinces (autonomous regions and municipalities) in China, with Beijing undertaking the largest number of projects, at 116 or 19% of the total. Heilongjiang and Shandong provinces came 2nd and 3rd, respectively, making up 12 and 7.5% of the total number of projects respectively. The amount of funding was basically positively correlated to the number of projects, with Beijing receiving RMB464.235 million or 17% of the total funding. As for project fields, the TOP3 fields in terms of the number of projects and the amount of funding were respectively: material science and technology, engineering and technology, and information science and technology. The material science and technology field won 167 projects and RMB790.75 million in funding, accounting for 28% of the total project funding. The engineering and technology field attracted 100 projects and RMB480.03 million in funding, making up 17% of the total project funds. The information science and technology field had 79 projects and received RMB373.34 million in funding, making up 14% of the total project funds. As for the nature of project implementation organizations, the TOP3 organizations were enterprises, universities, and research institutions respectively. Enterprises undertook 311 projects, accounting for 51% of the total projects, and received RMB119.035 million or 54% of the total funding. Universities implemented 161 projects, accounting for 26% of the total projects and received RMB671.86 million in funding, or 24% of the total funding. Research institutions carried out 102 projects, making up 17% of the total projects and won RMB479.21 million in funding, or 17% of the total funding.

The implemented bilateral science and technology cooperation projects are oriented towards the science and technology frontiers of the world and China's major needs for science and technology cooperation, covering high and new technologies in a number of fields including energy conservation and environmental protection, new energy, biomedicine, new materials, electronic information, achieving a large number of internationally advanced technological results, breaking some technological bottlenecks in corresponding industries in China and establishing smooth channels for Chinese scientific research institutions and enterprises to enter the international cooperation and competition platform. The project implementation results have played an irreplaceable and important role in pushing forward the development of strategic emerging industries and produced an effect of promoting the efficient transformation and upgrade of traditional industries in China.

The implemented bilateral science and technology cooperation projects have taken information technology, material technology, energy and other technologies as the breakthrough points, given priority to the overcoming of a number of key technological difficulties that hamper the development of the defense industry, the adjustment of economic growth mode and the restructuring and upgrading of industries in China, greatly improved the technology innovation abilities of the relevant research institutions and enterprises in their respective fields, enhanced China's national ability of science and technology innovation, narrowed the gap between China and countries of the internationally advanced level, and made great strides in several technological fields.

2.3 *Difficulties and Obstacles in China-Russia Cooperation in Science, Technology and Innovation*

Although China-Russia cooperation in science, technology and innovation has made great achievements in recent years, it should nevertheless be clear that in addition to the fact that science and technology development has its own laws, differences between the two sides in technology level, market environment, institutional mechanisms and thinking also make bilateral cooperation face big challenges for sustainable development at a high level.

(1) **China and Russia Need to Further Strengthen Their Cooperation in Major Projects of Science, Technology and Innovation**

In recent years, intergovernmental mechanisms for cooperation in science, technology and innovation between the two countries have been continuously enriched, with ever-strengthening government support and guidance for cooperation projects of science, technology and innovation. A number of typical cases of successful China-Russia cooperation in science, technology and innovation have emerged. But on the whole, the two countries have yet to fully tap their potential of cooperation in basic research, hi-tech R&D and other fields. There is still much room for improvement in the major science, technology and innovation projects and the influential results of such projects undertaken by the two sides in recent years in the priority development fields of new materials, life sciences, energy conservation and environmental protection, with a sight on long-term economic and social development needs.

(2) **China and Russia Still Need to Fully Establish Efficient Channels of Technology Transfer**

Affected by a planned economic system of the former Soviet Union, research is still seriously separated from the economy in Russia. The country has yet to fundamentally overcome the shortcoming of low transformation rate of research results, the low development of institutions that promote R&D and technology transfer, and the underdevelopment of specialized markets. Moreover, there are few international technology transfer service agencies in Russia. Although China has a relatively mature international technology transfer system, there are various obstacles for it to directly connect with Russian enterprises. The lack of effective information communication channels and specialized service support between enterprises and research institutions of the two countries has to some extent affected the healthy development of China-Russia cooperation in science, technology and innovation.

Chapter 5
Overview of Science, Technology and Innovation Development in Russia

Leonid Gokhberg, Tatiana Kuznetsova, Anna Pikalova and Alexander Sokolov

1 Introduction

1.1 Problems of Long-Term Resource-Led Growth

The Russian Federation faces a variety of challenges in securing adequate investment in new knowledge and technologies and deriving socio-economic benefit from them. The global financial crisis of 2008 and the ensuing stagnation were exacerbating domestic weaknesses, such as the limited market competition and persistent barriers to entrepreneurship, which were hampering the growth of the Russian economy. Despite some reforms since, these challenges have intensified since mid-2014.

The rapid growth of the Russian economy since the turn of the century had been largely fuelled by oil, natural gas and other primary products. Oil and gas alone account for more than two-thirds of Russia's exports and 10% of its GDP. High oil prices have helped Russia to improve the standard of living and accumulate large financial reserves. The growth rate slowed, however, in the aftermath of the global crisis in 2008, particularly after 2012 (Table 1). It has deteriorated farther since mid-2014, driven by a drop in global oil prices, combined with the economic,

---

The paper is based on the materials and results of studies obtained in the framework of implementation of the project UNESCO Science Report 2015 (Gokhberg and Kuznetsova 2015), as well as author's works in the field of international science and technology cooperation.

The original version of this chapter was revised: The affiliation of authors "L. Gokhberg, T. Kuznetsova, and A. Pikalova" was changed. The correction to this chapter is available at https://doi.org/10.1007/978-981-10-8078-4_13

L. Gokhberg · T. Kuznetsova · A. Pikalova · A. Sokolov (✉)
Institute for Statistical Studies and Economics of Knowledge, National Research University Higher School of Economics, Moscow, Russia
e-mail: sokolov@hse.ru

X. Zhao et al. (eds.), *BRICS Innovative Competitiveness Report 2017*,
Research Series on the Chinese Dream and China's Development Path,
https://doi.org/10.1007/978-981-10-8078-4_5

Table 1 Economic indicators for Russia, 2008–2013

Indicators	2000–2007[a]	2008	2009	2010	2011	2012	2013	2014
GDP	7.2	5.2	−7.8	4.5	4.3	3.4	1.3	0.6
Consumer price index	65,614	13.3	8.8	8.8	6.1	6.6	6.5	11.4
Industrial production index	6.2	0.6	−10.7	7.3	5.0	3.4	0.4	1.7
Capital investment	14.0	9.5	−13.5	6.3	10.8	6.8	0.8	−2.7
Exports	21.0	34.6	−36.3	32.1	31.3	2.3	−0.8	−5.1
Imports	24.2	29.4	−36.3	33.6	29.7	5.4	1.7	−9.8
Consolidated public sector balance (% of GDP)	–	4.8	−6.3	−3.4	1.5	0.4	−1.3	−1.2
Public external debt (% of GDP)	–	2.1	2.9	2.6	2.1	2.5	2.7	–

Percentage change over previous year, unless otherwise stated
[a]Annual average growth rate
Source Rosstat (2015), Ministry of Finance (2014)

financial and political sanctions imposed on Russia by the European Union (EU), USA and several other countries in response to events in the Ukraine and accession of the Crimea. This has fostered inflation and currency depreciation while curbing consumer spending. Capital outflows have become a major concern: the latest estimates for outflows are US$57 billion in 2015 and about 16 billion in 2016. The GDP contraction in 2015 exceeded 3%. A return to growth began in 2016 and continued in 2017.

The government has been obliged to cut back on spending and to use accumulated reserves to prop up the economy, in accordance with its anti-crisis plan adopted in 2015 and later, within the process of developing a strategy for socio-economic development of the country. The difficult economic and geopolitical situation has also prompted the government to implement vital structural and institutional reforms to revitalize and diversify the economy.

1.2 The Task of Transition to the Innovative Growth Pattern

Paradoxically, the rapid economic growth fuelled by the commodities boom between 2000 and 2008 actually weakened the motivation of enterprises to modernize and innovate. Another factor—the preservation of not very favorable conditions for entrepreneurship and for innovative entrepreneurship in particular. In the sphere of science, technology, and innovation (STI), this manifested itself in a boom in imports of advanced technologies and a growing technological dependence on developed countries in certain areas, such as in pharmaceuticals and high-tech medical equipment.

In the past few years, the government has sought to reverse this trend by encouraging companies, public research institutes and universities to innovate. Some 60 state-owned companies were obliged to implement special programmes to boost innovation. As a result, their investment in R&D doubled, and reached preliminary 2% of sales (on average). The share of innovative products in the total sales of state-owned companies had grown 27% in just a few years. Income from technology exports increased in 2013–2015 almost 4 times. Exports of innovative products also progressed, particularly in the aircraft industry, shipbuilding and chemicals. Central to the national strategy was the decision to enlarge the government's arsenal of competitive research funding for leading federal and national research universities. Public institutes and universities also received grants to commercialize new technologies and create small innovative firms (start-ups). In parallel, the government introduced schemes to foster academic mobility and expose scientists and engineers to the best training that money could buy. For instance, public research institutes and universities received grants to enable them to invite top Russian and foreign professionals to work on their campuses.

1.3 The Course Towards a New Economy

The domestic weaknesses observed in recent years include inadequate intellectual property protection, the obsolete institutional structure of the R&D sector, the lack of autonomy of universities and the relatively weak infrastructure for research and innovation. These chronic weaknesses augment the risk of Russia falling further behind the leading countries in global development. It is this concern, which has made national policy-makers particularly keen to galvanize STI-led recovery and development. Since 2010, the Russian authorities have adopted no fewer than 40 documents to regulate STI, including in the form of presidential decrees.

As early as 2012, President Putin acknowledged the need for a new economy. 'It is not acceptable for Russia to have an economy that guarantees neither stability, nor sovereignty, nor decent welfare,' he said. 'We need to create an effective mechanism to rebuild the economy and find and attract the necessary…material and human resources (Putin 2012).' More recently, he called for a widening of import-substitution programmes in May 2014, during a presentation to the St Petersburg International Economic Forum, and for active support of the national STI in 2016. 'Russia needs a real technological revolution,' he said, 'serious technological renewal, the most extensive in the last half-century, massive re-equipping of our enterprises'. In 2014 and 2015, action plans were launched in various industrial sectors, in order to produce cutting-edge technologies and reduce dependence on imports. Target products include high-tech machine tools, equipment for the oil and gas sectors, power engineering machinery, electronics, pharmaceuticals, chemicals and medical instruments. The federal Law on Industrial Policy adopted in 2014 provides a comprehensive package of supportive measures for companies, including investment contracts, R&D subsidies, preferential public procurement of the technologies produced, standardization, the creation of industrial parks and clusters and so on. A Fund for Industrial Development was established the same year to support highly promising investment projects initiated by companies. In 2016, the *Strategy of Science and Technology Development towards 2035* was developed and adopted. A new law regulating STI activities and policies is being finalised.

1.4 A New Agenda for Government Policy

In May 2012, the president approved several decrees proposing directives for STI development. These decrees fix qualitative objectives that are to be measured against quantitative targets to 2018 (Table 2). Although the potential for developing STI is relatively high, this potential is held back by weaknesses in private investment, low scientific productivity and incomplete institutional reforms. A fundamental lack of receptiveness to innovation and poor demand from many firms and organizations for scientific achievements and new technologies still hampers progress in this area. All stakeholders in the Russian innovation system, including economic actors, feel an urgent need for institutional change and more effective implementation of government policies. There are other bottlenecks too, which, if not overcome, could bring to naught state initiatives.

During last years, several policy documents[1] have identified the principal orientations of national S&T policies, as well as related implementation mechanisms. A wider format for promoting STI in Russia was provided by the report entitled *Strategy—2020: a New Framework for Innovation Policy*. It was drafted by leading Russian and international experts. Some of the ideas put forward in the report have since been transformed into official documents and are outlined below (Gokhberg and Kuznetsova 2011a).

1.5 Preservation of the Traditional Budget-Oriented Model of Science Funding

Gross domestic expenditure on research and development (GERD) in Russia rose by 2 times at constant prices from 2000 to 2015; federal budget allocations for civil R&D—even4 times. Nevertheless, Russia's R&D intensity remained relatively stable; in 2015, GERD accounted for 1.13% of GDP, compared to 1.25% in 2009 (Table 3). After rising steadily for years, state expenditure on R&D dropped slightly in 2010 because of the global financial crisis in 2008–2009 but has since

[1]Including the *Presidential Decree on the Approval of the Priority Areas for the Development of Science and Technology and the List of Critical Technologies* (2011), the *Strategy for Innovative Development* to 2020 (2012), the State Programme for Development of Science and Technology, 2013–2020 and the Federal Goal-oriented Programme on Research and Development in Priority Areas of Russia's Science and Technology Complex (2012).

Table 2 Objectives and quantitative targets to 2018 of the May 2012 presidential decrees

Decree	Objectives	Quantitative targets to 2018
On long-term economic policy (No. 596)	To increase the pace and sustain ability of economic growth and raise the real income of citizens	Labour productivity to grow by 150%
	To achieve technological leadership	Increase the share of high-tech industries in GDP by 130%
On measures to implement state social policy (No. 597)	To improve the conditions of employees in social sectors and science	Increase the average salary of researchers to double that of the average salary in the region
On measures to implement state policy in the field of education and science (No. 599)	To improve state policy in education and science and the training of qualified professionals to meet the requirements of the innovation economy To improve the efficiency and performance of R&D sector	Increase total funding of public scientific foundations to 25 billion roubles Raise the GERD/GDP ratio to 1.77% (by 2015) Increase the share of GERD performed by universities to 11.4% Boost Russia's world share of publications in the Web of Science to 2.44% (by 2015)

recovered (Fig. 1). The government fixed a target in 2012 of raising GERD to 1.77% of GDP by the end of 2015 (Table 2), which would bring it closer to the average for the European Union: 1.92% in 2012. Unfortunately, so far the goal has not been achieved.

In absolute terms, government funding of R&D amounted to PPP$35 billion in 2015, on a par with that of Germany and Japan (about US$33 billion) (HSE 2017a).

The low share of industry-financed R&D is a perennial concern. Despite government efforts, the contribution of industry to GERD in Russia actually fell from 32.9 to 26.5% between 2000 and 2015 (Table 4). This trend, unfortunately, is quite stable (unlike many other countries). This sector, which encompasses privately and publicly owned companies and large-scale industrial R&D institutes, nevertheless performs the bulk of GERD: 59.2% in 2015, compared to 31% for the government sector, 9.6% for higher education and just 0.1% for the private non-private sector (HSE 2017a).

The low propensity of companies to finance research is reflected in the modest place occupied by R&D in total expenditure on innovation: 22.9% overall in industry; 38.6% in high-tech sectors. On average, significantly less is spent on R&D than on the acquisition of machinery and equipment (48.2, 28.7%, respectively). In EU countries, the situation is diametrically the opposite. In Sweden, the ratio is even 5:1 and, in Austria and France, about 4:1. In the Russian industry, a low proportion of investment goes on acquiring new technologies (1.6%), including

Table 3 GERD/GDP ratio in the Russian Federation, 2003–2015

	Russia	USA	EU average	Germany	UK	Japan	Brazil	China	India	South Africa	Turkey
2003	1.29	2.55	1.70	2.46	1.67	3.14	0.96	1.13	0.71	0.79	0.48
2004	1.15	2.49	1.67	2.42	1.61	3.13	0.90	1.22	0.74	0.85	0.52
2005	1.07	2.51	1.67	2.43	1.63	3.31	0.97	1.31	0.81	0.90	0.59
2006	1.07	2.55	1.70	2.46	1.65	3.41	1.01	1.35	0.80	0.93	0.58
2007	1.12	2.63	1.70	2.45	1.69	3.46	1.10	1.39	0.79	0.92	0.72
2008	1.04	2.77	1.77	2.60	1.69	3.47	1.11	1.46	0.84	0.93	0.73
2009	1.25	2.82	1.84	2.73	1.75	3.36	1.17	1.66	0.82	0.87	0.85
2010	1.13	2.74	1.84	2.72	1.68	3.25	1.16	1.71	0.80	0.76	0.84
2011	1.09	2.77	1.88	2.80	1.68	3.38	1.21	1.78	0.81	0.76	0.86
2012	1.05	2.71	1.92	2.87	1.61	3.34		1.91		0.76	0.92
2013	1.06	2.74	1.93	2.82	1.66	3.48		1.99			0.94
2014	1.09	2.76	1.95	2.89	1.68	3.59		2.02			1.01
2015	1.13	2.79	1.95	2.87	1.7	3.49		2.07			

Source HSE (2017a); OECD's Main Science and Technology Indicators database, 2016/2; for Brazil and India: UNESCO Institute for Statistics

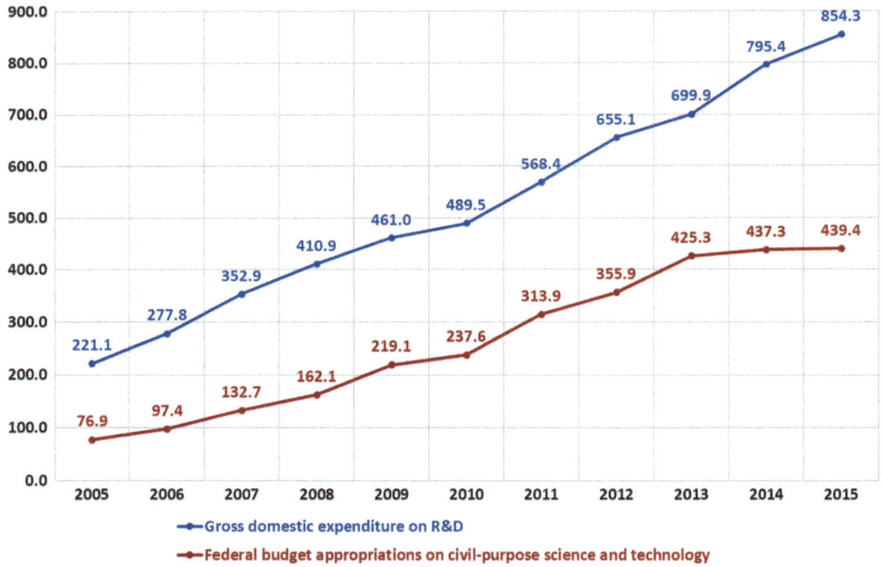

Fig. 1 GERD and Federal budget allocations for civil science and technology in Russia, 2005–2015. *In billions of roubles at current prices. Source* HSE (2017a)

patent rights and licenses (1.1%). This phenomenon is characteristic of all types of economic activity and limits both Russia's technological potential and its capacity to produce ground-breaking inventions (HSE 2017b). Normally, the generation of new knowledge and technologies would be expected to be driven by

Table 4 Industry-financed GERD in Russia as a share of GDP, 2007–2015 (%)

	Russia	USA	Germany	UK	Japan	China	South Africa	Turkey
2007	0.68	1.70	1.72	0.80	2.69	0.98	0.39	0.35
2008	0.69	1.70	1.75	0.77	2.71	1.05	0.40	0.34
2009	0.68	1.72	1.80	0.78	2.53	1.14	0.37	0.35
2010	0.64	1.86	1.82	1.02	2.49	1.26	0.30	0.36
2011	0.62	1.90	1.89	1.078	2.60	1.34	0.30	0.37
2012	0.61	1.87	1.95	1.02	2.56	1.45	0.29	0.42
2013	0.64	1.93	1.90	1.06	2.65	1.52	–	0.45
2014	0.65	1.96	1.95	1.09	2.79	1.56		0.50
2015	0.67	1.99	1.95	1.12	2.74	1.59		

Source OECD (2014, 2016)

technology-based start-ups and fast-growing innovative companies, including small and medium-sized enterprises (SMEs). However, this type of company is still uncommon in Russia (Table 4).

Table 5 depicts a growing orientation of R&D towards the needs of industry since 2008 and a drop in non-targeted (basic) research, referred to in official statistics as the general advancement of research. The share of R&D allocated to societal issues has risen somewhat but remains modest. The thin slice of the pie directly devoted to environmental issues has shrunk farther and that for energy-related research has stagnated; this is disappointing, given the growing interest globally in environmentally sustainable technologies. It also comes somewhat as a surprise, since the government has adopted a number of policies in recent

Table 5 GERD in the Russian Federation by socio-economic objective, 2009 and 2015 (%)

Socio-economic objectives	2008	2013	2015
Agriculture	2.6	2.4	2.2
Energy	4.3	4.4	3.1
Industry	25.1	28.2	27.0
Other economic objectives	4.4	5.4	4.3
Human health	2.3	3.0	3.0
Control and care of the environment	1.0	0.8	0.8
Social development	0.8	1.5	1.3
General advancement of research[a]	25.7	17.4	15.9
Earth and exploration and exploitation of the atmosphere	2.9	4.4	4.7
Civil space	3.2	6.9	6.3
Other	27.6	25.6	31.3
Total	100	100	100

[a]Refers to basic research
Source HSE (2017a)

years as part of an action plan for sustainable green growth that is aligned with the *Green Growth Strategy* of the Organisation for Economic Co-operation and Development (OECD 2011).

In 2009, the government adopted *State Policy Priorities to Raise Energy Efficiency in the Electric Power Engineering Sector based on the Use of Renewable Energy Sources*, covering the period to 2020. In 2012, it adopted *Principles of the State Policy on the Ecological Development of the Russian Federation*, which is valid to 2030. The problem of green growth and social progress is addressed by four Russian technology platforms: Environmentally Clean Efficient Fuel; Technologies for Ecological Development; Biotech 2030; and Bio-energy. These platforms co-ordinate the activities of industrial companies, research centres and universities to promote R&D and technology in related areas. Collectively, these measures represent only the first leg of the journey towards sustainable growth, of course.

The modest investment so far in sustainable technologies can largely be explained by the business sector's tepid interest in green growth. Empirical data show that 60–90% of Russian companies do not use advanced general-purpose and resource-saving technologies, or alternative energy-generating technologies and have no plans to do so in the near future. Only one in four (26%) innovative enterprises are producing inventions in the environmental field. Even when companies do have recourse to environmentally friendly inventions like energy-saving technologies, this gives them virtually no competitive advantage in the domestic market. Most companies are focusing their efforts on reducing environmental pollution, in order to comply with government standards. Very few are engaged in waste recycling or in substituting raw and other materials for more environmentally friendly ones. For instance, only 17% of companies use environmental pollution control systems (HSE estimates; HSE 2017b). This state of affairs prompted the government to adopt a series of regulations in 2012–2014 which encourage the usage of the best available technologies for reducing environmental waste, saving energy and upgrading technologies through a series of positive incentives (such as tax exemptions, certification and standardization) and negative ones, such as fines for environmental damage or higher energy tariffs.

1.6 Priorities of Innovation Activity

In the course of its transition to a market economy Russia has become an attractive destination for foreign technologies. Between 2009 and 2015, the number of patent applications submitted in Russia by foreign applicants increased 3 times—16,248 units (HSE 2017a, b). Patent activity by Russian applicants grew more slowly. As a result, the coefficient of technological dependence increased: the ratio of foreign to domestic patent applications submitted in Russia went from 0.23 in 2000 to 0.56 in 2015. If one takes into consideration the low patenting activity by Russian applicants abroad, this sends a negative signal to national policy-makers about the actions that are to be taken to ensure competitiveness of domestic technologies in the global market.

Intellectual property titles represent only roughly 6% of technology exports and only few companies engaged in R&D got revenue from exports of technology. The latter generated just US$1.7 billion in 2015, virtually the same as in previous years, compared to US$2.6 billion for Canada, US$6.6 billion for the Republic of Korea and US$136 billion for the USA (HSE 2017a). The Russian Federation's membership of the World Trade Organization since 2012 should help to boost technology transfer through exports and related revenue.

2 Human Resources

2.1 The Structure of S&T Personnel

Although Russia ranks 43th in the latest Global Innovation Index and 23th in the sub-index for human capital development (GII 2016), international competition for talent is intensifying. The issue of developing skills and behavioural patterns in line with the country's development strategy has never been more pressing in Russia. Policies introduced in recent years have addressed this urgent question.

In 2015, there were 738,857 people engaged in R&D in Russia, a group encompassing researchers, technicians and support staff. In absolute numbers, Russia figures among the world leaders for R&D personnel, coming only after the USA, Japan and China. However, there is an imbalance in the dynamics and structure of R&D personnel. Researchers (by head count) account for little more than half of R&D personnel (379,411) and support staff 41%, compared to just 8.4% for technicians. The large share of support staff can be explained by the dominance of R&D institutes, which have traditionally tended to function in isolation from both universities and enterprises and required labour intensive services for maintaining premises and finance. Russia lags behind many countries in terms of the number of people engaged in R&D per 10,000 employees, and the relative indicator for researchers. Two-thirds of R&D personnel are employed by state-owned organizations (HSE 2017a).

Between 2010 and 2015, there were some signs of improvement in the age pyramid. The proportion of researchers under the age of 40 rose to more than 40% and has since stabilized at this level. This trend reflects absolute growth in two age groups, those of scientists under the age of 30 and those aged between 30 and 39 years. After a long period of growth, the share of researchers over the age of 60 has at last stabilized in recent years at roughly 25% of the total (HSE 2017a).

2.2 A Serious Change in the Remuneration System of Researchers

In 2012–2013, several roadmaps were adopted to improve the attractiveness of careers in research, in order to stimulate productivity, redress the age pyramid and give research a greater economic impact. These documents introduced a new remuneration system primarily for researchers employed by public research institutes and universities. The corresponding target indicators were established by the *Presidential Decree on Measures to Implement State Social Policy* (2012) and the implementation schedule is controlled by the government.

The action plan fixes the target of raising researchers' salaries to at least 200% of the average wage in the region where the researcher is based by 2018. There are also similar plans to raise the salaries of teachers in universities and other institutions offering higher education programmes. Currently, research institutes and universities receive annual subsidies from the federal budget to enable them to increase salaries, as happens also for secondary schools, hospitals and agencies managing social security. The average salary of researchers tends to be rather high in Russia's research hubs like the Moscow region,[2] thereby contributing to the unequal distribution of R&D potential across the country. Reaching the aforementioned target in these research hubs may turn out to be problematic, as raising salaries that are already fairly generous will mean allocating substantial additional funding to R&D. Whatever their status, all regions may find it hard to reach the '200%' target, on account of budget shortfalls and the slowdown in the pace at which institutional reform is being implemented in the R&D sector.

In order to prevent the rise in researchers' salaries from becoming a goal in itself without any strong connection to their performance and the socio-economic impact of their work, the action plan also introduces performance-related pay mechanisms, implying that researchers will be regularly evaluated on their productivity (Gershman and Kuznetsova 2013, 2014).

2.3 Holders of University Degrees

Russia has long had a relatively high level of education. In recent years, interest in pursuing higher education has not waned. On the contrary, a Russian could expect to spend 15.7 years in the education system in 2013, up from 13.9 years in 2000. According to the 2010 population census, more than 27 million people over the age

[2]Roughly 60% of Russian researchers work in Moscow, the Moscow Region and St Petersburg. Six other regions together account for about 20%: Nizhny Novgorod, Ekaterinburg, Novosibirsk, Rostov, Tyumen and Krasnodar.

of 15 years hold university degrees in Russia, up from 19 million in 2002. This represents about 23% of the adult population, compared to 16% in 2002. In the 20–29-year age group, the percentage is as high as 28%, although this is down from 32% in 2002. At 55%, the overall proportion of the population with some form of tertiary education—including those with non-degree qualifications—is well above that of any member of the Organisation for Economic Co-operation and Development (OECD). Moreover, the number of people enrolled in higher education per 1000 inhabitants has risen sharply in the past decade from 162 in 2002 to 234 in 2010.

The rise in student rolls can partly be attributed to the hike in government spending on education in recent years. Federal expenditure on higher education now amounts to 0.7% of GDP and 3.7% of overall federal budget appropriations. This compares with 4.3% of GDP for public expenditure on education as a whole, or 11.4% of the consolidated budget (federal and regional levels). Spending per tertiary student has doubled since 2005 (HSE 2014a, c).

2.4 Training Scientists Becoming a Core Mission of Research Universities

As of the 2013/2014 academic year, 5.6 million students were enrolled in the country's tertiary institutions, 84% of which were state-owned: 2.8% of students were studying natural sciences, physics and mathematics; more than 20% engineering; 31% economics and management; and a further 20% humanities.

In Russia, postgraduate programmes that confer a Candidate of Science degree (equivalent to a Ph.D.) lead to the highest scientific degree in Russia, the Doctor of Science. In 2013, some 1557 institutions offered postgraduate programmes in science and engineering, almost half of which (724) were universities and other tertiary institutions and the remainder research institutes. Some 38% of these institutions (585) also hosted doctoral courses, including 398 universities. Women made up just under half (48%) of the 132,002 postgraduate and 4572 doctoral students in science and engineering. Most of the postgraduates (89%) and Doctor of Science candidates (94%) specializing in scientific disciplines are on the university payroll. The dominance of universities in this regard has been in force for decades, though the proportion of postgraduate students trained by research institutes was nearly threefold higher in early 1990s (36.4% in 1991) compared to its current level. This means that the education of highly qualified scientists in Russia, like elsewhere, is increasingly becoming a core mission of universities and a top priority. Engineering, economics, law, medicine and pedagogy are the preferred broad disciplines for postgraduate study.

2.5 High Level of Ng University Research—A Top Priority the Government

Russia's higher education sector has a long-standing research tradition that dates back to the Soviet Union. About seven out of ten universities perform R&D today, compared to half in 1995 and four out of ten in 2000. However, universities still occupy a fairly lowly position when it comes to the generation of new knowledge: in 2013, they performed just 9% of GERD. Although this is up from 7% in 2009 and on a par with China (8%), it remains less than in either the USA (14%) or Germany (18%). Although university staff are still insufficiently engaged in R&D, the situation has improved in recent years: the proportion of professors and teaching staff conducting research rose from 19 to 23% between 2010 and 2013 (HSE 2014a).

Boosting support for university research has become one of the most important strategic orientations of STI and education policies in Russia. This process has been under way for almost a decade. One of the first steps was the National Priority Project for Education, initiated in 2006. Over the next two years, 57 higher education institutions received competitive grants from the federal budget for the purposes of implementing innovative educational programmes and high-quality research projects, or acquiring research equipment.

Between 2008 and 2010, 29 institutions received the coveted label of national research university. The aim is to turn these 29 national research universities into centres of excellence. In parallel, eight federal universities are being turned into 'umbrella' institutions for regional education systems. This status entitles them to large-scale government support but there are strings attached—in return, they are expected to produce high-quality research, education and innovation.

Currently, the magnitude of support given to higher education and its main orientations are determined by the *Presidential Decree on Measures to Implement State Policy in the Field of Education and Science* (2012) and the *State Programme for the Development of Education*[3] (2013–2020). The presidential decree anticipates that universities will be performing 11.4% of GERD by 2015 and 13.5% by 2018 (Table 2). Moreover, the level of engagement of university staff in R&D has become a major criterion for proficiency testing and professional advancement.

[3]This programme provides schools, colleges and universities with full-scale financing for equipment procurement, offers subsidies to the best secondary schools and technical colleges, finances advanced teachers' training, etc.

3 STI Governance

3.1 Higher Education Must Adapt to Economic Needs

Despite undeniable success in boosting university research in recent years, one urgent problem remains: the discrepancy between the structure and quality of professional training, on the one hand, and current economic needs, on the other (Gokhberg and Kuznetsova 2011b; Kuznetsova 2013). This is reflected not only in the composition of educational programmes, graduate specializations and diplomas but also in the relatively small scale and low level of applied research, experimental development and innovation performed by universities.

In recent years, one of the most important steps towards modernizing higher education has been the adoption of the Federal Law on Education in 2012; it outlined the contours of a modern system respectful of international practices and standards, new developments in educational programmes and technologies, as well as new teaching methods and approaches to conducting experimental development and innovation.

3.2 Joining the Bologna Process

In accordance with the Bologna Declaration (1999), which launched the process of developing a European Higher Education Area, the various echelons of the Russian higher education system have been aligned with the International Standard Classification of Education to give:

- At the undergraduate level, the Bachelor's degree;
- At the postgraduate level, specialist training leading to a diploma or a Master's degree;
- Postgraduate study for academic staff leading to a Candidate of Science degree, equivalent to a Ph.D.

New legislation has raised the standards for a Ph.D. and made the process more transparent. University consortia and networking have been introduced into educational curricula and universities have been given the right to set up small innovative firms to commercialize their intellectual property. Students may also apply for scholarships or earmarked loans to cover the costs of their education.

3.3 New Funding Mechanisms

The 5/100 Programme was adopted in 2013 to raise the global competitiveness of Russian universities to the point where five of them figure in the top 100 (hence the programme's name) and the remainder in the top 200 of global university rankings. In 2013–2015, 15 leading universities[4] were selected on a competitive basis to receive earmarked subsidies to help raise their global competitiveness in both science and education. To this end, over 10 billion roubles were earmarked for 2013–2014 and 40 billion roubles for 2014–2016. The selection criteria included the university's publication output, international research collaboration, academic mobility and the quality of strategic programmes. These 15 universities are subject to a performance evaluation each year.

The *Presidential Programme for Advanced Training of Engineers* was launched in 2012. It offers training programmes and internships in leading research and engineering centres in Russia and abroad, with a focus on strategic industries. Between 2012 and 2014, the programme enabled 16,600 engineers to obtain higher qualifications and 2100 to train abroad; the programme involved 96 tertiary institutions located in 47 regions. The 'customers' of this programme were 1361 industrial companies which seized this opportunity to develop their long-term partnerships with tertiary institutions.[5]

The Russian Science Foundation[6] is a non-profit organization set up in 2013 to expand the spectrum of competitive funding mechanisms for research in Russia. The foundation received 48 billion roubles in state funding for 2013–2016. R&D-performing institutions may apply for grants to fund their large-scale projects in basic or applied research. To obtain a regular grant, applicants must include young scientists in their project team and guarantee that at least 25% of the grant will be spent on the salaries of young researchers. In 2015, the Russian Science Foundation launched a special grants programme to support post docs and introduced short- to medium-term internships to increase academic mobility (Schiermeier 2015). A total of 1100 projects received funding in 2014, one-third of which were in life sciences. Among the thematic priorities announced for the next call for proposals in 2015 weare: new approaches to identifying the mechanisms of infectious diseases, advanced industrial biotechnologies, neuro technologies and neuro cognitive research.

In recent years, the government has augmented its arsenal for stimulating research funding. A special government programme has been offering 'megagrants' to universities and research centres since 2010 to help them attract leading scientists. So far, the programme has seduced 144 world-class researchers, half of them foreigners, including several Nobel laureates. All the invitees have been selected to lead new laboratories with a total staff of more than 4000 scientists at 50 top Russian universities; this has led to the publication of 1825 scientific papers, more than 800 of which have appeared in scientific journals indexed by the Web of

---

[4]Including St Petersburg Polytechnic, the Far-East Federal University and three national research universities: the Higher School of Economics; Moscow Institute of Physics and Technology; and Moscow Institute of Engineering and Physics.

[5]See: http://engineer-cadry.ru.

[6]Not to be confused with the Russian Foundation for Basic Research, set up in 1993 to issue grants for basic research.

Science. Just 5% of applications were submitted by women, which explains why only 4 of the 144 megagrants went to principal investigators who were women (Schiermeier 2015). A total of 27 billion roubles in public funding has been allocated to the megagrants programme over 2010–2016, with recipient universities contributing about 20% of the budget.

In parallel, the government has increased funding for 'old' state foundations,[7] which focus on basic research and humanities, as well as for innovative SMEs (Gokhberg et al. 2011). It has also introduced grants to develop research networks and co-operation between universities, the national academies of science and industry, within the framework of the State Programme for the Development of Science and Technology for 2013–2020. Leading universities participating in this programme are expected to raise the share of their budget devoted to technology transfer from 18 to 25% between 2012 and 2020.

A Basic Research Programme has been designed for 2013–2020 to co-ordinate national efforts. It is part of the overarching State Programme for the Development of Science and Technology and contains specific provisions for selecting priorities in basic research and for an open public evaluation of scientific achievements. These provisions include the presentation of the programme's results in a freely accessible database and the mandatory publication of open-access articles on the Internet.

3.4 Incentives for Business

Since 2010, the government has also introduced a number of schemes to stimulate innovation in the business sector. These include:

- Programmes that make it mandatory for state-owned enterprises to develop innovation strategies and co-operate with universities, research institutes and small innovative businesses; to qualify for this programme, state-owned enterprises must raise their spending on R&D and actively produce innovative products, processes or services;
- a *Federal Law on Public Procurement* (2013) providing for the purchase of high-tech and innovative products by the state and promoting state procurement of goods and services from SMEs;
- state technology-oriented programmes supporting particular industrial sectors (aircraft, shipbuilding, electronics, pharmaceuticals, etc.) and overarching areas, such as biotechnology, composite materials, photonics, industrial design and engineering; and the Small and Medium-sized Enterprise Development Programme covering 2013–2020, which includes the distribution of federal

[7]The Russian Foundation for Basic Research, the Russian Foundation for Humanities and the Foundation for Assistance to Small Innovative Enterprises were all set up in the early 1990s.

budget subsidies to co finance regional SME development, support local clusters of engineering and prototyping centres and provide credit guarantees through the national system of guarantor institutions, the core of which is the new *Credit Guarantee Agency* (est. 2014).[8]

In 2015, two schemes were announced to drive technological development. The first is the *National Technology Initiative*; it introduces a new long-term model for achieving technological leadership by creating novel technology-based markets, such as in non-piloted drones and automobiles for the industrial and services sectors, neuro technological products, network-based solutions for customized food delivery and so on; technological projects will be coupled with support for the training of schoolchildren and students in these promising areas. The second scheme targets major traditional sectors and consists in funding a series of national technological projects with a high innovation component through public–private partnerships, with a focus on smart power engineering, agriculture, transport systems and health services, among other areas.

A key issue for businesses concerns how to demonstrate tangible results from their research. One possible mechanism would be for the state to allocate budgetary funds to businesses on the condition that expenses be cofinanced by interested companies and that effective partnerships be established between research institutes, universities and business enterprises (Gokhberg and Kuznetsova 2011a; Kuznetsova et al. 2014). It is also important to ensure co-ordination between government programmes targeting STI and programmes implemented by institutions oriented towards development, in order to build the so-called 'innovation lift' intended to pursue novel technologies, products and services along the overall innovation chain—from an idea to the market. It goes without saying that it would be vital to monitor the performance of these programmes to make timely adjustments.

3.5 The Problems in the Field of Patent Activities

The national intellectual property market is still at the developmental stage, with research output taking years to influence the economy: only 2–3% of all current patents are in use and patenting tends to be done more intensively than licensing of intellectual property. It is precisely during commercialization that the real competitive advantages emerge, such as income from the use of protected inventions and the accumulation of know-how. In Russia, however, the development of intellectual property is often disconnected from specific consumer needs and industrial demand.

[8]In 2015, it was renamed the Federal Corporation for the Development of Small and Medium Enterprises, a public company with 100% state ownership.

This makes it important to improve a legislative framework for intellectual property. The main regulation in this area comes from Section VI of the Civil Code, which is specifically devoted to issues related to intellectual property and the enactment of legislation. New norms developed in this area over the period 2009–2014 include:

- Assigning intellectual property rights generated by public research to the Russian Federation and establishing the principle of the free transfer of intellectual property from the public sector to industry and society, making it easier for research centres and universities to deal with licenses or other forms of commercialization of intellectual property;
- Regulating the conditions, amount and procedures relative to the payment of fees to authors for creating and commercializing in-service research results and technologies;
- Establishing an exhaustive list of the conditions under which the state may obtain exclusive rights to the fruit of intellectual creativity.

An action plan adopted by the government in 2014 contains additional measures for protecting intellectual property rights at the 'pre-patent' stage and on internet and introduces specialized patent courts, as well as better professional training in this area. Gradual steps are also being taken to improve the conditions under which R&D is capitalized upon, including by placing intellectual property on company balance sheets. This is particularly important for SMEs, as it allows the mto increase their balance sheet value, for example, or to attract investment and use their exclusive rights as a pledge to obtain credits.

3.6 Tax Instruments for R&D and Innovation

All fiscal affairs have been governed by a single document since 2008, the Russian Tax Code. The most important amendments in recent years concern new rules for calculating R&D expenditure and classifying certain specific types of spending by organizations as R&D expenditure, along with new regulations on the creation of reserves for forthcoming expenditure.

New tax incentives have been introduced since 2011 in favour of innovative SMEs, start-ups and spin-off companies, in particular:

- Zero tax (for three years) on profits channelled into developing intellectual property; in parallel, taxes on transactions involving intellectual property have been removed;
- Benefits and extensions to patent duty payment deadlines are offered to SMEs, as well as to individual inventors (enterprises);
- Residents of Skolkovo Innovation Centre have been given a 'tax holiday' for up to ten years (Box 2).

In the near future, there are plans to introduce tax incentives for individuals, such as business agents, inventors or entrepreneurs, who invest in projects developing innovation (or innovative companies) and for companies desirous to expand their intangible assets.

3.7 Institutional Reforms

The institutional structure of the Russian R&D sector is not yet fully adapted to the market economy. In the Soviet era, basic research was conducted predominantly by the research institutes of the state academies of science and major universities, whereas applied research and experimental development were concentrated mostly in branch institutions, design bureaux and specialized units of industrial enterprises. All R&D organizations were state-owned. Nowadays, most of the so-called industrial R&D in Russia is performed by large companies or legally independent research institutes. Industrial enterprises and design bureaux are mostly privately owned or semi-private organizations. This said, seven out of ten R&D-performing institutions are still state-owned, including universities and enterprises in which the government has a share of the capital. As already noted, small companies in the R&D sector are underrepresented, especially in comparison with other industrial nations (HSE 2017a).

Unaffiliated research institutes and design bureaux tend to dominate institutions of higher education and enterprises when it comes to R&D: they represented 48 and 9% of all R&D units respectively and employed three-quarters of all R&D personnel in 2015 (Table 6). Industrial enterprises account for just 7.4% of all R&D units, compared to 18% for institutions offering higher education (HSE 2017a). The government's desire to optimize the institutional structure of research triggered a long-awaited reform of the state academies of science[9] in 2013 that will have far-reaching consequences for Russian science (Box 1).

In parallel, the government is pursuing its plans to expand the network of state research centres (they now number 48) and to create a new network of large-scale national research centres (Gokhberg and Kuznetsova 2010). The first of these national research centres resulted, in 2009, from the subordination of three R&D institutes to the Kurchatov Research Centre, which specializes in nuclear energy and a broader spectrum of convergent[10] technologies. The second centre on a similar scale was established in the aircraft sector in 2014 by attaching several R&D institutes to the Central Aero-hydrodynamic Institute, renowned for aeronautic research. The Krylov Research Centre for Shipbuilding and the Research Institute

[9]Prior to the reform of 2013, there were six Russian academies: the Academies of Sciences; Medical Sciences; Agricultural Sciences; Education; the Arts; and Architecture and Construction Services.

[10]Such as bionanotechnology, neurobiology, bioinformatics, etc.

Table 6 Breakdown of R&D units in Russia by type and personnel, 2015

Types of R&D organisations	R&D organisations by type, %	R&D personnel by type of R&D organisations, %
Research institutes	40.9	58.9
Design organisations	7.7	18.4
Construction project and exploration organisations	0.7	0.4
Experimental enterprises	1.5	0.4
Higher education institutions	24.9	8.1
Industrial enterprises	8.9	7.3
Others	15.4	6.4

Source HSE (2017a)

for Aviation Materials are the next candidates on the list. To monitor the efficiency of national research infrastructure and identify avenues for targeted support, new arrangements were introduced in 2014 to assess the performance of public research institutions in the civil sector regularly.

Box 1: Reform of the Academy of Sciences

The reform of the Russian Academy of Sciences had been debated for over a decade. Since the late 1990s, the academy had functioned as a quasi-ministry, managing federal property and overseeing the network of institutions, which carried out the bulk of basic research in Russia. In 2013, the six academies comprising this sector accounted for 24% of Russia's research institutions, about one-fifth of R&D personnel, 36% of researchers and 43% of all researchers with Candidate and Doctor of Science degrees. They thus grouped a highly qualified labour force.

However, many of the institutions attached to the academy had developed a top-heavy age pyramid, with about one-third of researchers being over the age of 60 (34% in 2013), including about 14% over 70. The academies were also accused of low productivity—they received 20–25% of government research funding—and a lack of transparency. There was certainly a conflict of interest, in so far as some of those in charge of the academy and the distribution of resources among subsidiary institutes also happened to head these same institutes. Critics also reproached the academies for a lack of prioritization and weak ties to universities and industry.

The Russian Academies of Sciences, Agricultural Sciences and Medical Sciences attracted the most criticism, as they grouped about 96% of the research institutes placed under the academies, 99% of the academies' funding and 98% of their researchers in 2013. A series of 'soft' reforms in recent years had ironed out some problems, such as the introduction of rotation for management posts, greater internal mobility, a mandatory retirement age and teaching requirements and the expansion of competitive grants.

In September 2013, the government's long-awaited reform got under way with the adoption of a law stipulating the merger of the Russian Academy of Sciences with the two smaller academies for medical and agricultural sciences. The Russian Academy of Sciences was entitled to keep its name. A month later, the government passed a law establishing the Federal Agency for Research Organizations, with direct reporting lines to the government. These two laws served the immediate objective of establishing a system with two nodes of power divided between the Russian Academy of Sciences, on the one hand, and the Federal Agency for Research Organizations, on the other. The functions of co-ordinating basic research, evaluating research results across the entire public research sector and providing expert advice remain the preserve of the Russian Academy of Science, whereas the management of the academy's finances, property and infrastructure now falls to the Federal Agency for Research Organizations.

The more than 800 institutes that used to belong to the three academies of science are now formally the property of the Federal Agency for Research Organizations, even though they may still bear the label of one of the academies. This network remains extensive: the 800 institutes employ about 17% of researchers and produce nearly half of Russia's international scientific publications.

Source: Gokhberg et al. (2011), HSE (2017a), Stone (2014), authors' estimates.

3.8 R&D Priorities and Critical Technologies

Russia has an established system for identifying priorities so that resources can be distributed effectively to a limited number of fields, taking into account national objectives and both internal and external challenges. The current list encompasses eight priority areas and 27 critical technologies based on the results of a foresight exercise conducted in 2007–2010 (Shashnov and Grebenyuk 2011). This list was approved by the president in 2011. These research priorities address global challenges, ensure national competitiveness and promote innovation in key areas; they are being used to design governmental programmes for R&D and to streamline funding for other policy initiatives (see Sokolov 2013). Six priority areas for civil-purpose science and technology and their share of total funding structure are:

- Transport systems and space (37.7%);
- Safe and efficient energy systems (15.6%);
- ICTs (12.2%);
- Environmental management (6.8%);
- Life sciences (6.0%); and
- Nanotechnology (3.8%).

Two other priority areas refer to defence and national security.

In 2014, work began on updating this list, once the government had approved the findings of the most recent foresight exercise, *S&T Foresight—2030*, conducted between 2012 and 2014 (Gokhberg 2016; Sokolov and Chulok 2016). The report's recommendations are intended to serve as early-warning signals for the strategic planning of enterprises, universities, research institutes and government agencies.

Box 2: Skolkovo Innovation Centre: A Temporary Tax Haven Near Moscow

The Skolkovo Innovation Centre is currently under construction in the city of Skolkovo, near Moscow. This high-tech business complex has been designed to attract innovative companies and nurture start-ups in five priority areas: energy efficiency and energy saving; nuclear technologies; space technologies; biomedicine; and strategic computer technologies and software.

The complex was announced by the president in November 2009. It consists mainly of a technological university and a tech no park and is headed by Russian oligarch Viktor Vekselberg and co-chaired by former Intel head Craig Barrett. In order to woo potential residents, a bill according the residents of Skolkovo special legal, administrative and fiscal privileges was adopted by the State Duma (parliament) in September 2010. The law granted residents substantial benefits for up to ten years, including exemption from income tax, value-added tax and property taxes, as well as reduced insurance premiums of 14% rather than the going rate of 34%.

The law also made provision for the establishment of the Skolkovo Fund to support development of the university and thereby give personnel the skills that companies need. One of the centre's biggest partners is the Massachusetts Institute of Technology in the USA.

Once corporations and individuals become 'residents' of the city, they are entitled to apply for grants from the fund. Residents also have access to the centre's legal and financial infrastructure. In 2010, the government published a decree granting highly skilled foreign nationals who secured employment at Skolkovo a three-year work visa.

The Skolkovo Innovation Centre is financed primarily from the Russian federal budget. Its budget has increased steadily since 2010 and amounted to 17.3 billion roubles in 2013. A brand new motorway has been built linking Skolkovo to Moscow.

Today, more than 1000 companies from 40 Russian regions have set up shop in Skolkovo. In 2013, 35 agreements were signed with major global and domestic companies, including Cisco, Lukoil, Microsoft Nokia, Rosatom and Siemens. Industrial partners plan to open 30 R&D centres in Skolkovo, which would create more than 3000 jobs.

Source: compiled by authors

See also: http://economy.gov.ru/minec/press/interview/201412244.

3.9 Developing Technologies to 'Shrink' Distances

The development of transport systems has two key motivations: to strengthen the global reach of domestic technologies and ensure continuity across Russia's vast territory through the development of regional aviation hubs and high-speed railways.

S&T Foresight—2030 suggests some orientations for specific transport sectors. It recommends that the aircraft industry focus its technological portfolio on reducing the weight of planes, on the use of alternative fuels (bio fuel, condensed and cryogenic fuel), the development of 'smart' cabins for pilots with front windshield-based information panels and new composite (non-metal) materials, coatings and constructions (Gokhberg 2016). The Sukhoi Super jet 100 (SSJ) is one example of recent technological progress; this new-generation regional aircraft is equipped with advanced technologies and meets the demand of both domestic and global civil aviation markets. A novel integrated power system for regional and long-haul aircraft is also being developed by Snecma (the French Safran Group) and Saturn (Russia).

The state programme for the shipbuilding industry was adopted in 2013. This sector is experiencing a renaissance. More than 200 enterprises are engaged in manufacturing vehicles for maritime and inland cargo shipping, equipment for exploiting oil and gas reserves on the continental shelf, commercial and scientific shipping. The United Shipbuilding Corporation (est. 2007) is Russia's largest company in this sector; this fully state-owned company encompasses 60 enterprises and accounts for about 80% of the domestic shipbuilding industry's turnover, with exports to 20 countries.

According to Foresight—2030 and a special report on Foresight for Shipbuilding (Dekhtyaruk et al. 2014), research objectives for this industry principally concern the following areas: the development of composite materials based on nanotechnologies, organic and non-organic synthesis, metallurgy and thermal treatment; construction using novel materials and coatings; techniques to maximize the economic performance of vehicles; the construction of high-performance propulsion systems for small vessels based on the novel principles of energy generation, storage and conversion; high-performance tools and systems for ensuring the safety and durability of ships and vessels, including modern radio-electronic equipment based on nanotechnologies; and the design of highly automated smart adjustable systems for industrial production.

3.10 Alternative Energy and Energy Efficiency

Given the energy sector's key contribution to GDP and exports, any changes have an immediate impact on national competitiveness. You could say that, when the energy sector sneezes, the Russian economy catches a cold. In 2014, the government launched the Energy Efficiency and Development programme to tackle the challenges facing the sector, including low energy efficiency, high extraction costs for fuel and the predominant orientation towards traditional sources of energy. Within this programme, funds have been earmarked for the development of electric power engineering and the oil, gas and coal industries—but also alternative energy sources. Since 2010, four technological platforms have been in place for an Intellectual Energy System, Environmentally Neutral and Efficient Heat and Power Engineering, Advanced Technologies for Renewable Energy and Small Distributed Generation Systems.

There have been some noteworthy achievements in the field of alternative energy in recent years. High-performance separators, turbines and allied equipment are being used in the construction of new geothermal power stations in Kamchatka and Kurils, for instance. Mini-power plants using biogas generated from waste have also been built in many regions. Engines are also being produced for wind farms and small hydropower plants. In 2013, a complex engineering project got under way to develop the Prirazlomnaya ice-strengthened platform, offering a strong impetus for the exploitation of the Arctic shelf.

A cluster of projects are developing energy-efficient technologies at Skolkovo (Box 13.2). These focus on reducing energy consumption in industry, housing and municipal infrastructure. For example, the New Energy Technologies company is developing efficient thermos-electric generators for the direct conversion of thermal energy into electricity, based on nano structured membranes and highly efficient solar converters based on organic polymers. Meanwhile, the Wormholes Implementation company is creating intelligent systems for the monitoring and optimal exploitation of wells, in order to increase the efficiency of oil extraction and oil field development.

Foresight—2030 identifies 14 thematic areas for highly-promising applied R&D related to energy. These include specific technologies for the efficient prospecting and extraction of fossil fuels, effective energy consumption, bio-energy, storage of electric and thermal energy, hydrogen-based power generation, deep processing of organic fuels, smart energy systems, high-power fourth-generation water-cooled nuclear reactors and optimizing energy and fuel transportation (HSE 2014c).

3.11 Innovative Territorial Clusters

In the past five years, the government has taken steps to strengthen institutional infrastructure for the commercialization and transfer of technology. In 2012, Russia launched a series of pilot innovative territorial clusters to promote value-added production chains and drive growth in the regions. Initially, 25 clusters were selected on a competitive basis out of nearly a hundred applications. The applicants were cluster consortia grouping industry, research institutes and universities supported by local administrations. The clusters represent a variety of regions stretching from Moscow to the Far East; they specialize in areas ranging from high-tech (ICTs, biotechnology, nuclear energy, etc.) to the more traditional manufacturing sectors of the automotive, shipbuilding, aircraft and chemical industries.

In 2013, the 14 best-prepared clusters received funding from federal and regional authorities on a 50:50 basis (matching principle); in 2014, a further 11 clusters were earmarked for support. The next stage of the national cluster policy will involve creating broader regional cluster programmes and cluster development centres to ensure co-ordination and networking.

3.12 Technology Platforms and Engineering Centres

The first technology platforms were set up in Russia in 2010. They serve as a communication tool to unite the efforts by the state, businesses and the scientific communities to identify challenges, develop strategic research programmes and implementation mechanisms and encourage promising commercial technologies, new goods and services in specific economic sectors. There are currently 34 technology platforms across the country involving over 3000 organizations: 38% concern businesses, 18% universities, 21% research institutes and the remainder NGOs, business associations and so on. In many cases, the platforms' strategic research programmes have been inspired by the recommendations of *S&T Foresight—2030* (Gokhberg 2016).

Two key tools used to regulate the activity of these platforms are the co-ordination with government technology-oriented programmes and the provision of interest-free loans for innovative projects from the Russian Technology Development Fund, which was renamed the Foundation for Industrial Development in 2014.

Among the best-performing platforms are Medicine of the Future; Bio-industry and Bioresources—BioTech2030; Bio-energy; Environmentally Neutral and Efficient Heat and Power Engineering; Advanced Technologies for Renewable Energy; Technologies for Hydrocarbon Extraction and Use; Hydrocarbon Deep Processing; Photonics; and Aviation Mobility.

All 34 platforms will be evaluated to assess their level of support for industry; the list of platforms will then be adjusted accordingly. State support will only be renewed for those platforms that have demonstrated a high potential and tangible results.

Research and federal universities, state research centres and academic institutes form the core of the country's federal centres for collaborative use of scientific equipment, the first of which appeared in the mid-1990s. Since 2013, these centres have been brought together in a network of 357 entities to improve their effectiveness. Their funding comes from the Federal Goal-oriented Programme for

Research and Development in Priority Areas. Centres can obtain annual subsidies of up to 100 million roubles for a maximum of three years for a specific project.

In 2013, a related pilot project started to create engineering centres at leading technological universities. Its objective is to advance university-led development and the provision of engineering and training services. Support comes from budgetary subsidies that offset some of the expenses incurred in carrying out projects in engineering and industrial design: in 2013, each centre received 40–50 million roubles, for a total of 500 million roubles in subsidies.

Technoparks

There are currently 88 technoparks in Russia. The main tools of public support for these are the programme for The Creation of High-Tech Technoparks in the Russian Federation (2006) and, since 2009, an annual competitive programme for SMEs. Technoparks mostly specialize in ICTs, medicine, biotechnology, instrument-making and mechanical engineering but one-third (36%) exhibit a cross-sectorial specialization.

Technopark policies are fraught with problems, owing to some 'grey areas' in legislation and organizational procedures. According to the Russian Association of Technoparks in High-Tech Sectors, only 15 technoparks are actually effective.[11] The remainder are in the planning, construction or winding-up stages. The main reason for this is the excessive length of time taken by regional authorities to establish the titles to plots of land and give town-planning permission, or to render decisions on funding.

Special Zones

Special economic zones date back to 2005, when the government decided to instigate a favourable regime for innovative entrepreneurship at the local level. Certain locations were identified specifically to encourage the development of new high-tech businesses and high-tech exports.

By 2014, five such zones were in operation in St Petersburg, Dubna, Zelenograd, Tomsk and the Republic of Tatarstan. These five zones host a total of 214 organizations. Each one benefits from a preferential regulatory environment, such as a zero property tax for the first ten years or other tax benefits, free customs regimes, preferential leasing terms, the opportunity to buy plots of land and state investment in the development of innovation, engineering, transport and social infrastructure. In order to increase the efficiency of these policy instruments, particular attention should be paid to arriving at a critical mass of organizations and to strengthening linkages between residents and the external environment.

---

[11]Some technoparks were not able to fulfil their mission and achieve the prescribed objectives (measured in terms of highly skilled jobs created, turnover of goods produced and services rendered to resident businesses, completed projects, etc.). See: http://nptechnopark.ru/upload/spravka.pdf.

4 Trends in International Co-operation in

4.1 *Political Tensions Are Affecting Some Areas of Co-operation*

Economic sanctions imposed on Russia by the EU in 2014 are limiting co-operation in certain areas, such as dual-use military technologies, energy-related equipment and technologies, services related to deep-water exploration and Arctic or shale oil exploration. The sanctions may ultimately affect broader S&T cooperation.[12]

Over the past 20–25 years, there has also been significant co-operation with the USA in key areas such as space research, nuclear energy, ICTs, controlled thermonuclear fusion, plasma physics and the fundamental properties of matter. This co-operation has involved leading universities and research organizations on both sides, including Moscow State University and Saint Petersburg University, Brookhaven and Fermi national laboratories and Stanford University. The level of mutual trust was such that the USA even relied on Russian spacecraft to transport its astronauts to the International Space Station after its own space shuttle programme was wound up in 2011.

However, these contacts with the USA are now being affected by the recent political tensions over Ukraine. For example, joint efforts to secure nuclear materials actually ceased when the US Department of Energy announced the termination of co-operation in April 2014. For the time being, co-operation between Russia and the USA is being maintained at the level of particular research centres and universities. This approach was approved, for example, by a meeting of the Skolkovo Scientific Advisory Council in November 2014 in Stanford (USA). At this meeting, several areas were selected for joint activities, namely brain and other bioscience research, molecular diagnostics, environmental monitoring and the forecasting of natural and technogenic emergencies.

4.2 *STI Cooperation with BRICS Countries*

The reforms implemented in Russia include a serious ‘rationale’ for partnerships with foreign countries, such as with the fellow BRICS countries, Brazil, India, China and South Africa, as well as other rapidly developing nations.

In 2013, the RF President Vladimir Putin approved the Concept of participation of the Russian Federation in the BRICS association. The Concept sets the main goals of Russia’s cooperation with BRICS countries in the science, technology, and innovation sphere:

---

[12]See: http://europa.eu/newsroom/highlights/special-coverage/eu_sanctions/index_en.htm#5.

- exchanging information on science and technology policy and programmes, and formulating joint long-term problem-oriented cooperation programmes on this basis;
- encouraging research in the areas of primary interest for the Russian Federation and other BRICS states, such as aeronautics, high-speed transportation vehicles, microelectronics and information technology, nanotechnology, food security and sustainable agriculture, biotechnology and veterinary science, medical science, basic research, search for and exploration of mineral resources, remote sensing of the Earth, climate change, water resources and water treatment technology;
- cooperating in the field of space research and the use of space technology;
- providing organisational, legal, financial and personnel support for cooperation in the field of science, technology and innovation in the BRICS framework, including creating high-tech zones (science parks) and incubators, forming common technology platforms, promoting joint investment in development of high technologies, research and innovation centres such as Skolkovo in Russia and similar centres in other BRICS states;
- increasing interaction in the area of education, training of scientific personnel, and implementation of joint research programs;

At the sixth BRICS summit in Brazil in 2014, the five partners established a New Development Bank hosted by China, and a Contingency Reserve Agreement (CRA) to provide them with alternatives to the World Bank and International Monetary Fund in times of economic hardship, protect their national economies and strengthen their global position. Russia is contributing US$18 billion to the CRA, which will be credited by the five partners with a total of over US$100 billion. The CRA is already operational. Currently, work is under way to develop financing mechanisms for innovative projects with the new bank's resources.

On 18 March 2015, in Brasilia governments of the BRICS countries signed international agreement "Memorandum of Understanding on Cooperation in Science, Technology and Innovation". The document aims at strengthening cooperation in the following areas:

- Exchange of information on policies and programmes and promotion of innovation and technology transfer;
- Food security and sustainable agriculture;
- Natural disasters;
- New and renewable energy, energy efficiency;
- Nanotechnology;
- High performance computing;
- Basic research;
- Space research and exploration, aeronautics, astronomy and earth observation;
- Medicine and biotechnology;
- Biomedicine and life sciences (biomedical engineering, bioinformatics, biomaterials);

- Water resources and pollution treatment;
- High tech zones/science parks and incubators;
- Technology transfer;
- Science popularization;
- Information and communication technology;
- Clean coal technologies;
- Natural gas and non-conventional gases;
- Ocean and polar sciences;
- Geospatial technologies and its applications.

The Brazilian Declaration's Work Plan made each country responsible for coordinating cooperation in one of five priority areas:

- Prevention and management of natural disasters (Brazil);
- Water resources, and preventing pollution (Russia);
- Geospatial technologies and their application (India);
- New and renewable energy sources, energy efficiency (China);
- Astronomy (SAR).

Between 1 April 2015 and 15 February, 2016 Russia for the second time in history acted as BRICS chairman. Russia's and partner countries' long-term goal is turning BRICS into a full-fledged mechanism for strategic cooperation on key issues.

The Moscow Declaration signed by Brazilian, Russian, Indian, Chinese, and South African education and science ministers in October, 2015 established new working groups on major research infrastructures, funding multilateral research projects, technology commercialisation, and innovation. The following new priorities were added to the list of BRICS countries' cooperation areas:

- Establishing BRICS Young Scientists Forum (coordinated by India);
- Biotechnology and biomedicine, including healthcare and neuroscience (coordinated by Russia and Brazil);
- Information technology and high-performance computing (coordinated by China and SAR);
- Marine and polar studies, and relevant technologies (coordinated by Russia and Brazil);
- Materials science, including nanotechnology (coordinated by Russia and India);
- Photonics (coordinated by Russia and India).

Memorandum of Understanding for Establishment of the BRICS Network University signed in 2015 has started off a joint research and education project aimed at creating an integrated education environment, promoting academic mobility, and training highly qualified professionals specialising in areas seen as priorities by the BRICS countries. In April, 2016 in Yekaterinburg, at the First Forum of the BRICS Network University Rectors, work plans on developing joint curricula were approved. According to these plans, the first summer and winter schools will be held in the next year already, along with exchange programmes for participating universities; a year later joint Masters' and Ph.D. programmes are planned to be launched. The BRICS University thematic priorities include energy, informatics and information security, BRICS countries' research, ecology and climate change, water resources and pollution prevention, and economics.

The BRICS STI Framework Programme aims to support excellent research on priority areas, which can best be addressed by a multinational approach. The initiative should facilitate cooperation among the researchers and institutions in the consortia, which consist of partners from at least three of the BRICS countries.

As part of the initiative, the following research funding organizations from the BRICS countries have agreed to jointly establish a new scheme for funding multilateral cooperative activities:

Brazil: National Council for Scientific and Technological Development (CNPq);
Russia: Foundation for Assistance to Small Innovative Enterprises (FASIE); Ministry of Education and Science (MON); Russian Foundation for Basic Research (RFBR);
India: Department of Science and Technology (DST);
China: Ministry of Science and Technology (MOST); National Natural Science Foundation of China (NSFC);
South Africa: Department of Science and Technology (DST); National Research Foundation (NRF).

Collaborative multilateral basic, applied and innovation research projects in the following thematic areas can be submitted in response to the call:

- Prevention and monitoring of natural disasters
- Water resources and pollution treatment
- Geospatial technology and its applications
- New and renewable energy, and energy efficiency
- Astronomy
- Biotechnology and biomedicine including human health and neuroscience
- Information technologies and high performance computing
- Ocean and polar science and technology
- Material science including nanotechnology
- Photonics

In 2016, in the scope of the BRICS Framework Science, Technology and Innovation Programme, coordinated multilateral calls for project proposals were held jointly by the Russian Ministry of Education and Science, the Brazilian National Council for Scientific and Technological Development, the Russian Fund for Assistance to Small Innovative Enterprises, the Russian Foundation for Basic Research, the Indian Department of Science and Technology, the Chinese Ministry of Science and Technology, the National Natural Science Foundation of China, and the South African National Research Foundation, in the following subject areas.[13]

- Prevention and management of natural disasters;
- Water resources, and preventing pollution;
- Geospatial technologies, and their application;
- New and renewable energy sources, energy efficiency;
- Astronomy;
- Biotechnology and biomedicine, including healthcare and neuroscience;
- Information technology, and high-performance computing;
- Marine and polar studies, and relevant technologies;
- Materials science, including nanotechnology;
- Photonics.

The Jaipur (India) Declaration, and BRICS countries' Work Plan on Cooperation in Science, Technology and Innovation signed by the BRICS countries' ministers in October, 2016 extended cooperation prospects in the following areas:

- Establishing the Network Innovation Platform, as a mechanism for coordinating cooperation and involving businesses and academic communities, promoting technology transfer, and involving small and medium enterprises, innovative and technology clusters, science parks and incubators, and R&D and innovation centres from BRICS countries in innovation and technology development;
- Establishing the BRICS Global Research Infrastructures Network, to support joint mega-science projects. In 2015 Russia proposed an initiative to set up the Global Research Advanced Infrastructure (GRAIN) in the BRICS format, to provide access to mega-projects' infrastructure to all BRICS countries. Implementation of these plans would allow countries to avoid duplicating similar projects on their own territory, sparing them unnecessary expense and allowing to make more efficient use of huge resources invested in development of research infrastructures.

The main goals of cooperation in the sphere of Research Infrastructure are:

- to promote cooperation within large-scale research infrastructure;
- to support initiatives leading to efficient use and development of mega-science projects in the BRICS countries thus contributing to implementation of the BRICS Research and Innovation Initiative;
- to create dynamically developing complex of Research Infrastructures amongst BRICS member countries for providing fundamental and applied tasks solution on a cutting-edge of science;
- to engage Global research community to the BRICS RI.

[13]The 2017 call for proposals of basic research projects held by the Russian Foundation for Basic Research jointly with organisations participating in the BRICS Framework Science, Technology and Innovation Programme.

These goals will be achieved through:

(i) Coordination of existing large-scale national programs of the BRICS countries and harmonization their national strategies,
(ii) Establishment of the BRICS Global Research Advanced Infrastructure Network (BRICS GRAIN).

Collaboration on geospatial technologies and their application is also strengthening, specifically for sustainable development purposes in the food security context; to assess and reduce risks of accidents and disasters; to study Earth, natural, and anthropogenic phenomena. BRICS countries show significant interest in stepping up cooperation in the area of physical and satellite geodesy.

The project "Joint National Telemedicine Systems—an Efficient Way to Improve the Quality of Medical Assistance and Social Services in BRICS Countries" approved by professionals from all BRICS countries and proposed for funding by the BRICS New Development Bank is aimed at putting in place an effective and cost-efficient advanced medical infrastructure, and supporting sustainable development of BRICS countries and developing nations on the basis of cutting-edge approaches and solutions. In June, 2016 representatives of BRICS countries signed an agreement on establishing International Telemedicine Society which would coordinate joint work on development and application of integrated telemedicine systems.

Current agreements on Russian-Chinese S&T cooperation:

- Agreement between the RF government and the PRC Government on S&T cooperation (18 December, 1992)
- Protocol to the Agreement on S&T cooperation (of 18 December, 1992) on the principles of protection and allocation of intellectual property rights, signed by the RF government and the PRC Government on 25 February, 1993
- Inter-government international agreement "Memorandum of Understanding on Cooperation in Science, Technology and Innovation", signed by the BRICS countries on 18 March 2015.

4.3 STI Cooperation with China

Russia's dynamic collaboration with China stems from the Treaty on Good Neighbourliness, Friendship and Co-operation signed by the two countries in 2001, which has given rise to regular four-year plans for its implementation. The treaty provides the basis for about 40 collaborative projects, as well as student exchanges at the secondary and tertiary levels and the joint organization of conferences and symposiums, among other forms of co-operation. Dozens of joint large-scale projects are carried out. They concern the construction of the first super-high-voltage electricity transmission line in China; the development of an experimental fast neutron reactor; geological prospecting in Russia and China; and joint research in optics, metal processing, hydraulics, aerodynamics and solid fuel cells. Other priority areas for co-operation include industrial and medical lasers, computer technology, energy, the environment and chemistry: geochemistry, catalytic processes, new materials, including polymers, pigments, etc. One new priority theme for high-tech cooperation concerns the joint development of a new long-range civil aircraft. To date, the aircraft's basic parameters have been elaborated, as well as a list of key technologies and a business plan, which has been submitted for approval.

Russia and China are also co-operating in the field of satellite navigation, through a project involving Glonass (the Russian equivalent of GPS) and Beidou (the regional Chinese satellite navigation system). They have also embarked on a joint study of the planets of our Solar System. A resident company of Skolkovo, Optogard Nanotech (Russian) and the Chinese Shandong Trustpipe Industry Group signed a long-term deal in 2014 to promote Russian technologies in China. In 2014, Moscow State University, the Russian Venture Company and the China Construction Investment Corporation (Chzhoda) also signed an agreement to upscale co-operation in developing technologies for 'smart homes' and 'smart cities'.

Russo–Chinese collaboration shifts from knowledge and project exchanges to joint work. Joint technoparks have been operated in the Chinese cities of Harbin, Changchun, Yantai and others since 2003. Within these technoparks, there are plans to manufacture civilian and military aircraft, space vehicles, gas turbines and other large equipment using cutting-edge innovation, as well as to mass-produce Russian technologies developed by the Siberian Branch of the Russian Academy of Sciences.

In the past few years, the Russian government has removed a number of administrative barriers to closer international co-operation with its partners. For example, the visa application process has been simplified, along with labour and customs regulations, to promote academic mobility and flows of research equipment and materials related to collaborative projects.

An important mechanism for implementing cooperation projects and programmes in the S&T area which has proved its efficiency is the Chinese-Russian Sub-commission for S&T Cooperation of the Russian-Chinese Commission on Preparation of Meetings of Heads of Governments.

Mutually beneficial and efficient cooperation between the PRC and the RF goes on in numerous S&T areas, including the following priority fields:

- Nanotechnology;
- Information and telecommunication systems;
- Life sciences;
- Efficient environment management;
- Energy efficiency and energy saving;
- Nuclear energy, etc.

Russia's main partners in terms of the number of joint publications in the 2000s were (in descending order) Germany, the US, France, the UK, Italy, and Japan; note that relevant figures (absolute and relative ones alike) are growing. For the first time China joined the first twenty of Russia's S&T cooperation partners. This is evidence of gradual re-orientation of Russia's scientific links towards China, which rapidly improve sits positions on the global academic arena. Results of bibliometric analysis of joint publications indexed in the Web of Science database in 2003–2014 show that the number of joint Russian-Chinese publications was 6.6 thousand. Main subject areas of S&T cooperation include the following:

- Elementary particles physics;
- Quantum field theory;
- Nuclear physics;
- Applied physics;
- Interdisciplinary studies in the materials science field;
- Earth sciences;
- Optics;
- Condensed state physics;
- Physical chemistry;
- Biochemistry and molecular biology;
- Instrumentation;
- Atomic, molecular, and physical chemistry;
- Nuclear science and technology;
- Spectroscopy;
- Metallurgy and metal lography;
- Interdisciplinary studies in the field of chemistry;
- Meteorology, and atmospheric sciences.

Among China's international initiatives, the Daya Bay Reactor Neutrino Experiment in the elementary particles physics aimed at studying neutrino oscillations[14] is considered to have the best prospects. The international partnership includes researchers from China, Russia, the US, Taiwan, and the Czech Republic. The experiment is conducted at an installation comprising three detectors; anti-neutrinos are generated by six nuclear reactors located 500 metres away from the detectors.

Together with Russia and a number of other countries, China participates in the International Thermonuclear Experimental Reactor (ITER) project.

One of the priority areas for Russian-Chinese cooperation is stepping up joint innovation activities, and extending the range of joint initiatives by R&D centres, academic institutes, and science parks. More than 30 academic institutes cooperate with various Chinese R&D centres in the scope of inter-institute agreements.

In July 2016, the Declaration on Establishing the Association of the Russian Federation and the People's Republic of China Universities was signed in Moscow.

[14]Daya Bay neutrino oscillation facility's official website [Electronic resource]. Access mode: http://dayabay.ihep.ac.cn/twiki/bin/view/Public/.

200 Russian and 600 Chinese universities have already established partnerships in nationally important strategic areas, signing 900 direct agreements. Eight associations of Russian and Chinese universities were created.

The Russian Foundation for Basic Research actively cooperates with the National Natural Science Foundation of China. The partners hold calls for proposals of basic research projects implemented jointly by Russian and Chinese scientists in the following areas:

- Mathematics, mechanics, and informatics;
- Physics and astronomy;
- Chemistry, and materials science;
- Biology, and medical sciences;
- Earth sciences;
- Application of natural science methodologies in humanities;
- Information and communication technologies, computing systems;
- Basic foundation of engineering sciences.

5 Conclusion

5.1 A Need for Longer-Term Horizons in Policy-Making

Despite the current complex economic and geopolitical situation, Russia has the firm intention of consolidating its national innovation system and pursuing international co-operation. In January 2015, the Minister of Education and Science, Dmitry Livanov, told Nature magazine as much. ‘There will be no substantial reductions in the level of science funding caused by the current economic situation,’ he said. ‘I strongly believe that scientific co-operation should not depend on temporary changes in the economic and political situation. After all, the generation of new knowledge and technologies is a mutually beneficial process’ (Schiermeier 2015).

The rapidly changing landscape of science and technology—with supply and demand for innovation shifting incessantly—is obliging policy-makers to address longer-term horizons and tackle emerging challenges. In a context of rapidly evolving global economic and geopolitical climates, coupled with growing international competition, both the government and public and private companies need to adopt more active investment strategies. To this end, future policy reforms in Russia should incorporate:

- preferential support for competitive centres of excellence, taking into account international quality standards for research and the centres’ potential for involvement in global networks; research priorities should be influenced by the recommendations of Foresight—2030;

- better strategic planning and long-term technology foresight exercises; an important task for the near future will be to ensure the consistency of foresight studies, strategic planning and policy-making at the national, regional and sectorial levels and that national priorities are translated into targeted action plans;
- greater financial support for the research of leading universities and research institutes, together with incentives for them to collaborate with businesses and investment bodies;
- further development of competitive research funding, coupled with a regular assessment of the effectiveness of budget spending in this area;
- stimuli for technological and organizational innovation in industry and the services sector, including subsidies for innovative companies—particularly those engaged in import substitution–tax deductions for companies investing in high-tech companies, a wider range of incentives for companies to invest in R&D, such as tax rebates and corporate venture funds; and
- regular appraisals of specific institutional mechanisms to support innovation, such as the technology platforms, and monitoring of their funding levels and performance.

STI will obviously develop most intensively in those sectors where resources are concentrated, such as in fuel and energy, traditional high-tech manufacturing and so on. At the same time, we expect to see future STI intensity around newly emerging competitive industries where the conditions for global competition have already been met, such as in advanced manufacturing, nanotechnology, software engineering and neuro technology.

In order to strengthen domestic STI in a globally competitive environment, Russia needs to establish a climate conducive to investment, innovation, trade and business, including through the introduction of tax incentives and lighter customs regulations. The National Technology Initiative adopted in 2015 has been devised to ensure that Russian companies capture their share of future emerging markets.

It is of vital importance that administrative barriers blocking the entry to markets and the development of start-ups be removed; the intellectual property market must also be further liberalized by gradually reducing the role of the state in managing intellectual property and enlarging the class of owners, with the introduction of support measures to raise demand for innovation. Some of these issues have been addressed in the action plan adopted in 2015 to implement *Russia's Strategy for Innovative Development to 2020.*

References

Dekhtyaruk, Y., Karyshev, I., Korableva, M., Velikanova, N., Edelkina, A., Karasev, O., et al. (2014). Foresight in civil shipbuilding—2030. *Foresight—Russia, 8*(2), 30–45.

Gershman, M., & Kuznetsova, T. (2013). The 'effective' contract in science: the model's parameters. *Foresight—Russia, 7*(3), 26–36.

Gershman, M., & Kuznetsova, T. (2014). Performance-related pay in the Russian R&D sector. *Foresight—Russia, 8*(3), 58–69.

GII. (2016). *The Global Innovation Index 2016: Winning with Global Innovation*. Ithaca/Fontainebleau/Geneva: Cornell University/INSEAD/WIPO.

Gokhberg, L. (Ed.). (2016). *Russia 2030: science and technology foresight*. Ministry of Education and Science of the Russian Federation, National Research University Higher School of Economics, Moscow.

Gokhberg, L., Kitova, G., Kuznetsova, T., & Zaichenko, S. (2011). *Science policy: A global context and Russian practice*. Moscow: Higher School of Economics.

Gokhberg, L., & Kuznetsova, T. (2010). Russian Federation. In *UNESCO science report 2010.*

Gokhberg, L., & Kuznetsova, T. (2011a). Strategy2020: A new framework for innovation policy. *Foresight—Russia, 5*(4), 40–46.

Gokhberg, L., & Kuznetsova, T. (2011b). S&T and innovation in Russia: Key challenges of the post-crisis period. *Journal of East-West Business, 17*(2–3), 73–89.

Gokhberg, L., & Kuznetsova, T. (2015). Russian Federation. *The UNESCO science report: Towards 2030* (pp. 342–363). Paris: UNESCO.

HSE. (2014a). *Education in figures: 2014. Brief data book.* Moscow: Higher School of Economics.

HSE. (2014b). *Science. Innovation. Information society: 2014. Brief data book.* Moscow: Higher School of Economics.

HSE. (2014c). *Education in the Russian Federation: 2014. Data book.* Moscow: Higher School of Economics.

HSE. (2017a). *Science indicators: 2017. Data book.* Moscow: Higher School of Economics.

HSE. (2017b). *Indicators of innovation activities: 2017. Data book.* Uses OECD data. Moscow: Higher School of Economics.

Kuznetsova, T. (2013). Russia. In V. Scerri & H. M. M. Lastres (Eds.), *BRICS national system of innovation. The role of the state*. Routledge.

Kuznetsova, T., Roud, V., & Zaichenko, S. (2014) Interaction between Russian enterprises and scientific organizations in the field of innovation. *Foresight—Russia, 8*(1), 2–17.

Ministry of Finance. (2014). *Execution of the federal budget and budget system of the Russian Federation*. Moscow.

OECD. (2011). *Towards green growth*. Paris.

OECD. (2014). *Main science and technology indicators* (Vol. 2014, Issue 1). Paris.

OECD. (2016). *Main science and technology indicators* (Vol. 2014, Issue 2). Paris.

Putin, V. (2012). We need a new economy. *Vedomosty*. January 30, 2012.

Rosstat—Russian Federal Statistical Service. (2015). *Russian Statistical Yearbook: 2015*. Moscow.

Schiermeier, Q. (2015). Russian science minister explains radical restructure. *Nature*.

Shashnov, S., & Grebenyuk, A. (2011). Science and technology priorities for modernization of Russian Economy. *Foresight—Russia, 5*(2), 48–56.

Sokolov, A. (2013). Foresight in Russia: Implications for policy making. In D. Meissner, L. Gokhberg & A. Sokolov (Eds.), *Science, technology and innovation policy for the future: Potential and limits of foresight studies*. Springer.

Sokolov, A., & Chulok, A. (2016). Priorities for future innovation: Russian S&T Foresight 2030. *Futures, 80*, 17–32.

Stone, R. (2014). Embattled president seeks new path for Russian academy. *Science*. http://news.sciencemag.org.

Chapter 6
India Report on Science, Technology and Innovation

Zuhua Shan, Liangliang Bi and Bingqing Xin

1 Basic Situation

India is a country with a long history of science and technology development. Since ancient times, the country has accomplished many outstanding achievements. However, when India became independent, its science and technology infrastructure was fairly weak and its organizational structure was far from adequate. As a result, the country relied heavily on foreign support for science and technology. After decades of efforts guided by national needs, India has strengthened its science and technology infrastructure and capability, continuously reducing its dependence on foreign countries and growing from small to big, weak to strong. Furthermore, India made major progress in the areas of institutions, facilities, service and product, etc. In both basic and applied research fields, India commands considerable strength, which provides powerful support and motivation for national economic construction and development.

1.1 Evaluation and Analysis of India's National Innovative Competitiveness

For a comprehensive analysis of India's national innovative competitiveness, China Science and Technology Exchange Center and other organizations have excerpted the innovation scores of the BRICS countries for 2001–2015 from 5 well-known domestic and foreign national innovation index reports including The World Economic Forum's *The Global Competitiveness Report*, INSEAD's *Global Innovation Index*, *The World Innovative Competitiveness Development Report*

Z. Shan (✉) · L. Bi · B. Xin
China Science and Technology Exchange Center, Beijing, China

X. Zhao et al. (eds.), *BRICS Innovative Competitiveness Report 2017*,
Research Series on the Chinese Dream and China's Development Path,
https://doi.org/10.1007/978-981-10-8078-4_6

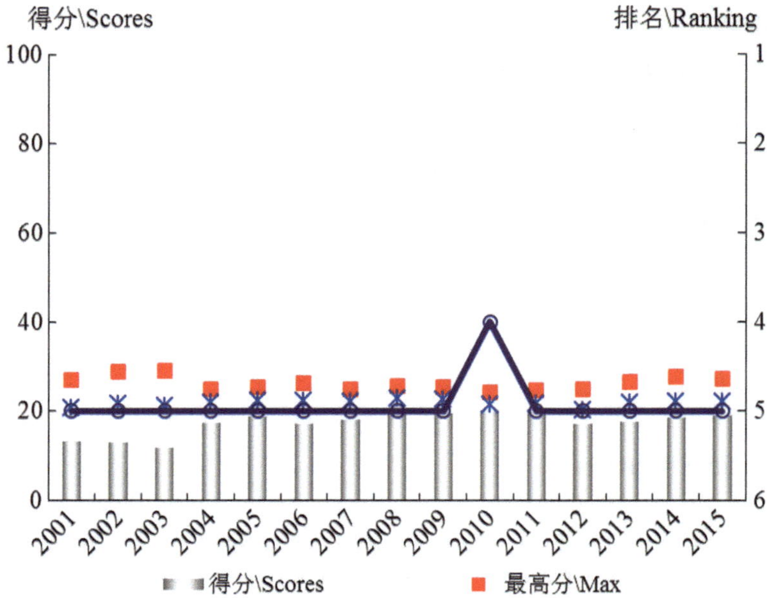

Fig. 1 India's national innovative competitiveness score and ranking 2001–2015

issued by the Chinese Academy of Social Sciences and Fujian Normal University, *Bloomberg Innovation Index* and China Academy of Science and Technology for Development's *National Innovation Index*. A portfolio evaluation method is adopted to calculate the national integrated Innovative competitiveness index and the rankings of the BRICS countries for the past 15 years and forecast their Innovative competitiveness for the next 5 years.

Figure 1 shows changes in India's position and score on national Innovative competitiveness among the BRICS countries.

The innovation index forecast results for India in the next 5 years are given in Table 1.

(1) As for the change in overall rankings, India came 5th among the BRICS countries in terms of national Innovative competitiveness in 2015, same as its ranking in 2001. In 2010, India rose to 4th place but later fell back to 5th position. Overall, the rankings fluctuated during the evaluation period.
(2) As for the scores, India scored 19.10 points on the national Innovative competitiveness in 2015, 8.21 points lower than the highest score and 2.99 points lower than the average score of the BRICS countries. Compared with 2001,

Table 1 Innovation index forecast of India in the next 5 years

Year	2016	2017	2018	2019	2020
Innovation index	24.97	24.90	25.40	26.02	26.79

India saw a rise of 5.90 points in Innovative competitiveness in 2015, narrowing the gap from both the highest score in 2001 by 5.59 points and the BRICS average score by 4.61 points.

(3) According to the forecast, India's innovation index score will rise year by year, expecting to reach 26.79 points by 2020.

1.2 STI Policy, Strategy and Development Plans

1.2.1 S&T Management System

The science and technology management system of India mainly consists 2 levels, the central government level and the state government level:

1. **The science and technology department of the central government**. The Ministry of Science and Technology (MOST) is established by the central government of India as the administrative department for science and technology work in India, also the main undertaker and consulting agency for the formulation of major science and technology policies and plans. Its main responsibilities involve promoting national development of science and technology, coordinating and guiding related activities. MOST not only performs the administrative functions of the government but also directly undertakes scientific research and technology development activities. However, MOST has no independent administrative body; instead, it implements science and technology policies and plans through its science and technology management departments. Currently, the 3 vice-ministerial science and technology management departments implement most of the tasks:

   (1) **Department of Science and Technology** (DST): Responsible for promoting national development of science and technology, coordinating and guiding related activities. Its functions include: formulate national science and technology policy and plans, coordinate and support research institutions' research activities, support national technology development projects and their industrialization, coordinate and manage affiliated research institutions to carry out science and technology work, support basic and applied research, support science and technology infrastructure construction, support technology development and commercialization, engage in international cooperation in science and technology, and popularize science and technology.

   (2) **Department of Scientific and Industrial Research** (DSIR): Responsible for part of the policy formulation process and technology import management of R&D institutions inside enterprises. Its goal is to make Indian industries develop timely innovation abilities and competitiveness through research and technical means.

(3) **Department of Biotechnology** (DBT): Responsible for biological research policy and technology development management, supporting biotechnology R&D and production, promoting the large-scale use of biotechnology, and formulating biosecurity guidelines.

The highest-ranking officer of these 3 departments is Secretary, which is a vice-ministerial post. The Secretary of DST is responsible for the specific science and technology management affairs of MOST. The Secretary is the highest post a civil official can hold in India and also the highest level in the public service system. The office above is the Minister, which is generally appointed by the Prime Minister of India after a general election.

In addition, the central government of India also includes the following full-time departments for science and technology work: The Ministry of Earth Sciences (MOES), the Department of Atomic Energy (DAE) and the Department of Space (DOS).

2. Social, economic and other departments of the central government. Some industrials and economic departments in the central government of India are also involved in science and technology work. These departments mainly include: The Office of the Principal Scientific Advisor, the Planning Commission, the Ministry of Agriculture, the Ministry of Communications and Information Technology, the Ministry of Environment and Forests, the Ministry of Food Processing Industries, the Ministry of Health and Family Welfare, the Ministry of Heavy Industries and Public Enterprises and the Ministry of New and Renewable Energy. These departments administer and guide science and technology work in their respective industries.
3. The science and technology department of states. According to the Indian Constitution, each state is responsible for the establishment and the authority of the local governments. Some states have a department of science and technology, whereas some do not. However, all states have a relevant department in charge of planning, organizing, implementing and promoting science and technology work. The science and technology department of each state coordinates and manages their science and technology funds, institutions, human resources and projects based on their specific local situation as well as development plans to promote the development of the state.

1.2.2 STI Policy

Revitalizing the country through science and technology has been a policy India has long adhered to as the Government of India attaches great importance to science and technology innovation. The Indian Constitution stipulates that science must penetrate every aspect of national life and every fields Indians striving to achieve. Since India became independent, it has issued a total of 4 national science policy resolutions. As early as in 1958, the Indian Parliament issued a *Science Policy Resolution* (1958). At the time, the science and technology policy mainly focused on scientific research, while treating technology promotion as a byproduct of science and technology activities. In 1983, a *Technology Policy Statement* (1983) was established to advocate independent technological capability in India. During this period, the emphasis was mainly on the leading role of the government in economic planning, the protection and support of domestic enterprises, the promotion of an "import alternative" policy and the encouragement of enterprises to carry out independent R&D activities. A primary direction of achieving self-sufficiency and building India into a big power was put forward. Through science and technology legislations and the establishment of a science and technology decision-making system, India promoted the development of science and technology, set up a series of special R&D institutions, and focused on developing hi-tech industries. The government paid attention to the training of science and engineering talents and encouraged them to engage in R&D activities. Active government intervention enabled India to establish a complete industrial system within a short period of time. Innovation activities during this stage mainly centered on national research institutions. Universities and research institutions worked closely together on basic research. Enterprise R&D activities were primarily conducted by large-sized State-owned enterprises and mainly funded by the government. Such R&D activities included experimental R&D, as well as R&D to localize imported technologies. However, enterprise R&D activities were almost detached from the R&D activities undertaken by universities and national research institutions.

Entering the 21st century, the Indian government has successively issued a *Science and Technology Policy* (2003) and a *Science, Technology and Innovation Policy* (2013) and launched *India's Ten-Year Innovation Roadmap* (2010–2020). The *Science and Technology Policy* of 2003 combined science policy with technology policy and stressed the importance of R&D inputs, while the *Science, Technology and Innovation Policy* of 2013 took science, technology and innovation as a united framework and put forward the idea of establishing a national science and technology innovation (STI) system to open up a high technology-led development path for India, achieve faster, sustainable and inclusive growth and put India among the world's Top 5 science and technology powers by 2020. The new science and technology policy focused on the following 4 aspects:

(1) Increase R&D inputs. It plans to raise the R&D inputs/GDP ratio to 2% within 5 years and double the number of scientific publications. Through improving the investment environment of the private sector, stimulate enthusiasm among the private sectors and guide them to increase research inputs. In the next 5 years, the investment ratio of the public sector over the private sector should be changed to within 1:1 from 3:1 to achieve a doubled growth in science and technology inputs.
(2) Nurture innovation talent. Increase the full-time equivalent of R&D personnel in the whole country by 66% in the next 5 years. Nurture and recruit 4 types of talent: young talent, female talent, science and engineering talent and overseas talent with Indian origin. In the teaching stage, strengthen education reform, pay attention to science and engineering discipline building, and set up

cross-university research centers; at the research stage, actively participate in global R&D infrastructure construction and mega science projects, implement result-oriented personnel incentive policies, and attract more talents to the research field.

(3) Key fields of focus. Pay priority attention to the fields of agriculture, communications, energy, water management, healthcare, pharmaceuticals, materials, environment, climate diversity and change, etc. Boost R&D intensity through major strategic tasks; stimulate the export of hi-tech products and the development of the innovation industry.

(4) Support innovation and entrepreneurship. Energetically support innovation and entrepreneurship activities in the whole society by establishing a "Venture Creativity Fund", exploring "small creativity, small funds" mechanisms, strengthening entrepreneurship incubation services and encouraging business model innovation. Regarding innovation and entrepreneurship risks, design preventive mechanisms with the public sector as the leading sector and the private sector as participating sector, continue to explore new ways of financial investment and compensation, promote the policy of "first purchase first use" of innovative products, implement a legal framework of intellectual property rights sharing between inventors and investors, and protect against the risk of innovation and entrepreneurship failures from both the supply and the demand side.

(5) Create a sound atmosphere of innovation. On one hand, increase public awareness on science, release new white papers on science and engineering, publicize scientific knowledge, nurture a respectful attitude towards science among the public and improve the practical application ability of science among young people of all social status. On the other hand, raise the level of technology decision making, establish an independent, autonomous evaluation and indicators system, and shift from subjective intuition to reliance on evidence and methodology in science and technology decision making.

1.2.3 Science and Technology Development Plans

Since 1957, India has been formulating its national development strategies taking China's Five-Year plan as a reference. India's current Twelfth Five-Year Plan (2012–2017) has set specific development goals for the science and technology sector, specifying R&D inputs, science and technology output, talent training and other indicators, as shown in detail in Table 2.

In addition to national development plans, MOST has also formulated science and technology output indicators of its own department for the Twelfth Five-Year period. In 2011, DST issued a Working Group Report of the *Twelfth Five-Year Plan (2012–2017)*, planning the science and technology development goals for the Twelfth Five-Year period and setting specific goals for every important development field, as shown in detail in Table 3.

Table 2 Science and technology goals of India's Twelfth Five-Year Plan

Category	Goal	Category	Goal
R&D/GDP ratio	>2%	Share of global publications	>5%
Global ranking of SCI papers published	Enter the TOP6	Global ranking of Patent Cooperation Treaty (PCT)	Enter the TOP10
Number of full-time equivalent R&D personnel	250,000	Number of outbound Ph.D. students in all scientific fields	12,500/year
Public/private sector investment ratio	1:1	Relative global ranking of patent portfolios	Enter the TOP9
Commercialization percentage of patents	>5%	Percentage of high technology in exports	>20%
Global ranking of innovation index	Enter the TOP25	Gender equality (PI) of EMR research funding	>60:40

1.2.4 Science and Technology Programs and Funds

Research activities in India are mainly concentrated in research institutions affiliated to central and state governments and enterprises' internal research institutions. Generally, India's science and technology programs and projects are only undertaken by the research institutions of the concerned department, whose influence is also only limited to one field and one department.

Currently, MOST has 4 major types of science and technology programs: science and engineering research program, technology development program, social and economic development program, and international cooperation in science and technology program. In addition, there are also some specialized science and technology programs, such as the New Millennium Science and Technology Development Program, the Innovation Incentive Research Program designed to attract outstanding students and talents, the Women Scientists Program, and National Task Programs-Solar Energy and Water Resources.

The Science and Engineering Research Program is oriented to basic research and mainly concentrates on the fields of chemistry, earth and atmospheric sciences, engineering, life sciences, mathematics and physics. The Program aims to promote high and new technology research, strengthen the research ability of concerned institutions, and encourage young scientists to engage in R&D activities. Moreover, there are also sub-basic research programs such as Nano Technology Application and Development Program, Water Resources Program and so on.

The Technology Development Program is open to applied research and mainly covers research on drugs and pharmaceuticals as well as the research and development of instruments and other infrastructure. All Indian companies can apply for the drug and pharmaceutical research projects but they must do so in collaboration with an Indian university or research institution.

Table 3 Science and technology output goals set by DST for the Twelfth Five-Year period

Category	Goals
Overall human resources capability building	Fulfill all the targets of the INSPIRE Program, and expand the scope of sponsorship of the award; provide scholarships to 1000 students pursuing a Ph.D. degree and 250 Ph.D. holders pursuing postdoctoral research overseas; recruit personnel and female scientists with promising R&D capabilities from college and universities to increase the full-time equivalent of R&D personnel by 40–50%
Research institutions' capacity building	Double the basic research output of India's TOP10 research institution; double the R&D output of India's TOP10 universities; greatly improve the R&D output of the Top40 universities in the country; significantly strengthen the research abilities of five states in the country; markedly improve the research abilities of ten states; 5 research institutions of the country enter the World's TOP 300 Research Institutions, 10 research institutions and 15 universities of the country increase their research output by 50%; 100 universities increase their research results by 25%
Important technology developments and plans	Provide and implement technical solutions for 10 districts; meet the technology demands in 5 key fields related to national development, namely: solar energy, water technology, homeland security, Nano and biomedicine technology; establish "input-output" linkage among technological measures; in technical fields relating to public interests (water, energy, telecommunication and computing materials technology development), establish at least 4 R&D (technology) platforms of the PPP model
International cooperation and technology alliances	Strengthen 5 major international partnership and alliance relationships; in line with the principle of mutual benefit, establish 5 new global alliances and partnerships; establish at least 5 PPP-based partnerships and 5 state–central technology cooperation partnerships
Science and technology that supports social development	Establish and improve implementation mechanisms, devote 5% of the fiscal budget to the poor and development-related evaluation systems; Central Silk Technological Research Institute (CSTRI) launches 50 rural model demonstration projects and implement women and children health R&D programs

The Social and Economic Development Program consists of R&D projects on regional livelihood. It aims to raise the standard of living in economically underdeveloped regions through strengthening technology development. The project's main applicants are research personnel in universities or research institutions.

The International Science and Technology Cooperation Program is designed to promote bilateral, multilateral and regional science and technology cooperation between India and other countries and regions. It is a way for India to open up its national science and technology programs. India mainly engages in international cooperation in science and technology by organizing joint work workshops, supporting mutual visits between scientists, establishing joint research centers, and undertaking joint R&D projects. Today, India has established several themed R&D centers, as well as international cooperation promotion centers with France, the United States, Uzbekistan and some non-aligned countries.

Inaugurated in 2000, the New Millennium Science and Technology Development Program is the largest public-private collaborative science and technology R&D program in India. Focusing on the future of science and technology development, the Program aspires to leverage the R&D abilities of national research institutions, the financial strengths and innovation needs of private companies to establish India's as a major power of science and technology in the world. The financial inputs from the government play the role of guiding funds in the Program, supporting selected promising applied science and technology R&D projects. The project funding and policy support run throughout the stages of R&D, results commercialization and industrialization. For over 10 years since its implementation, the Program has played an excellent role, supporting a total of 57 major joint research projects in the fields of agriculture, biotechnology, bioinformatics, pharmaceuticals, chemicals, materials, information communications and energy. These projects involve 80 industrial enterprises and 270 R&D institutions, with the participation of more than 1700 research personnel. So far, these 57 research projects have generated an output value of over 5 billion rupees.

1.3 Innovation and Entrepreneurship Support Policy and Practice

To address the problem of the commercialization of science and technology results obtained by State research institutions and universities, MOST, the Ministry of New and Renewable Energy, and the Ministry of Commerce and Industry of China have gradually formulated guiding advices on the research results of public-funded science and technology programs and research projects. Meanwhile, research institutions and universities affiliated to various ministries and commissions have also introduced corresponding detailed implementation rules based on their specific situation.

1. **MOST Has Introduced Several Policies and Provisions on the Industrialization of Publicly Financed Science and Technology Results**

The *Science and Technology Policy (2003)* and the *Science, Technology and Innovation Policy (2013)* have both set out guiding and programmatic provisions on the industrialization of science and technology results. The *Science and Technology Policy (2003)* called for efforts to encourage public-private partnership in the economic and social R&D field, develop flexible mechanisms to help scientists and technical personnel transfer their professional skills to the actual production and become partners in technology industrialization, ensure legislation for all intellectual property rights in India and provide maximum possible incentives for independent innovation technologies that have achieved large-scale and rapid commercialization. The *Science, Technology and Innovation Policy (2013)* made the following recommendations: speed up the social and commercial application of R&D results, reproduce successful business operation models and establish new commercialization models of a PPP structure, establish PPP models for large-scale R&D facilities, formulate and improve provisions on benefits sharing, revise intellectual property rights policy, and under the PPP model, provide publicly financed science and technology results with intellectual property rights sharing mechanisms that accord to public interests and benefits sharing.

In addition, MOST issued *Guiding Opinions on Technology Transfer and Intellectual Property Rights* in 2000 with the aim of mobilizing the enthusiasm for science and technology innovation among scientists, research institutions and universities. The main policy items include: encourage research institutions to protect the intellectual property rights of their research results, push forward the commercialized development of their patent results, and allow patent-holding institutions to keep the proceeds and earnings of their intellectual property. In addition, research institutions should allocate 25% of their intellectual property rights proceeds to set up a Patent Promotion Fund.

2. **The Ministry of Commerce and Industry Has Submitted for Deliberation *A Motion on The Protection and Application of Publicly Financed Intellectual Property Rights*, the Indian Version of the Bayh-Dole Act**

Making reference to the *Bayh-Dole Act* of the United States, the Ministry of Commerce and Industry of India in conjunction with other ministries and commissions submitted in 2008 *A Motion on The Protection and Application of Publicly Financed Intellectual Property Rights* to the Upper House of the Indian Parliament for deliberation. The Motion was intended to establish applicable rules and code of conduct by formulating specific policies, promote the commercialization of the intellectual property rights of publicly financed R&D institutions, and prevent potential risks and uncertainties". In addition, the Motion allows Indian R&D institutions to apply for intellectual property rights protection in other countries with a punitive clause put forward to address cases of infringement. Through future legislation, the Motion attempts to address the difficulty in the commercialization of R&D activities and the lack of connection between universities and enterprises. However, because of the added punitive clause (such as the provision to prevent science and technology personnel from abusing intellectual property rights), the Motion has attracted attention from the government, academia and industries, and remained a controversial topic for debate until today.

3. **The Ministry of New and Renewable Energy Has Set Out Provisions on the Intellectual Property Rights Protection Obligations and Income Distribution of Research Institutions**

The Ministry of New and Renewable Energy of India has laid down provisions for its affiliated research institutions on intellectual property rights protection and income distribution, indicating that project transferees or inventors are under obligation to apply for intellectual property rights protection for technologies related to the research project and share the proceeds from the intellectual property rights in accordance with the following rules: the government has the right use intellectual property for administrative purposes for free; if the transferee is unable to apply for patents, the Ministry of New and Renewable Energy will be responsible for the patent application; proceeds from technology transfer and commercialization will be shared between the Ministry of New and Renewable Energy, research institutions and intellectual property rights inventors at a 40, 40, and 20 respectively; consent of the Ministry of New and Renewable Energy is needed when research institutions and enterprises wish to transfer their intellectual property rights to any other company for further development. However, such transfer must be "non-exclusive proprietary right" development.

4. **"Startup India, Standup India" Has Made Achievements**

In January 2016, India launched the "Startup India, Standup India" Initiative ("Entrepreneurial India") to encourage entrepreneurial spirit of the whole nation and implement preferential policies to push forward technology innovation and mass entrepreneurship in India to boost economic growth. The policies provide facilitative measures to entrepreneurs, offering start-up companies with a 3-year exemption from labor and environmental inspection, shortening patent application approval time and cutting the application fee down by 80%. Funds support is also provided, with a $1.47 billion fund set up to support innovation projects in manufacturing, agriculture, healthcare and education sectors. A $300 million credit security fund is created to help start-up companies obtain credit loan from financial institutions. Entrepreneurship India Centers" are also established to provide institutional mechanism guarantee.

Remarkable achievements made. First, the scales of entrepreneurship investment and the number of enterprises have risen significantly. On average, 3–4 startup companies are established each day, and the amount of entrepreneurship investment increased to $7 billion in 2015 from $2.2 billion in 2014. Second, there is a wide range of entrepreneurship fields which are mostly concentrated in e-commerce, digital advertising, Big Data, data analysis, cloud computing, hardware, education and healthcare sectors. Third, many young and highly educated entrepreneurial personnel have been attracted to the scheme. In 2015, entrepreneurship stimulated employment for 80,000–85,000 people, while 72% of the entrepreneurs were under 35 years of age and female entrepreneurs accounted for 50%. Fourth, the initiative has promoted the formation of Delhi economic zone as well as Mumbai and Bangalore entrepreneurship agglomerations.

1.4 R&D Expenditure, Output and Personnel

1.4.1 R&D Expenditure

In recent years, India has seen a rapid growth in its influence on the global economy and scientific research. It has made significant progress in biopharmaceuticals, material chemistry and other key fields and has gained its place in related fields in the world. India's R&D expenditure has the following characteristics:

1. **Financial expenditure on R&D is less than 1% of GDP**. Since 2001, India's gross expenditure on R&D kept rising from 170.4 billion rupees in 2001 to 853.2 billion rupees[1] ($13.3 billion) in 2015 at a growth rate of 11.3%. However, the R&D expenditure/GDP ratio was only 0.63%, far below the standard value[2] advocated by the European Union. Though India has announced to dramatically increase its R&D expenditure, aiming for a R&D expenditure/GDP ratio of 2% by 2020; nonetheless, as far as the current situation is concerned, the level of difficulty remains unneglectable (Fig. 2).
2. India's R&D expenditure mainly comes from government inputs, with 50% of the expenditure contributed by government funding. Out of India's R&D expenditure, 52.5% of the funds come from the central government, and 43.6% from the business sector. R&D spending by businesses has risen in recent years, up to 43.6% in 2015 from 19.3% in 2001. In addition, on average, each enterprise spends 0.61% of its sales revenue on R&D: state-owned enterprises spend 0.27% of its revenue on R&D while private businesses spend 0.82% (Figs. 3 and 4).
3. Drugs and pharmaceuticals, transportation and IT are the top 3 recipients of R&D expenditure. As for the sector-by-sector distribution of research expenses, the top 10 are drugs and pharmaceuticals, transportation, IT, defense industries, fuels, chemicals (excluding fertilizers), biotechnology, electrical and electronics, metallurgical industries, and telecommunications. The percentage of government spending by industry to total R&D expenditure in descending order is as follows: national defense industries (basically all government invested), fuels (approximately 80% invested by government), metallurgical industries (approximately 60% invested by government), telecommunications (approximately 15% invested by government), and electrical and electronics industry (approximately 10% invested by government). However, in the drugs and pharmaceuticals and biotechnology industry which cost the most R&D expenditure, all funds come from private investment. Similarly, in chemicals (other than fertilizers), transportation and IT industries, most R&D expenditure also comes from private investment (Fig. 5).

---

[1]Source: UNESCO Data Center.

[2]The European Commission believes that for a mature economy, the appropriate percentage of R&D expenditure to GDP (R&D/GDP) should be around 2%.

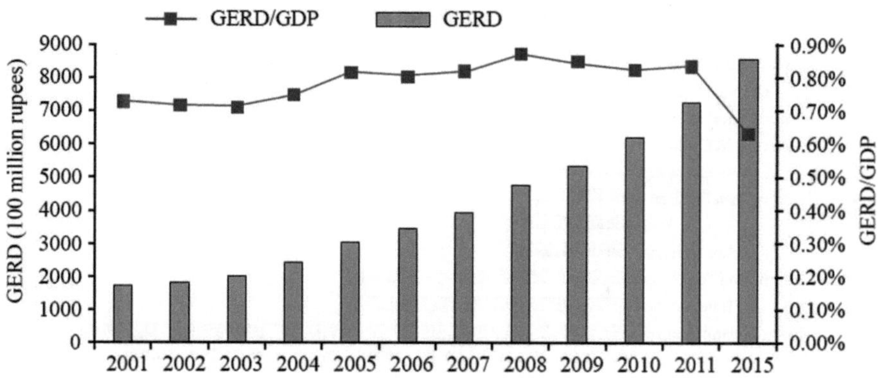

Fig. 2 National R&D expenditure and its percentage in GDP of India (2001–2015)

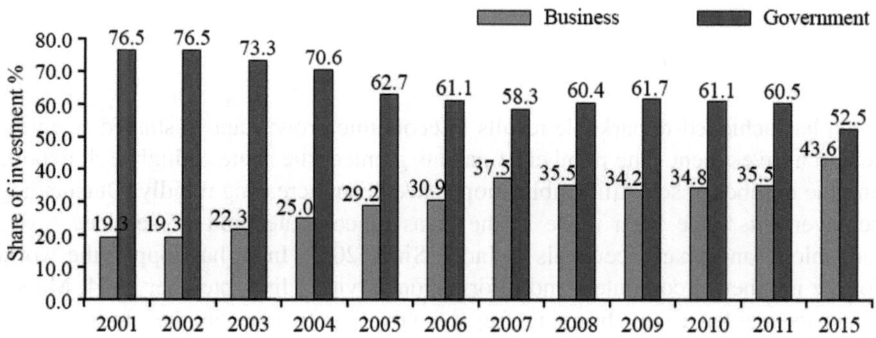

Fig. 3 Percentage share of government and business sector in GERD (2001–2015)

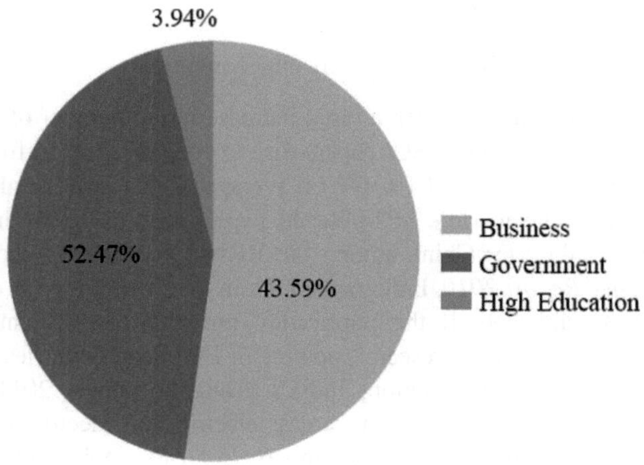

Fig. 4 Distribution of R&D expenditure in India (2015)

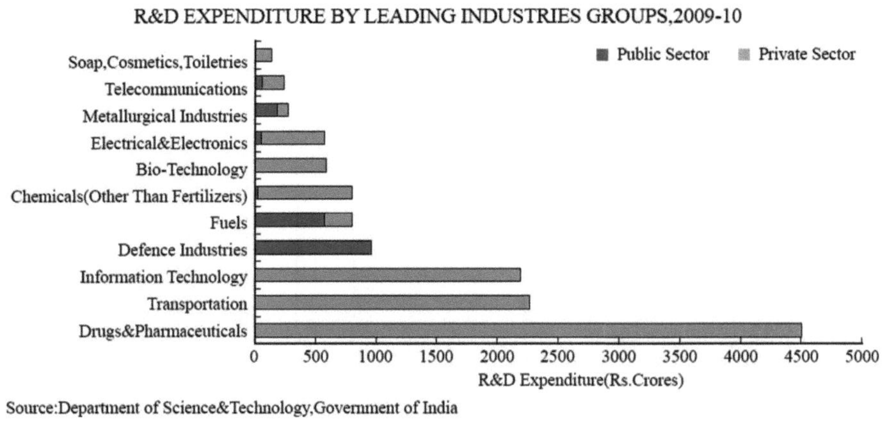

Fig. 5 Distribution of R&D expenditure by leading industries groups in India (2009)

1.4.2 Research Output

India has achieved remarkable results in economic growth and sustained scientific research investment. The number of patents granted, the share of high-tech exports and the number of scientific publications have been increasing rapidly. Outstanding achievements have been made in the areas of computer and IT services, space technology and pharmaceuticals in India. Since 2005, India has topped the world for the number of computers and information services. In September 2014, Mars 6 made its maiden flight, bringing frugal innovation to new heights. India is also actively engaged in international mega science projects, such as participation in the construction of the European Organization for Nuclear Research (CERN), the Large Hadron Collider (LHC), and the "Facility for Antiproton and Ion Research in Europe GmbH" (FAIR GmbH).

(1) Number of Academic Papers and Advantageous Fields

In the past 10 years, India has seen a rapid increase in the number of its published research papers. The number of SCI papers rose to 95,300[3] in 2016 from 35,000 in 2006 at an annual growth rate of 10.56% on average, way above the global average of 4%. Its share of the world's SCI-indexed papers rose to 4.39% in 2016 from 2.35% in 2006, only after China among the BRICS countries. According to the *UNESCO Science Report 2010*, India ranked 9th in the world in terms of number of scientific papers published. If the publication growth trend continues, India is expected to catch up with the research powers in European countries (Fig. 6).

As for research field distribution, India's academic papers 2011–2016 were mainly concentrated on the following areas: electric and electronic engineering (14.12%), material science (7.13%), chemistry (5.56%), applied physics (5.4%),

[3]Source: Web of Science Core Collection, July 7, 2017.

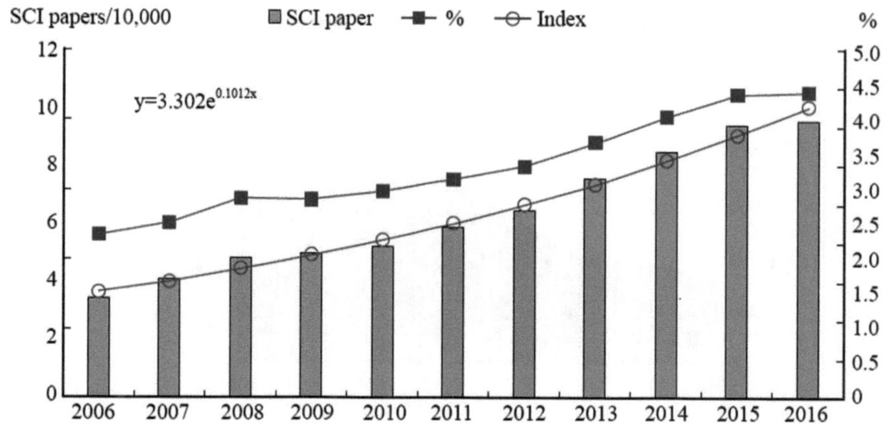

Fig. 6 India's research publication trend

and computer science (5.36%). The top five contributors of SCI papers are: Indian Institute of Technology (IIT), Council of Scientific Industrial Research (CSIR), Indian Institute of Science Bangalore, Indian Institute of Technology Kharagpur and Bhabha Atomic Research Center.

(2) Research Influence—Academic Paper Quality and Advantageous Fields

Data estimates based on the above paper citation database show that if the average global citation influence value is 1%, India had 235 frequently cited papers in 2011, accounting for 0.52% of the country's total science and technology paper output (China had 1131 frequently cited papers, accounting for 0.72% of its total research paper publications, ranking at the top among the BRICS countries. The percentage for Britain was 1.4%).

The top 10 research fields of India's frequently cited papers are: engineering science, physics, computer science, material science, social sciences, mathematics, space science, environmental science and ecology, chemistry, psychiatry, and psychology. Among them, in the 3 fields of engineering science, physics and computer science, India's paper commands an influence above world average. These fields are the future R&D directions for India with the most promising prospects.

(3) Economic Influence—Number of Patents and Related Areas

According to the WIPO statistics, India had 43,031 patents in 2010. Most of the patent applications were concentrated in the 3 fields of computer/electronics, machinery and chemicals. In 2010, 60% of the patent applications were from foreign applicants, most of whom are from the United States (33.5%), Japan (13.2%), and Germany (11.8%). Since 2007, there has been a decrease in the percentage of patent applications by Indians to India's total number of annual patent applications.

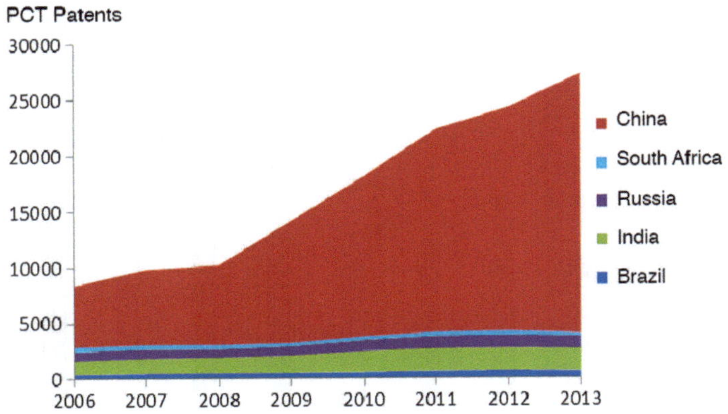

Fig. 7 Increase in patent applications: India versus other countries (2006–2013)

In terms of the growth trend of patent application, India is performing well with an annual increase of approximately 16.6%. This is expected to put India in 3rd place among the BRICS countries, after China and Russia. In comparison, China's patent applications have increased about 6 times in the past 10 years (Fig. 7).

Based on the patent application data analysis of the 35 technology fields defined by the World Intellectual Property Office, India enjoys a high share of patent applications in pharmaceutical and refined organic chemistry. In comparison, China performs outstandingly well in its patent applications in the fields of power machinery, devices and energy, digital switching and computer technology.

Compared to global patent applications activities of various fields, India performs better in pharmaceuticals, agricultural chemicals, food detergents, water treatment, biotechnology, and general chemicals than in machinery and electronics technology fields. This is directly related to the fact that India is home to many large-sized and superior quality pharmaceutical and chemical companies. In contrast, China does not have a particular technology field in which it performs exceptionally well.

1.4.3 Number of Research Personnel and Human Resources

According to UNESCO statistics, India had a total of 528,200 R&D personnel in 2015 next only to China and Russian among BRICS countries. Among them, there were 283,000 (53.6%) researchers and 125,200 (23.7%) technicians. A total number of 376,000 researchers and technicians (71.2%) were employed by public research institutions and universities, 127,000 (24%) were employed by businesses. 14.7% of India's researchers and technicians were female, reaching 77,700. The number of R&D personnel per million people in India increased from 341 in 2005 to 402.9 in 2015, which is still lagging behind China (2731 R&D personnel per million people) (Table 4).

Table 4 R&D personnel of BRICS countries (10,000 persons)

Year	2000	2005	2010	2015
China	92.21	136.48	255.38	375.88
Russia	100.73	91.97	84.00	83.37
India	31.84	39.11	44.11	52.82
Brazil	13.30	19.63	26.67	–
South Africa	2.12 (2011)	2.88	2.95	3.80 (2013)

Source UNESCO Data Center

India faces a serious problem of brain drain. The Indian government estimates that about 30 million Indians have emigrated overseas, with 2.5 million to the United States; moreover, 80% of India's lost high-end talents are in the United States. Among adult Americans of Indian descent, 74.1% have a Bachelor's degree or higher, while 40% hold Masters' degrees, Ph.D. or other professional titles. Among these people, 68.9% have an ideal job, joining management posts or conducting research with their professional skills. They achieve outstanding performance in almost all important field including academic research, industries, finance, science and technology, healthcare, and business management. Out of the 2000-plus new immigrant hi-tech companies in the Silicon Valley of the United States, 40% were founded by Indians. Statistics from the US Department of Homeland Security, Immigration Service and other departments show that among foreigners holding an H-1B visa in the United States, 46.9% were born in India. In 2010, 98,000 Indians obtained an H-1B visa, whereas only 14,700 Chinese.

1.4.4 India's STI Strength and Characteristics of Its Advantageous Fields

(1) **Diversified Research Work and Advantages, and a Wide-Range of Research Focuses**

India has a relatively balanced research output. Its research capacity (academic papers) advantages are distributed in the fields of chemistry, pharmacology and virology, agricultural science, materials science, microbiology, physics, engineering science, plant and animal science, earth science, biology and biochemistry, showing diversified research work and advantages and a wide-range of research focus. By comparison, the top 10 advantageous fields of China's academic papers are material science, chemistry, physics, mathematics, engineering science, computer science, earth science, pharmacology and virology, environmental science and ecology, biology and biochemistry.

(2) **There Is Discrepancy Between Advantageous Fields of Research Output and Advantageous Fields of Economic Influence**

There is discrepancy between India's advantageous fields of research output and its advantageous fields of economic influence. With the exception of the pharmaceuticals industry being an advantageous field of both research output and patents, **India's patent applications are mainly concentrated in the 3 fields of computer/electronics, machinery and chemicals. Compared with overall global patent applications in various fields, India has a fairly big share of the pharmaceuticals and refined organic chemistry fields**, performing much better than it does in the computer/electronics and machinery fields. By comparison, there is a basic match between China's advantageous fields of research output (e.g. material science, chemistry, physics, mathematics, engineering science, computer science, earth science, etc.) and its advantageous fields of economic influence (power machinery, devices and energy, digital interaction and computer technology, etc.).

(3) **Advantageous Fields of Research Influence Are Not High-Yield Research Fields**

India's advantageous fields of research output (number of academic paper) are chemistry, pharmacology and virology, agricultural science and material science, while its advantageous fields of research influence (quality of academic paper) being engineering science, physics and computer science, with a mismatch between the two advantageous fields. The difference also reflects the insufficient ability of the Indian government when guiding the direction of economic and social development through research spending. In comparison, China's advantageous fields of both research output (**number of academic paper**) and research influence (**quality of academic paper**) are both concentrated in material science, mathematics, engineering science, and computer science, with a basic match between research output and research influence.

(4) **Low R&D Expenditure Results in Low Quality Patent Applications**

India's top 5 R&D expenditure consuming fields are drugs and pharmaceuticals, transportation, IT, defense industries, and fuels. But, its gross R&D expenditure is less than 1% of GDP and the number of patent applications is also relatively small. In terms of the number of patent applications, India lags far behind China (in 2010, China received 391,200 patent applications, the 2nd largest in the world, only after the United States; in 2011, China had 526,412 invention patent applications, exceeding the 503,582 applications in the United States), which conforms to the empirical research results from European countries and the United States that "low-level R&D investment results in low-level patent applications". *The Global TOP100 Innovative Institutions* report shows that though China has an outstanding number of patents and achieves rich returns from research investment, poor patent quality has cost China its position among the Global TOP100 innovative institutions. Thus, for China, the most urgent task is to improve the quality of its patents.

(5) **The Ability to Produce Research Output Outperforms Patent Application Growth**

There is a gap between the trends in patent application and academic publication in India. While its paper publications and patent applications both increase, the slow growth of patents to a certain extent shows India's low efficiency in turning research results into the productive force. This is proven by the Indian Patent Office's work performance. The number of patent applications to be reviewed grew by 47.3% from March 2011 to April 2012, standing as many as 123,255. By comparison, China has shown consistent performance in patent application activities and paper output data.

(6) **There Is a Basic Match Between Advantageous Fields of Economic Influence and Advantageous Fields of Industry Development**

India's global advantageous fields of patents include drugs and pharmaceuticals, agricultural chemicals, food detergents, water treatment, biotechnology, and general chemicals. This is closely related to India's heavy R&D spending on drugs and pharmaceuticals, its pillar industry. By comparison, China does not have obvious advantageous fields of invention patents advantages when compared with the rest of the world. On the one hand, this shows a balance between China's invention activities in various fields. On the other hand, it indicates that China should strengthen the support of invention patents for current industry development and boost the linkage between research inventions and industry technology applications.

1.5 International Cooperation

India's strategy for international cooperation in science and technology is to strategically select major powers of science and technology output, focus on international alliance and partnership building and fully leverage advantages of international cooperation. In *the 2010–2015 Science and Technology Development Strategy*, MOST set out the key items for international cooperation in science and technology, including: establish platforms for international cooperation in science and technology and for collaboration between institutions; establish strategic science and technology partnerships with selected countries; attract foreign companies to set up industry bases and research institutions in India, such as those on computer software, photoelectric systems, medical diagnosis equipment, auto parts and so on; expand the recruitment of international experts to join in important national science and technology R&D projects and advanced research facilities construction, set up state-of-art research centers and train research talents in particular research field; make effort to join in international mega science projects, and participate in the research and development of international advanced facilities and international science databases.

India attaches importance to expanding channels of science and technology cooperation and has greatly increased its international bilateral science and technology cooperation activities. While engaging in a broad-range and in-depth cooperation with almost 40 countries at present, India also actively establish contacts with international institutions and has formulated and expanded multilateral cooperation mechanisms. Based on projects, it has engaged in many international cooperation endeavors, expanded the influence of international cooperation and raised its overall level of research level and strengths of innovation. India has its own focus in science and technology cooperation. Its international cooperation is mainly with major powers of science and technology, such as Russia, Britain, the United States, Japan and Germany. The fields of cooperation cover almost all frontier and high-end fields of science and technology with different focus on different countries. In addition, it has also established bilateral R&D programs and set up joint R&D centers with 13 countries including Russia and France. India is involved in the establishment of the ASEAN Science and Technology Fund, the India-US Endowment Fund and the New Africa Science and Technology Program.

India has widely participated in international organizations, collaborating with more than 60 international science and technology organizations. It aims at raising international reputation and having a greater say in international affairs through participating in international science and technology projects. Firstly, it has joined the various science and technology-related organizations and agencies of the UN system and played an active role in their activities. Secondly, by becoming members or signing agreements, India has maintained ties with international and regional science and technology organizations and organized multilateral activities. Thirdly, it attaches importance to initiating and creating inter-governmental organizations for international cooperation in science and technology. Currently, India has established 4 such organizations, including the International Commission on Irrigation and Drainage and the International Center for Genetic Engineering and Biotechnology.

1.6 Key STI Fields

In its *Science and Technology Innovation Policy (2013)*, India would prioritize its focus on agriculture, communications, energy, water management, health, pharmaceuticals, materials, environment, climate diversity and change, green manufacturing. Through major strategic tasks, India will step up R&D intensity, stimulate the export of hi-tech products and the development of the innovation industry development. Its support of and investment in key fields will cover all links of the innovation chain, including basic research, technology development, and research results transformation and industrialization. Table 5 shows the fields of science and technology development which the Modi administration pays attention to.

Table 5 Key fields of science and technology development in India

Key areas	Development goals
New energy policy	Produce and provide low-cost energy, and raise the development goals of the National Solar Energy Program
"Strategic Uranium Reserves" program	Increase nuclear R&D spending by 7%
Made in India	Raise the percentage of the manufacturing industry in India's GDP to 25% from the current 15% and strive to build India into a competitive global manufacturing industry center that can rival China
Digital India	To provide the public with better network connection and services and push forward the development of India's electronics manufacturing industry, plan to significantly cut down electronics product imports by 2020 and by establishing a network of "public service centers", provide the public with e-healthcare, education, banking, insurance, endowment pension and agricultural services. Such centers are scheduled for opening to 250,000 villages across the country by 2019
Smart City	Build 100 Smart Cities, establish e-government and expand business promotion. Build inhabitable cities: reduce pollution and develop clean energy, Internet medical care, electronic vehicles and clean water
Bioagriculture (transgenosis)	Encourage scientists to step up research and pilot plantation, encourage farmers to grow genetically modified crops and speed up the commercialization of genetically modified crops
Mars and Venus Exploration Tasks	India launched its first low-cost Mars orbit space-craft, becoming the world's first to succeed in the first attempt of Mars exploration. India has announced that its next target of space exploration is Venus
Commercial Satellite Launches	Since 1993, India has used its polar orbital launch vehicle to launch nearly 80 satellites for 19 countries
National Supercomputing Program	Plan to become a world power in supercomputing science and technology. India's Centre for Development of Advanced Computing (C-DAC) and Indian Institute of Science school (IISc) will cay out joint R&D work, with an expected investment of $72 million in the next 7 years
Deep-sea Exploration and Marine Science Expedition	Conduct research in the areas of monsoon, polar expedition, ocean observation, disaster prevention and mitigation, and climate change, and strengthen its position in the World Climate Research Program (WCRP) and other international projects

2 China-India STI Cooperation

2.1 History

Both China and India are developing countries and important neighbors to each other. There has been a long history of cooperation and exchanges in science and technology between the two countries. Since the reform and opening-up of China, the bilateral cooperation in science and technology has been advancing.

In 1988, Prime Minister Rajiv Gandhi of India visited China, marking the start of China-India relations gradually coming out of a low ebb. The two sides decided to set up a China-India Joint Commission on Science and Technology (JCSTC). The two countries also signed an agreement on science and technology cooperation, which laid a foundation for bilateral cooperation in science and technology.

In 1991, Premier Li Peng of China visited India. The two sides signed a Memorandum of Understanding on Science and Technology Cooperation for the Peaceful Use of Outer Space. Subsequently, the President, Prime Minister and Speaker of Parliament of India successively visited China.

In 2002, Premier Zhu Rongji of China visited India. The two sides signed 6 cooperation documents on science and technology, water conservancy, space and other fields, pushing forward the further development of China-India cooperation in science and technology.

In 2005, Premier Wen Jiabao of China visited India. The two countries signed a *Joint Declaration of China and India*, announcing the establishment of a strategic cooperative partnership for peace and prosperity between them.

In 2006, the Fifth Session of the China-India Joint Commission on Science and Technology Cooperation was held in Beijing. The two countries signed a *Memorandum of Understanding on Science and Technology Cooperation* and set up a ministerial Steering Committee for Science and Technology Cooperation between China and India.

In 2013, the Sixth Session of the China-India Joint Commission on Science and Technology Cooperation was held in Beijing. The two sides agreed to focus their science and technology cooperation in earthquake and natural disaster mitigation and management, astronomy and astrophysics, climate change, technology research, traditional knowledge, medicine and pharmaceutical field, while encouraging scientists from both countries to exchange, hold workshops, establish joint laboratories and joint research centers, and engage in other forms of cooperation.

In September 2014, during his visit to India, President Xi Jinping of China announced that China would work with South Asian countries to implement the China-South Asia Science and Technology Partnership Program. The China-South Asia Science and Technology Partnership Program will support the co-building of national joint laboratories so as to establish stable cooperative relationships between research institutions, conduct high-level joint research and development, organize the implementation of the program for outstanding young scientists in South Asian countries to come and work in China, sponsor outstanding talents in South Asian countries to come and work in China for a period of 6 months to 1 year, and help various countries to train their science and technology talents; organize the implementation of major technology demonstration projects and build technology demonstration and promotion bases; establish China-South Asia technology transfer centers, cooperate in technology transfers cooperation, jointly organize enterprise connectivity and technology demonstration and training, and serve the development of enterprises in different countries. India is an important member of the China-South Asia Science and Technology Partnership Program.

In May 2016, the State Oceanic Administration of China and the Ministry of Earth Sciences of India signed a *Memorandum of Understanding on Strengthening Cooperation in Marine Science, Marine Technology, Climate Change, Polar Science and Cryosphere*. The two sides have successfully held the First China-India Workshop on Marine Science and Technology Cooperation and reached consensus on cooperation in Southwest India Ocean monsoon research and forecast, Arctic and Antarctic scientific expedition, and biogeochemical process research. In May 2016, the China-India Joint Commission on Marine Science and Technology Cooperation held its first meeting in Beijing, and deliberated on and adopted 8 collaborative projects on Indian ocean circulation and monsoon climate research.

Since the two countries signed their first intergovernmental science and technology cooperation agreement in 1988, 6 sessions of the China-India Joint Commission on Science and Technology Cooperation have been held. Science and technology cooperation between China and India has covered many fields, including agriculture, biotechnology, chemicals, medicine, electronics and new materials, achieving considerable breadth and depth. In recent years, China and India have made frequent high-level contacts, ushering China-India relations into a new stage of all-round development. Consequently, exchanges and cooperation between the two countries in the field of science and technology have also been comprehensive and developed fast.

2.2 Current Status

(1) Mechanisms and Platforms Preliminarily Established

Intergovernmental cooperation in science and technology between China and India now has a history of almost 30 years. Since the two countries signed their first intergovernmental science and technology cooperation agreement, the China-India Joint Commission on Science and Technology Cooperation has held 6 sessions and approved dozens of projects on science and technology exchanges and cooperation projects. The concerned departments of the two countries have also more than 20 inter-departmental science and technology cooperation agreements and memorandums of understanding, involving the Ministry of Education, the Ministry of Agriculture, the Ministry of Water Resources, the Ministry of Health, the State Forestry Administration, the China Earthquake Administration, the State Oceanic

Administration, the China Meteorological Administration, the China State Space Administration, the Chinese Academy of Sciences and the National Natural Sciences Foundation of China. Some provinces, autonomous regions, municipalities and universities in China have also reached an intent of cooperation in various forms with their counterparts in India, involving a wide range of cooperation fields.

In 2006, China and India issued a Joint Declaration. Science and Technology cooperation once again became a key focus for the two countries. In September 2006, the ministries of science and technology of both countries also signed in Beijing a *Memorandum of Understanding on Science and Technology Cooperation* and set up a ministerial China-India Steering Committee on Science and Technology Cooperation. Since the global financial crisis broke out in 2008, science and technology cooperation and exchanges between the two countries have progressed further. All these measures have not only coordinated and solved strategic issues in bilateral cooperation and guided the development path of bilateral cooperation in science and technology but also established platforms for China-India cooperation in science and technology and provided operation support mechanisms for science and technology cooperation and exchanges between the two countries.

(2) **Fields of Cooperation and Exchanges Have Continued to Expand**

In the 1980s and 1990s, science and technology cooperation and exchanges between China and India were concentrated in agriculture, biotechnology, chemicals, medical science, electronics and new materials. In 1989, the First Session of the China-India Joint Commission on Science and Technology Cooperation made agriculture, forestry, remote sensing technology, medical science, biotechnology and science and technology policy as the fields of cooperation between the two countries. In 1991, the Second Session of the Joint Commission adjusted them into laser, materials, aerospace and remote sensing, agriculture and fishery, chemicals, electronics, biotechnology, and pharmaceuticals. The Third Session of the Joint Commission held in 1993 made almost no adjustment to the fields of cooperation. The Fourth Session of the Joint Commission held in 1999 did not designate any field of cooperation but only discussed modes of cooperation, funding arrangements and other matters.

Entering the new century, science and technology cooperation and exchanges between China and India have expanded to fields that include water conservancy, space, earthquake and Nano technology. China-India ties in science and technology cooperation have deepened and developed further. In 2002, the Fifth Session of the China-India Joint Commission on Science and Technology Cooperation decided to continue China-India cooperation in the fields of biotechnology, chemicals, new materials, earth science, disaster mitigation, medical science, aviation, electronics, information technology, and software development. It also called for further strengthening bilateral exchanges in the areas of traditional medicine, intellectual property rights, science and technology incubators, science and technology and rural social development. In 2006, China and India reached consensus on cooperation in earthquake engineering science, climate change and weather forecast,

Nano technology, biotechnology and pharmaceuticals. In the *Joint Declaration* issued during Premier Li Keqiang's visit to India in May 2014, India proposed to strengthen bilateral cooperation in the areas of earthquake and natural disaster early warnings and management, astronomy and astrophysics, climate change technology research, traditional knowledge and medicine, clean energy (civilian nuclear power), ocean research and environmental protection and cross-border rivers.

(3) **Multilateral Cooperation Under the BRICS Framework**

For a long time, developed countries have been the main targets for international cooperation in science and technology for both China and India. The two countries have respectively established wide-ranging and in-depth relations with the United States, Britain, Germany and other developed countries in science and technology cooperation. By comparison, bilateral science and technology cooperation relations between China and India are not deep and are mostly within the BRICS framework. For example, during the Davos Summer Forum in Dalian in September 2011, the BRICS countries discussed ways to promote cooperation in science and technology innovation among them and elaborated on the strategies, priority fields and work mechanisms of their cooperation in science and technology innovation. All the countries agreed to collaborate in key fields of science and technology innovation. Within the BRICS framework, India has cooperated with China's Taiwan province in the Nano technology field and supported 20 on-going projects and approved 10 new projects to develop advanced Nano materials needed for biosensors and energy storage equipment and flood forecast systems. In 2011 and 2012, the BRICS Working Group Meeting made some key suggestions for bilateral and multilateral cooperation and exchanges in science and technology. In 2013, the Durban Action Plan specifically proposed to hold the Meetings of Science and Technology Ministers and High-Level Science and Technology Officials of the BRICS countries. The BRICS cooperation framework provides a good opportunity for China-India cooperation in science and technology.

(4) **China-India Software Industry Park Has Become a Model of China-India STI Cooperation**

Located in the Linyi Economic and Technological Development Zone in Shandong Province, the China-India Software Industry Park is jointly built by Linyi Infotop Co., Ltd. and the internationally known India IT company SRM Group, with an investment of RMB560 million. As of June 2017, the Park had attracted 40 high-end Chinese and foreign software R&D companies and 7 leading international collaborative institutions, forming an internationalized industry cluster led by IBM, SRM Group, Bangalore-based IIHT Group and Ukraine State University Smart Robots. Through strong-strong collaboration between advantageous industries of both China and India, the Park has actively attracted international high-level talents, including 60 Indian experts and a number of international experts from Russia, Czech, Ukraine and New Zealand. The cumulative number of joint R&D projects by foreign experts has now reached 80. The China-India Software Industry Park has been designated by the Chinese Ministry of Science and Technology as China's first "National park of International Innovation" in the software industry between China and India. Shandong Province hopes to build the China-India Software Industry Park in Linyi into an India city of software industry cooperation in China. In addition, it also intends to attract Indian healthcare and automotive industries to develop there. Meanwhile, the Park has also expanded its cooperation to the United States, Israel, Germany and other countries. In their cooperation, Chinese and Indian partners are fully confident about pushing forward an all-round cooperation in IT and software industries and are actively working together to expand their comprehensive cooperation in software sourcing service, software talent exchanges and joint research projects.

2.3 *Difficulties and Obstacles*

(1) **That There Is a Lack of Trust Between China and India in Their Cooperation in Science and Technology Has Hampered the Progress of the Cooperation**

International political relations have an important influence on the development of international cooperation in science and technology. Though both China and India have an intention to promote communication as well as cooperation and there are also available spaces for them to do so, long years of isolation and alienation in bilateral relations make the two countries view each other as geopolitical rivals with a lack of political mutual trust between them, which had a negative influence on their science and technology cooperation. The real situation is that the Indian side has a strong attitude of defense, restriction, suspicion, doubt and rivalry towards China. In its advantageous fields, India is often arrogant and does not bother to cooperate with China. Instead, it is keen to collaborate with the stronger Western powers of science and technology. In China's advantageous fields, India often has a mentality of resistance and jealousy thus not willing to collaborate. Under such circumstances, it is often very difficult for China and India to find a suitable point to start their science and technology cooperation.

(2) **Mechanisms for Bilateral Cooperation in Science and Technology Are Not Adequate, Resulting in Mediocre Operational Performance**

Though the ministries of science and technology of China and India have regular meeting mechanisms between them and both sides hope to expand fields of science and technology cooperation, there still lack substantial cooperation contents. Though China and India have set up a Joint Commission on Science and Technology Cooperation and signed relevant science and technology exchange programs, relatively little substantial work got carried out. The Indian Ministry of Science and Technology is a weak department, with limited fields of jurisdiction and limited ability of coordination, thus can hardly provide sufficient support for the subsequent implementation of cooperative projects. This has caused many collaborative projects between China and India to remain on paper, and only a very few projects actually got implemented. Both the scope and depth of the China-India cooperation in science and technology have not met the expectation of the two countries.

(3) **Science and Technology Cooperation Has Not Produced Notable Results, and There Is Still Potential for Cooperation**

Firstly, there is limited cooperation opportunity between China and India in the research, development and application of science and technology projects. This is especially true for basic research and application in unknown fields and collaborative development, transfer and industrialization of certain technologies. There are relatively few substantive projects in which the two countries can conduct collaborative research or development. Secondly, the existing cooperation between China and India is rather small in scale in terms of mutual visits between science and technology personnel, technical, expert consultancy, information circulation, academic conferences and science and technology results exhibitions. Finally, though China and India have signed many inter-departmental science and technology cooperation agreements and memorandums of understanding, there is not much substantive science and technology cooperation. Moreover, China and India have only held 6 sessions of Joint Commission on Science and Technology Cooperation in the past 30 years. Thus, it is clear that when comparing to the 50 protocols signed between China and the United States on science and technology cooperation in the past 30 years, China-India cooperation in science and technology does not match their positions as leading developing countries. China-India cooperation in science and technology evidently lags behind, and both countries still need to expand their fields of cooperation.

(4) **Insufficient Information Exchanges Makes Projects Difficult to Implement**

The important body of international cooperation in science and technology are industrial and commercial enterprises. Industrial and commercial companies in the two countries only decide to enter each other's market after having gained sufficient information on the science and technology in the other country. However, due to the lack of political mutual trust and long years of isolation and alienation of bilateral ties, the Chinese and Indian industrial and commercial circles do not have a sufficient knowledge of each other's research market, research information, experimental technology and industrialization. They even lack effective channels to obtain science and technology information of the other side and thus find it difficult to find a suitable project or a reliable cooperation partner. The level of science and technology cooperation between China and India is not high enough, making it even more difficult to actually implement those potential science and technology cooperation projects.

Chapter 7
China Report on Science, Technology and Innovation

Hongwei Huo, Zhongcheng Wang and Wenjing Li

1 Evaluation of China's National Innovative Competiveness

Located in the east of the Asian continent and the west of the Pacific Ocean, China has a land area of about 9.6 million km^2, and the total length of its boundaries is 22,280 km. Its sea area is 4 million 730 thousand km^2, and the coastline of China is more than 18 thousand km long. China shares borders with 14 countries including the BRIC countries of Russia and India, and is a sea neighbor of 8 countries. At the provincial level there are 4 municipalities directly under the jurisdiction of the central government, 23 provinces, 5 autonomous regions, and 2 special administrative regions. Beijing is the capital. The total population is 1 billion 364 million 300 thousand (in year 2014). Since reform and opening-up, China's economy has maintained rapid growth over the long term. At the end of 2014, the total GDP reached USD 10.3601 trillion.[1] This section will focus on detailed analysis of China's national innovative competitiveness and the changes in the ranking of the BRICS in the 20 years from 2001 to 2020 (Fig. 1).
Changes of positions and scores of China's innovative competitiveness in the BRICS are shown in Fig. 2. The forecast of China's innovation index is shown in Table 1.

[1]Website of the Central Government of the People's Republic of China (http://www.gov.cn/guoqing/index.htm) on July 5, 2017.

H. Huo (✉) · Z. Wang · W. Li
China Science and Technology Exchange Center, Beijing, China

X. Zhao et al. (eds.), *BRICS Innovative Competitiveness Report 2017*,
Research Series on the Chinese Dream and China's Development Path,
https://doi.org/10.1007/978-981-10-8078-4_7

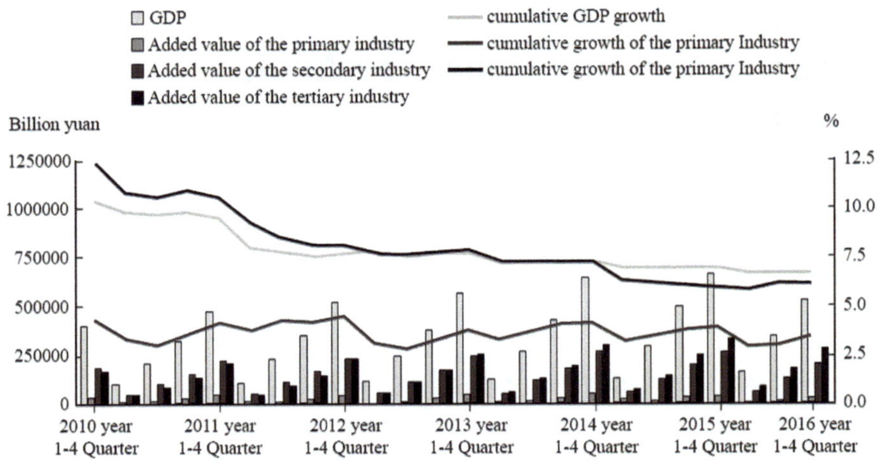

Fig. 1 Changes in China's GDP in the first three quarters from 2010 to 2016 (*Source* National Bureau of Statistics of the People's Republic of China)

Fig. 2 Trends of scores and rankings of China's innovative competitiveness from 2001 to 2015

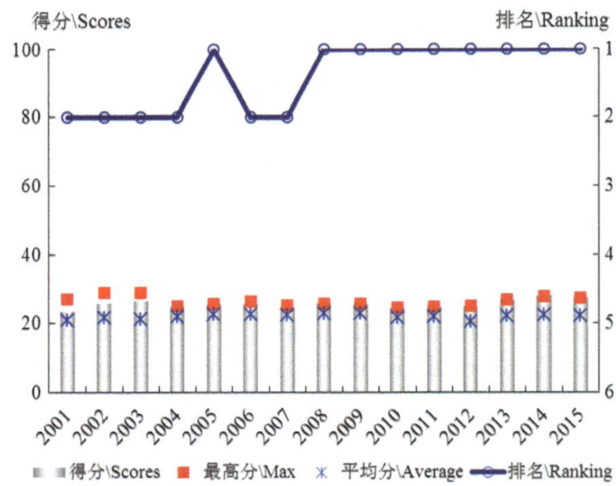

Table 1 Forecast of China's innovation index over the next 5 years

Year	2016	2017	2018	2019	2020	2021
Index	60.76	64.50	68.38	72.41	76.59	76.99

(1) From the comprehensive ranking, China's ranks first in terms of national innovative competitiveness in the BRICS in 2015, one place up compared with that of 2001. Overall speaking, the evaluation period showed an upward trend with volatility.

(2) In 2015 China's score of national innovative competitiveness reached 27.31 points, the highest among the BRICS, 5.22 higher than the average, 4.01 points higher compared with that of 2001, narrowing the gap between China and the country with the highest score in 2001 to 3.70 points, and the gap in average scores of the BRICS has increased by 2.72 points.
(3) In terms of forecasting, Chinas comprehensive innovation index is expected to achieve rapid growth from 2016 to 2021 with an increase by 16.43 points and the annual growth rate will exceed 5.1%.

Comprehensive research found that: first, China's rankings of innovative competitiveness has been on a rapid rise. Among BRICS, China has always been at the first place; among the G20, China's ranking rose from 12th in 2001 to 9th in 2015 and is the only developing country to enter the top 10. Second, it is expected that China's innovative competitiveness will steadily increase. In the timespan of the 13th Five-Year Plan, China will continue to take a leading position among the BRICS, and around 2020 enter the top 5 of the G20. The same will happen in the world rankings as well, by 2030 the country will have entered the global top 3.

2 Overview of China's Science, Technology and Innovation

2.1 Background

(1) **China Is About to Enter the Rank of Innovative Countries**

Since the 12th Five Years and the 18th CPC National Congress in particular, the central leadership has been attaching great importance to science, technology and innovation (STI). Important decisions were made on the implementation of the innovation-driven development strategy. The R&D investment in China has been increased significantly. The country has been improving the ability of making original innovations with a great number of R&D achievements. Innovative businesses are proactively competing with each other. Technological innovation provides a strong support for the economy which has maintained rapid growth towards the higher-end of the value chain. In the STI arena, China has entered a new stage of being a follower in some areas, a parallel runner in other areas and a frontrunner in still other areas. It is an important period from quantitative to qualitative changes, from breakthroughs in points to systemic growth of the overall capability.

The core position of STI in the overall national development has become more prominent. In the global landscape of innovation, China has further improved its position as a major country with significant international influence (Fig. 3).

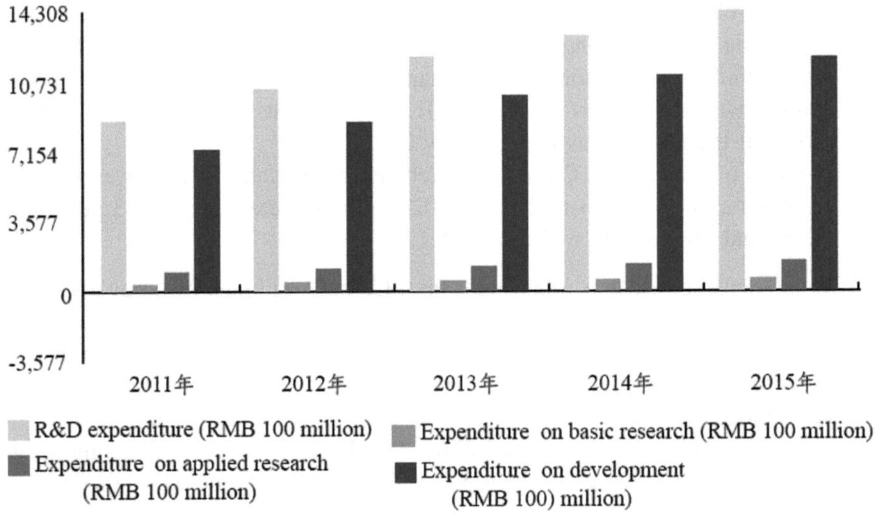

Fig. 3 R&D expenditure in China from 2011 to 2015 (*Source* National Bureau of Statistics of the People's Republic of China)

On the whole, the STI capability of China continues to grow; breakthroughs continue to be made in strategic high technologies. The country has also continued to enhance its international influence of basic research. Important innovation achievements have been made in manned space flight, lunar exploration, the manned deep submersible, deep drilling, supercomputing, the quantum anomalous Hall effect, quantum communications, neutrino oscillations, and induced pluripotent stem cells. Breakthroughs have been made in major equipment and products of high speed railways, hydropower, ultra-high-voltage (UHV) power transmission, hybrid rice, fourth generation mobile communications (4G), earth observation satellites, the Beidou navigation system, and electric vehicles among others. Some of these technologies and products have been well received in the international market. The management reform of national R&D programs has been further carried out and substantial progress has been made in the management of R&D programs (projects and funds) sponsored by the central government to further enhance the planning and coordination of resource allocation, enable the market-based mechanism of technological innovation to become more sophisticated, and have enterprises play a leadership role in technological innovation. The level of STI internationalization has been raised dramatically, with more in-depth international cooperation, and the accelerated agglomeration of high-end resources such as talent and research institutions. Science and technology diplomacy has been playing a more important role in the country's foreign affairs.

The climate for innovation and startups has become better and better. National innovation demonstration zones and high-tech zones are important carriers of innovation and entrepreneurship. Incentives such as super-deduction of business

R&D expenses have been implemented and proved to be quite effective. Technology and finance are connected more closely; the scientific literacy of citizens has been improved and the public are now more innovative with a growing awareness of innovation.[2]

(2) **A New Round of Global Scientific and Technological Revolution and Industrial Transformation Has Taken Shape**

It has become increasingly obvious that science and technology is evolving from the micro to macro scale, and there is a growing trend that breakthroughs are made through multi-disciplinary studies and crosscutting knowledge. New directions and frontiers are identified by original breakthroughs in major scientific issues of material science, space science, life science, and brain science among others. Generic breakthroughs have been made in artificial intelligence, biological technology, information network, clean energy, new materials, advanced manufacturing and other fields. Disruptive technologies keep emerging, which stimulates new economies, new industries, new business formats the new modalities with profound impacts on the way of production and lifestyles. STI are playing an increasingly important role in mankind's efforts to tackle common challenges and realize sustainable development. At the same time, innovation and entrepreneurship have become highly intensive and active across the globe. The speed, scope and scale of worldwide resource mobility have reached an unprecedented level. Like nerve before, knowledge exchanges and technology transfer have received robust momentum. The restructuring of the global innovation landscape is speeding up and STI are key for countries to achieve economic rebalancing and build up new competitiveness.

(3) **China's Economic Development Has Entered a New Normal**

The 13th Five-Year National Plan on STI notes that the supply-side reform should be promoted to enhance the quality and efficiency of economic transition and upgrade. That presents an urgent need to leverage STI to create new drivers for growth. It is urgent to rely on STI to break the resource and environmental constraints so as to coordinate and promote new industrialization, information technology, urbanization, the green modernization of agriculture, and ecological protection. In order to cope with an aging population, eliminate poverty, enhance people's health and innovate in social governance, it is urgent to use STI to improve people's livelihood. To implement the overall concepts of national security and safeguard national security and strategic interests, there is an urgent need to rely on the powerful guarantee of STI. In the meanwhile, along with growing national income, market demands have increased more rapidly, the industrial system has become more comprehensive, institutions have significantly enhanced their vitality, and the education level has been raised and quality of human capital improved. The economy is resilient with great potential for sustainable sound development and

[2]Contents from the 13th Five-Year National Plan on Science, Technology and Innovation.

room for maneuver. The comprehensive national strength is expected to reach a new level and provide solid economic and human resource bases for expediting STI breakthroughs.

2.2 *Status Quo of STI in China*

(1) R&D Intensity Has Again Reached a Record High. State Finance and Taxation Have Played a Bigger Role in Supporting STI

According to statistics, in 2015 the country's total R&D expenditure was RMB 1.4 trillion, an increase of 38.1% over 2012, with an average annual growth of 11.4%. According to exchange rates, China's R&D funding overtook that of Germany in 2010, and then Japan in 2013. At present, China has become the world's second largest R&D investing country. In 2015, China's R&D intensity (ratio between R&D funding and GDP) was 2.10%, 0.17 percentage points higher than that of 2012; has reached the level of moderately developed countries and ranked first in developing countries. The increase of R&D investment has created favorable conditions for China to become a "parallel runner" and "frontrunner" in some STI areas (Table 2).

In 2014, the R&D input of national financial expenditure was RMB 645 billion 450 million, up 15.3% over 2012, an average annual growth of 7.4%; R&D spending accounted for 4.25% of the country's total fiscal expenditure. In

Table 2 Main STI indicators of China at the end of 12th Five-Year Plan

Indicators		Value of indicators in 2015
1	World ranking of national comprehensive innovation capability	18
2	Contribution of scientific and technological progress to economic growth (%)	55.3
3	R&D Intensity (%)	2.1
4	R&D personnel (full-time equivalent) per 10,000 employees	48.5
5	Operating income of high-tech enterprises (RMB trillion)	22.2
6	Added value of knowledge-intensive service sector/GDP (%)	15.6
7	R&D expenditure of large industrial enterprises/main business income (%) (large industrial enterprises refer to those with an annual revenue of no less than RMB 20 million from main business operations)	0.9
8	World citation ranking of international scientific papers	4
9	PCT patent applications (10,000 pieces)	3.05
10	Invention patents per 10,000 people	6.3
11	Turnover of national technical contracts (RMB 100 million)	9835
12	Proportion of scientifically literate citizens (%)	6.2

particular, the R&D input from the central government was RMB 289 billion 920 million, with an average annual growth of 5.3%; the local fiscal expenditure on R&D was RMB 355 billion 540 million, with an average annual growth of 9.1%. In 2014 the super-deduction of R&D costs from taxable income of large industrial enterprises (with annual revenue of no less than RMB 20 million from their main business operations) and high-tech enterprises reached RMB 37 billion 980 million and RMB 61 billion 310 million respectively, with growth rates of 27.2 and 16.2% over those in 2012, and the average annual growth was 12.8 and 7.8% respectively.

(2) **Capabilities of Original Innovation Have Been Continuously Improved with Significant R&D Achievements**

According to statistics, in 2015 China's basic research funding was RMB 67 billion 60 million, an increase of 34.4% over 2012 with an average annual growth of 10.4%. In particular, the input of basic research from active knowledge innovators such as universities and research institutions reached RMB 34 billion 720 million and RMB 29 billion 500 million, up 26 and 49% respectively compared with those in 2012. In 2015, China's basic research funding accounted for 4.7% of the total R&D expenditure of the whole society.

In 2015 China's patent applications hit 2 million 799 thousand, an increase of 36.5% over 2012; the number of initial invention patent applications exceeded one million, reaching 1 million 102 thousand, an increase of 68.8% over 2012. For 5 consecutive years, the number of patent applications received in China ranked first in the world. In 2015, the number of patent grants in China was 1 million 718 thousand, up 36.9% in 2012; in particular, the number of invention patent grants was 359 thousand, an increase of 65.5% over 2012. Invention patent grants accounted for 20.9% of the total, 3.6 percentage points higher than that of 2012. As of the end of 2015, valid patent grants and valid invention patents were 5 million 478 thousand and 1 million 472 thousand respectively, an increase by 1 million 969 thousand and 597 thousand than those in 2012; the number of invention patents per 10,000 people was 6.3. In 2015, China accepted 30,548 Patent Cooperation Treaty (PCT) patent applications. From 2013 onwards, the country ranked third in the world for two consecutive years in terms of PCT patent application. The rapid growth of patents with an improved structure and upgrade of international rankings indicate the increased capability and efficiency of China's scientific and technological output (Fig. 4).

(3) **Innovation Resources Are Clustering Around Enterprises, Giving Rise to Innovation in Enterprises**

According to statistics, in 2015, China's business expenditure on R&D (BERD) exceeded RMB 1.1 trillion, an increase of 40.3% over 2012, at an average annual growth of 11.9%. BERD/GERD reached 77.4%, up by 1.2% over 2012. Among them, the R&D expenditure of full-scale industrial enterprises, for the first time, exceeded RMB 1 trillion, reaching RMB 1 trillion and 15.09 billion. Business R&D personnel totaled 4.25 million, an increase of 26.2% over 2012. By the end of 2015, the technology centers of state-level enterprises (groups) had reached 1187, an

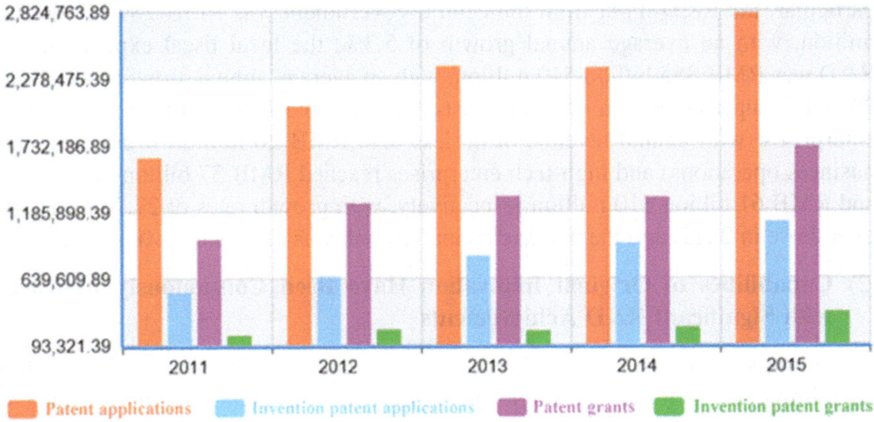

Fig. 4 Patent applications and grants in China from 2011 to 2015 (*Source* National Bureau of Statistics of the People's Republic of China)

increase of 300 compared with 2012. As of the end of 2015, the number of State Key Laboratory based in enterprises had been 177, accounting for 36.8% of the total. The national engineering centers built in companies amounted to 144, accounting for 41.6% of the total. In 2015, the gross funding for emerging industries was RMB 55.68 billion, an increase of 92% over 2012, and 1233 enterprises were invested, representing an increase of 995 compared with 2012 (Fig. 5).

According to the 2014 national survey on innovation by enterprise, during 2013–2014, 41.3% of China's 646,000 large-scale enterprises, 266,000 enterprises

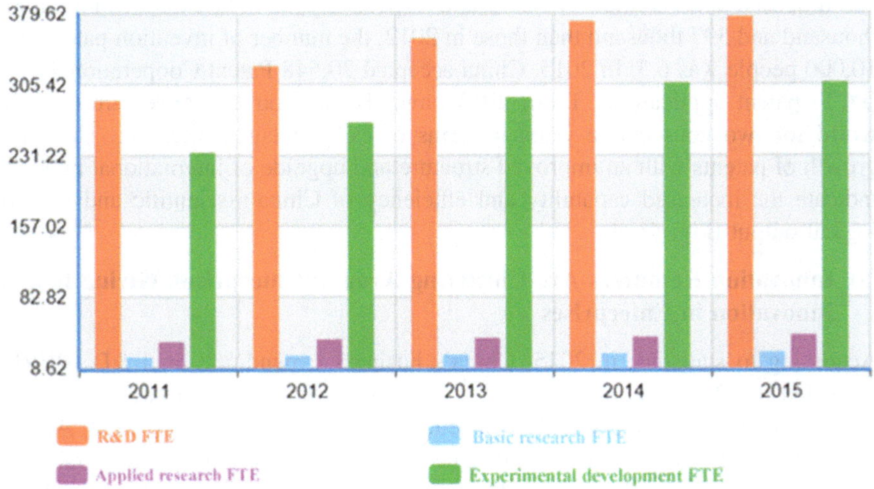

Fig. 5 FTE R&D personnel 2011–2015 (*Source* National Bureau of Statistics of the People's Republic of China)

carried out innovation activities. Among them, product innovation, process innovation, organizational innovation and marketing innovation accounted for 18.7, 20, 27.9 and 25.8%. By the industrial sector, there were 177,000 and 84,000 enterprises with innovative activities in industrial and service enterprises, accounting for 46.8 and 32.6% respectively. By scale, innovation activities of large and medium-sized and small enterprises in the industrial enterprises were 42,000 and 135,000, accounting for 64.8 and 43.1%. By type of registration, 148,000 domestically-funded enterprises and 29,000 enterprises funded by Hong Kong, Macao, Taiwan and foreign capital carried out innovative activities, accounting for 45.9 and 52.4% respectively. To a certain extent, innovation has become a common choice for the survival and development of different types of enterprises.

2.2.1 Science, Technology and Innovation Have Optimized the Industrial Structure and Supported Economic and Social Development

In 2014, the sales revenue of new products by large-scale enterprises arrived at RMB 14.3 trillion, an increase of 29.3% over 2012, at an average annual growth rate of 13.7%; new product sales revenue accounted for 12.9% of main operating income, 1% higher than that of 2012. During 2012–2014 years, the contribution of new product sales to the main operating income increments was 18.2%. During the shift of economic growth rate in China, high-tech manufacturing industry, e.g. electronic and communications and pharmaceutical, kept steady growth, laying foundation for the optimization of industrial structure in China. In 2015, the value added of high-tech manufacturing grew by 10.2% over the previous year, 4.1% higher than the growth rate of the value added of large-scale enterprises during the same period, accounting for 11.8% of the value added of large-scale enterprises. The main operating income totaled RMB 13.7 trillion, an increase of 33.9% over 2012, at an annual increase of 10.2, 3.9% higher than the annual growth rate of the large-scale industry. During 2012–2015, the high-tech manufacturing industry contributed 19.9% to the increment of the main operating income (Fig. 6).

Since the 18th Congress of the Communist Party of China (CPC), science, technology and innovation has given rise to rapid economic and social development. The industrial chain of TD-LTE has grown more mature. By the end of 2015, the number of 4G users had exceeded 380 million. A new generation of high-speed railway is leading the world and entering the overseas market, with a total mileage of high-speed rail reaching 19 thousand km, accounting for more than 60% of the world total mileage. The advanced regional jet (ARJ) has been sold and delivered for operation. C919, China's home-grown passenger jet, completed its maiden flight.[3] The solid-state lighting (SSL) technology has been widely promoted, with the overall size of the SSL industry in 2015 amounting to RMB 424.5 billion, a

[3]Xinhua News Agency, "China's home-grown passenger jet C919 completes maiden flight", http://news.xinhuanet.com/politics/2017-05/07/c_1120929083.htm.

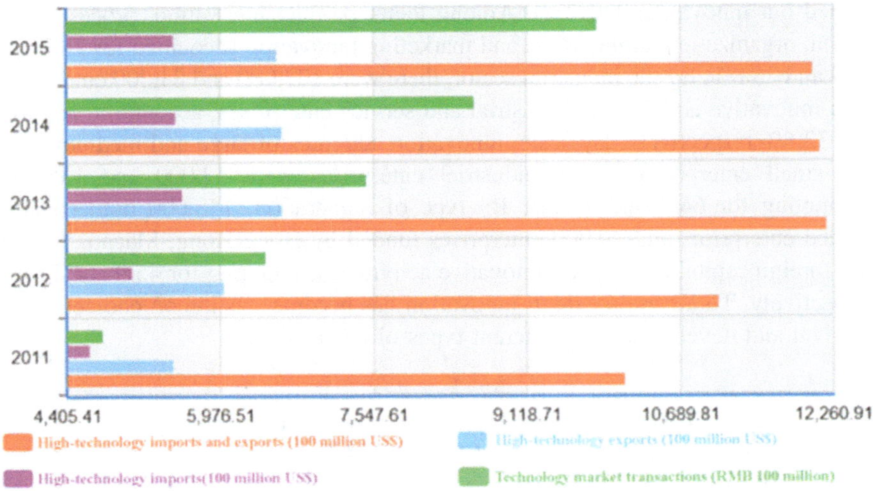

Fig. 6 Import and export of high-tech products and the growth of technology markets in China from 2011 to 2015 (*Source* National Bureau of Statistics of the People's Republic of China)

year-on-year increase of 21%. The fourth-generation super rice has created a new record with an average yield of 1026.7 kg/mu. In the field of health, the world's first artificial bioengineering cornea and anti-tumor drugs like Apatinib and chidamide have gone to the market. It is fair to say that science, technology and innovation have played an important role in improving people's life.

2.2.2 Regional Innovation Witnesses New Strides and Remarkable Progress in Some Areas

Since the 18th CPC Congress, efforts have been made in building Beijing-Tianjin-Hebei innovation community, transforming the Yangtze River Economic Belt, launching regional innovation and reform pilots in full swing. In 2015, the local expenditure on R&D in Jiangsu, Guangdong, Shandong and Beijing exceeded RMB 100 billion, with the R&D intensity of Beijing, Shanghai, Tianjin Jiangsu, Guangdong, Zhejiang, Shandong and Shaanxi reaching or exceeding the national average. During 2013–2014, businesses contributed 44, 39.8, 37.9 and 26.3% to innovation activities in eastern, central, western and northeastern parts of China. Tianjin, Jiangsu, Zhejiang, Guangdong and Shaanxi are at the forefront for business engagement in innovation. Beijing and Shanghai are actively building science and innovation centers with global influence. Innovative provinces and innovative urban construction projects have achieved initial success, and the national innovation demonstration zones and high-tech zones have developed rapidly. During the "12th Five-Year" period, the revenues of 146 high-tech zones have maintained an average annual growth rate of 17%, with the operating income reaching RMB 28 trillion in 2016, up by 11.5% year on year. The total industrial output has increased by 10.3%. High-tech zones have become models of regional innovation in China. The 17 national innovation demonstration zones have played a tremendous role in promoting the regional economic development. Thanks to the support of innovation demonstration zones, Donghu in Wuhan, Changsha-Zhuzhou-Xiangtan in Hunan, and Chengdu in Sichuan remained about 30% growth rate in recent years. Similarly, the Z-park contributed 24.7% of Beijing's GDP.[4]

3 China's Cooperation with BRICS Members in Science and Innovation

3.1 China's Engagement with International Partners in Science and Innovation

As the world economy undergoes deep adjustment and twisted recovery, a new round of scientific and technological revolution and industrial changes are in the making. The global governance system has undergone profound changes. It has become a global consensus to promote sustainable development through innovation. Information communication, biology, new materials and new energy technologies are widely applied, giving rise to technological breakthroughs that are green, intelligent and ubiquitous. Technological innovations continue to break geographical boundaries, with innovation resources flowing across the globe. Sustained, extensive and in-depth international cooperation in science and innovation has become a necessary way for China to meet global challenges and achieve economic growth and sustainable development.

Around the world, the developed countries are deepening international cooperation in science and innovation. Emerging economies have become major players in the global scientific and technological cooperation. International flows of research funding, technologies and researchers are picking up speed, with foreign capital accounting for a bigger share of R&D funding in a country. The major countries have formulated policies and measures to attract foreign talent, thus promoting technological innovation and economic growth. Closer business partnership has become the emphasis of countries, which have issued innovation strategies to create a sound policy environment for business collaboration in innovation. Open science and open innovation have become an important model for international innovation and development. The globalization of science and innovation calls for an open strategy of win-win cooperation, deep engagement in international economic cooperation and competition, and access to domestic and international resources.

[4]Vice Chairman of CPPCC and Minister of Science and Technology of China Dr. Wan Gang spoke at the news center of the fifth session of the 12th National People's Congress on March 11, 2017, http://scitech.people.com.cn/n1/2017/0311/c1007-29139006.html.

Domestically speaking, China is embracing opportunities, with a transition from factors-driven to innovation-driven development. In order to ensure that by 2020, China could join the rank of innovative countries, achieve the "two centennial goals", and balance short-term steady growth and long-term economic adjustment, international cooperation in science and innovation is an effective way to implement the Strategy of Innovation-driven Development, cluster global resources, and enable China to play a more important role in the global value chain. To follow up on the five concepts of "innovative, harmonious, green, open and sharing development" proposed at the fifth Plenary Session of the 18th CPC Central Committee and facilitate the "Belt and Road" Initiative, we need to have closer ties among each other. International cooperation in science and innovation is an important solution to create a community of shared future, reinforce partnership with developing countries, promote multilateral diplomacy, and facilitate the reform of the international system and global governance. As China's national strength continues to increase, we need to adopt a more proactive strategy of cooperation in science and innovation.

Since the beginning of the 12th Five-Year, China's scientific and technological innovation capability has achieved a historic leap, having bigger influence in international scientific and technological cooperation. An all-round, multi-level and multi-channel international cooperation system has taken shape. The investment has increased significantly, and so has the cooperation capability. A relatively complete network of international scientific and technological cooperation has gradually taken shape. International collaboration has played a key role in the three aspects. First, support the overall diplomacy of the country, develop new relations between major powers, and promote cooperation with China in developing domestic affairs. Two, through scientific and technological partnership, we should take part in the international science program, solve the bottleneck problems, and achieve leapfrogging development in related fields. Three, enhance mutual trust and create a sound international environment for economic restructuring.

By the end of the "12th Five-year period", 549 cooperation bases have been established including international innovation parks, international joint research centers, international technology transfer centers and pilot bases for international cooperation. China has established technological partnerships with 158 countries and regions, signed 111 inter-governmental agreements, joined more than 200 inter-governmental science and technology cooperation organizations, and sent science diplomats to 70 embassies and consulate generals stationed around the world. China also launched the mechanism of innovation dialogue with the US, the EU, Germany, France, Israel, Brazil, Russia and Canada, forming a stable inter-governmental cooperation mechanism with major countries, regions and international organizations. Meanwhile, China has forged science and technology partnerships with Africa, ASEAN, South Asia, SCO member countries, CELAC member countries and Arab nations, to carry out pragmatic cooperation in science and technology and support enterprises to go to international markets (Table 3).

Table 3 Main goals of international STI cooperation during the 13th Five-year Plan period

– Build an STI system oriented to the world, and support in meeting the demand of economic and social development. Set up an STI cooperation system based on the innovation cooperation mechanism and provide strong support to deal with major core S&T issues in light of the needs of industries, sectors and localities. Enhance capabilities of various innovation players, cultivate new strengths for international STI cooperation and competition, and work to promote STI, industrial transformation and livelihood improvement
– Build initially STI clusters of international impact and attractiveness. Further open national innovation system, better integrate into the global innovation network, and effectively pool innovation resources of talent, technology and funds. Expedite the cultivation and introduction of versatile talents with international vision, and make the research and innovation personnel more international. Support a number of international cooperation bases, and conduct international joint research and cooperation between enterprises, universities and research institutes through various channels. Facilitate the building of regional STI community of mutual benefit and cooperation
– Bring about new pattern of international STI cooperation featuring mutual benefit, win-win outcomes and joint development. Serve general diplomacy and introduce new mechanisms of inter-governmental cooperation to put in place a new pattern of international win-win STI cooperation. Advance the Belt and Road initiative, and facilitate interconnectivity. Take part in and when appropriate take the lead in implementing international mega-science programs and engineering projects. Fulfill the responsibility of a big country, and work with all countries against global challenges like climate change, human health, energy security, food security and environmental problems. Get involved in global STI cooperation governance through bilateral and multilateral cooperation mechanisms, and improve the say and impact in the formulation of cooperation rules. Enhance foreign assistance through science and technology, and deepen cooperation with developing countries
– Support enterprises in getting deeply involved in international STI cooperation, and promote mass entrepreneurship and innovation. Encourage and help enterprises to "go global" and "bring in", make them more international and enhance their global competitiveness by establishing professional service systems. Advance mass entrepreneurship and innovation through international S&T cooperation

3.2 Cooperation with BRICS Cooperation

With BRICS cooperation constantly furthered, a multi-domain and multi-tiered cooperation mechanism where BRICS Summit plays the leading role and hi-level meetings in relevant departments and areas serve as the supplement.

In terms of cooperation mechanism, STI cooperation under BRICS framework contains three tiers of working mechanisms, namely ministerial meeting, coordinators' meeting and working group meeting.

According to the MOU between BRICS Countries on Inter-governmental STI Cooperation signed at the 2nd Ministerial Meeting in March 2015, 19 areas were confirmed as the priority areas, such as new energy, renewable energy and energy efficiency, natural disaster management, water resource and pollution treatment, geospatial technologies and application, astronomy, high-performance computing and nano-technology. It was also decided that the countries continue pragmatic cooperation in 5 special working areas launched at the 1st BRICS STI Ministerial

Meeting, among which China leads in new energy, renewable energy and energy efficiency, Brazil natural disaster management, Russia water resource and pollution treatment, India geospatial technology and application, and South Africa astronomy.

At the 3rd BRICS STI Ministerial Meeting held in October 2015, *Moscow Ministerial Declaration* was issued, which increased the number of special areas to 10. Brazil and Russia lead in bio-tech and bio-medicine, including human health and neuroscience, China and South Africa IT and high-performance computing, Brazil and Russia ocean and polar S&T, India and Russia material science including nanotechnology, and India and Russia photo electricity.

In order to further STI cooperation, in 2016 BRICS set up a funders' working group, signed BRICS STI Framework Program and the Implementation Plan, and decided to launch joint call for multilateral R&D projects. This initiative is aimed at supporting and promoting cooperation between partners from at least three countries. At present, the following funders have participated in the initiative: CNPq from Brazil; FASIE, MON and RFBR from Russia; DST from India; MOST and NSFC from China; and DST and NRF from South Africa (Table 4).

In terms of the content of cooperation, extensive, pragmatic and fruitful cooperation has been conducted in recent years with regard to bilateral STI projects, bases and exchange of research personnel. The cooperation with Russia dates back long time and turns out to be fruitful. Over the past twenty years or so, China and Russia have conducted frequent cooperation in space, aeronautics, ocean exploration, nuclear energy, agriculture, biology and machine making, made marked progress in the cooperation of science parks, and drew upon each other's innovation experience. With an enduring history of S&T cooperation, China and Brazil have conducted in-depth and fruitful cooperation in space, new energy, agricultural S&T and nano-innovation research. China and India have carried out cooperation in software development, bio-tech and bio-medicine, and green and low-carbon industries. China and South Africa have collaborated on bio-tech (food processing, agriculture and medicine), new material and advanced manufacturing technologies, IT, environmental protection, mining and metallurgy, resource exploitation, space technology, transportation, paleoanthropology and local knowledge system.

Moreover, Russia has conducted close and successful hi-tech cooperation with India especially in areas of nuclear energy, worked closely with Brazil in aeronautics, and cooperated in an in-depth manner with South Africa in nuclear energy. In June 2003, South Africa, India, and Brazil have founded a tri-party forum of IBSA, under which extensive S&T cooperation has been conducted.

3.2.1 Bilateral STI Cooperation

China and the other four BRICS countries have conducted content-rich, fruitful and promising STI cooperation in international S&T cooperation project, bases and exchange of research personnel.

Table 4 Overview of special areas supported by BRICS countries

No.	Special areas	Brazil	Russia			India	China		South Africa	
		CNPq	FASIE	MON	RFBR	DST	MOST	NSFC	DST	NRF
1	Natural disaster management	▲	▲	▲	▲	▲	▲			▲
2	Water resource and pollution treatment	▲	▲	▲	▲	▲		▲		▲
3	Geospatial technology and its application	▲	▲	▲	▲	▲	▲			▲
4	New energy, renewable energy and energy efficiency	▲	▲	▲	▲	▲	▲			▲
5	Astronomy		▲	▲	▲	▲		▲		▲
6	Bio-tech and bio-medicine, including human health and neuroscience	▲	▲	▲	▲	▲		▲		▲
7	IT and high-performance computing	▲	▲	▲	▲	▲	▲			▲
8	Ocean and polar S&T	▲	▲	▲	▲	▲		▲		▲
9	Material sciences, including nanotechnology		▲	▲	▲	▲		▲		▲
10	Photo electricity		▲	▲	▲	▲	▲			▲

From 2007 to 2015, the Chinese government allocated RMB 2.729 billion of funding for the STI cooperation projects, with the funding volume and number of projects rising year-on-year. The areas of cooperation were mainly material, engineering, information and life sciences. From 2007 to 2016, China has set up 190 BRICS-related national-level bases of S&T cooperation concerning life sciences, advanced manufacturing, IT, material sciences, earth sciences, energy and environment, and agriculture. The bases are evenly distributed in all the areas.

The exchange of research personnel mainly involved technical training for foreign assistance and the Talented Young Scientist Exchange Program. Since 2006, MOST and its affiliated institutions have held 411 technical training workshops, among which BRICS attended 221. Among the 7885 participants, 566 come from BRICS countries, 7.18% of the total. And the number of those coming from BRICS countries are twice higher than that of the average. The workshops they attended mainly dealt with agriculture, information, manufacturing, response to climate change, resource and environment, new energy, medical health, and S&T policy and management.

(1) **Cooperation with Russia**

From 2007 to 2015, there were 609 joint projects with Russia with a total funding of RMB 2.626 billion, with the amount of projects and funding both accounting for over 90% of the total. The cooperation mainly focuses on areas of material, engineering & technology, and information. There are as many as 157 Russia-related cooperation bases, far exceeding that of other BRICS countries.

(2) **Cooperation with India**

Cooperation with India mainly dealt with people-to-people exchange. The number of the participants for technical training workshops was the highest among BRICS countries. Since 2006, there has been 256 Indian participants, accounting for 45% of the total. From 2007 to 2015, there were only 9 cooperation projects with India reaching a total funding of RMB 8.46 million, the least among BRICS countries. In the same period of time, the 8 joint projects with other South Asian nations totaled a funding of RMB 22.67 million, among which 5 Pakistan-related ones reached a funding of RMB 15.01 million.

(3) **Cooperation with South Africa**

From 2007 to 2015, there were 30 cooperation projects with South Africa with a funding of RMB 63.03 million, second to Russia in terms of both amount and volume. This shows that China has maintained sound STI relationship with South Africa. At the same time, there are 27 projects for Russia with a funding of RMB 107 million, among which 3 projects with Egypt reached a volume of RMB 15.7 million, 2 with Kenya reached RMB 11.53 million, and 2 with Algeria reached RMB 9.94 million.

(4) **Cooperation with Brazil**

From 2007 to 2015, there were 17 joint projects with a total fund of RMB 31.21 million.

3.2.2 Outcome of Cooperation with BRICS Countries

(1) **Tackle Key Technologies in Priority Areas and Enhance Innovation Capacity of BRICS Countries**

With information, material and energy as the breakthroughs, we work to bring about a batch of key technologies vital to the change of growth mode and upgrading of industrial structure, which greatly enhanced the technical innovation capability of relevant research institutes and enterprises and elevated the STI capacity of those involved in the cooperation. As an important component of the overall diplomatic strategy, BRICS STI cooperation has brought benefits to relevant countries by promoting a batch of advanced and applicable technologies.

(2) **Further BRICS Partnership Through Personnel Exchange**

Under the framework of training for outstanding talents from developing countries, we help produce top scientists and engineers for BRICS countries, establish long-term partnership between research institutes, universities and enterprises home and abroad, strengthen exchange of research personnel, build up closer bonds between the peoples, enhance capability building, lay a solid foundation for STI cooperation and play a supporting role in advancing mutual benefit and win-win outcomes.

(3) **BRICS STI Cooperation Becomes Role Model of Cooperation Between Developing Countries**

Governments of BRICS countries can play a dominant role in BRICS S&T cooperation in light of their strategic orientations, build platforms for S&T cooperation, set up joint labs and conduct joint research. The Chinese government has, within its capabilities, allocated a great deal of resources, played a leading role in BRICS S&T cooperation, act in line with its image as a big developing country, and build BRICS S&T cooperation into a role model of South-South S&T Cooperation.

Chapter 8
South Africa Report on Science, Technology and Innovation

Zhongyang Wang, Dong Zhang and Zongwen Ma

Located in the southernmost tip of the African continent, with an area of 1213,100 km^2, South Africa is renowned as the "Country of Rainbow". The country is surrounded by the Indian Ocean and the Atlantic Ocean in the east, south and west. At the end of 2015, South Africa had a population of 54.9569 million and gross domestic product (GDP) of 314.572 billion US dollars, the second largest economy in Africa (second only to Egypt, 330.779 billion US dollars). With per capita gross national income (GNI) of 6080 US dollars, South Africa is a developing country of medium to high income while keeping people's living standards among the best in Africa. Since the establishment of the new South Africa in 1994, it economic development has gone through three stages: 1994–2002, a slight decline in GDP which maintained around 100–150 billion US dollars; 2003–2011, rapid economic growth where GDP increased by about three times, with an annual growth rate of 16.37%; 2012–2015, the economy began to decline, down by 6.7% per year (Fig. 1).

With a relatively complete financial and legal system, South Africa is able to provide sound communication, transportation, energy and other infrastructure. South Africa holds an important political and economic status in Africa. Mining, manufacturing, agriculture and services are the four pillars of South Africa's economic development. Minerals are the main source of its economy. It is a leading player in deep mining and other technologies in the world.

Over the recent years, in order to tackle the economic slowdown and achieve sustainable development, South Africa has been working together with a new round of technological revolution and industrial revolution by introducing new policy initiatives to stimulate scientific and technological innovation that facilitate its industrialization and modernization process at home and by strengthening international cooperation that provides new driving force to its innovation.

Z. Wang (✉) · D. Zhang · Z. Ma
China Science and Technology Exchange Center, Beijing, China

X. Zhao et al. (eds.), *BRICS Innovative Competitiveness Report 2017*,
Research Series on the Chinese Dream and China's Development Path,
https://doi.org/10.1007/978-981-10-8078-4_8

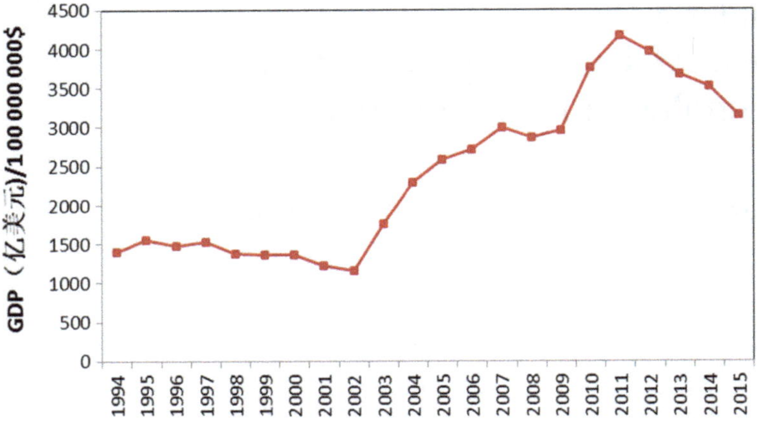

Fig. 1 GDP of South Africa from 1994 to 2015. *Source* OECD data base

1 The Landscape of Science, Technology and Innovation in South Africa

1.1 Evaluation and Analysis of Innovation Competitiveness of South Africa

According to this report, the ranking and score of the innovation competitiveness of South Africa among the "BRICS" countries are shown in Fig. 2. Forecast of South Africa's innovation index is displayed in Table 1.

Fig. 2 Trend forecast in score and ranking of innovative competitiveness of South Africa from 2001 to 2015

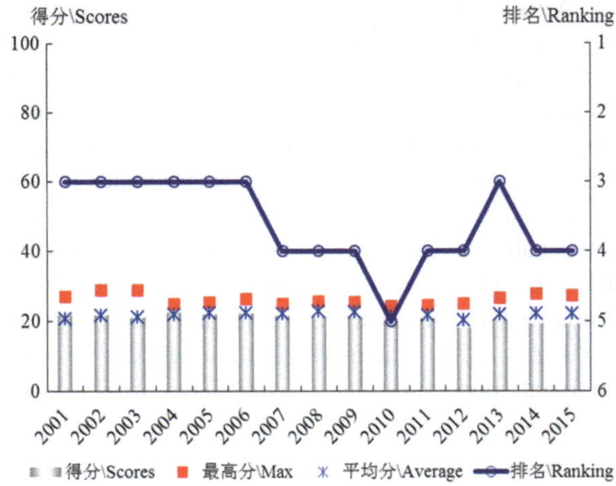

Table 1 Forecast of South Africa's innovation index in the next five years

Year	2016	2017	2018	2019	2020
Score	27.16	27.23	27.29	27.36	27.42

(1) In terms of overall ranking of national innovation competitiveness, South Africa ranked 4th in "BRICS" countries in 2015, down by one place compared with 2001. On the whole, the ranking showed a downward trend.
(2) In terms of scoring, South Africa's innovation competitiveness scored 19.38 in 2015, 7.93 points lower than the highest in the "BRICS", and 2.71 points lower than the average; compared with 2001, South Africa's national innovation competitiveness score fell by 2.52 points, the gap with the highest score widened by 2.83 points, and the gap with the "BRICS" average expanded by 3.81 points.
(3) In terms of the forecast, South Africa's Composite Innovation Index will score low in the long run, which is expected to increase by only 0.26 points from 2016 to 2020.

1.2 Science, Technology and Innovation Policy, Strategy and Development Plan

The overall goal of scientific and technological innovation in South Africa is to serve the economic and social development and build a knowledge-based economy and society. Since the plan to build a national science and technology innovation system was released in 1996, the South African government has given full play to the functions of the government to coordinate the national technological innovation activities, focusing on the construction of the national innovation system, concentrating on the development of science and technology resources in the national key priority areas, as well as promoting the synergy of business, academia, and research in order to ensure enterprises become major undertaker of innovation.

1.2.1 Value Science, Technology and Innovation, and Build a Long-Term Strategy of National Economic and Social Development

The South African Department of Science and Technology promulgated and implemented the *South African National Research and Development Strategy* in July 2002, proposing the strategic initiatives that should be taken to establish a national innovation system and laying the foundation for the development of the national innovation system. In 2007, the Department of Science and Technology developed the "*Embrace the Knowledge Economy—Ten-Year Innovation Plan*

(2008–2018)", which for the first time outlined a long-term blueprint for the future development of science and technology in South Africa, further clarifying the goal of building the national innovation system that leads the country to "a knowledge-based economy and society by 2018". The planning addressed five priority areas: medicine and bio-economy, space science and technology, energy security and global climate change. In June 2009, the Presidential Office of South Africa released the *Medium-Term Strategic Framework 2009–2014*, further emphasizing that science, technology and innovation are one of the key policy areas. It announced to focus on existing strategies to give special support to enterprise innovation and sector development programs. Focusing on building the national innovation system, the Department of Science and Technology introduced a national biotechnology strategy, advanced manufacturing technology strategy, technology transfer strategy, information and communication technology strategy, human resources development strategy, integrated manufacturing strategy, national nanotechnology strategy, national Antarctic Research strategy, the origin of Africa—ancient scientific research strategy and other related strategic planning.

1.2.2 Focus on Sharing Research Platforms in Order to Back up the Construction of National Innovation System

The South African government believes that research infrastructure is an important basis of the national innovation system. The South African Department of Science and Technology plans to build a world-class technology and innovation infrastructure, and strives to achieve the sharing of science and technology resources in 20 years. To this end, the government has adopted a series of measures: firstly, the construction and updating of national research infrastructure by implementing the National Equipment Program (NEP) and the National Nano Equipment Program (NNEP), with a view to set up advanced research equipment in South Africa to allow its scientists to carry out high-level research and innovation; secondly, implementing three pillar programs of information infrastructure, including the construction of a high-performance computing center, the national research network, and the national super data center.

1.2.3 Take Multiple Measures to Give Full Play to the Main Role of Enterprises in Innovation

The South African government has taken measures to create a favorable policy environment and put in place an innovation service system to promote enterprise development. Firstly, create a favorable policy environment to promote enterprise technological innovation. Major measures include the introduction of tax incentives to encourage enterprises to increase R&D investment; harness government procurement to support business growth; and strengthen intellectual property management to promote local transfer of research results. Secondly, establish technology plans and funds to directly support enterprise R&D and innovation. Major measures include Innovation Funds, Industrial Technology and Human Resources Development Program (THRIP), Industrial Innovation Support Program (SPII), Sector Partnership Fund, and Industrial Development Group Venture Capital Fund, etc. Thirdly, build the enterprise innovation. Major measures include the creating a small business incubation and technology transfer platform, setting up technical service stations in the University of Science and Technology to provide technical support to technical SMEs, and establishing a biotechnology innovation center.

1.2.4 Strengthen Industrial Technology Innovation to Promote Economic and Social Development

As a global power of mining, South Africa leads the world in coal chemical technology and isotope technology. In recent years, the South African government introduced an *Added Value in Mining* strategy, and implemented the Advanced Metal Program, focusing on supporting fluorine chemical research and development, titanium manufacturing technology to enhance the added value of mineral products. In addition, the South African government also actively promoted the application of deep processing technology of agricultural products, aquaculture technology and information and communication technology, which has generated sound economic and social benefits.

1.2.5 Introduce International Talent to Make up for Talent Shortage in Key Areas

In order to achieve the ambitious goal listed in its innovation planning, and to solve the bottleneck of talent shortage in South Africa, to reverse the situation of high-efficient brain drain, to attract world-class talents, to enhance competitiveness in key areas, and to revive national innovation, the Southern African Department of Science and Technology launched the Chief Scientist Program (SARChi) in 2006, and identified 62 SARChi seats in priority areas listed in the National R & D Strategy, Ten-Year Innovation Plan, and National Medium and Long Term Development Strategy and opened these seats to candidates for international pioneers in science and technology. Through the implementation of the Program, South Africa not only retained a number of the original science and technology elites, but also attracted more than 20 world-class scientists from the United States, Britain, Germany, Italy, Netherlands, Sweden and other developed countries.

1.3 *Strategy for the Development of Key Areas of Technological Innovation*

1.3.1 Introduce the *Research Infrastructure Roadmap* to Vigorously Develop the Research Infrastructure Focusing on SKA Mega-Science Projects

Over the past eight years, South Africa has invested as much as over seven billion rand in research infrastructure, of which 3.5 billion goes to the MeerKAT radio telescope and the SKA mega-science project, and 1.5 billion rand to the network infrastructure. In order to improve its overall scientific and technological competitiveness, as well as to promote open sharing of scientific research, cooperative research, personnel training, and enhance R&D capacity in key areas, South Africa officially released the *Research Infrastructure Roadmap* in October 2016. The strategic goal was to build first-class research infrastructure, and to lay the foundation for improving the competitiveness of scientific research and attracting world-class talent. The Roadmap focused on promoting mega-science projects such as SKA, including allocating 273 million rand to implement the national integrated network infrastructure system; establishing high-performance computing centers to meet the urgent needs of universities and research institutions for infrastructure; continuing to expand the construction of the national network covering all colleges, universities and major public research institutions in South Africa. The Roadmap also focused on funding 13 projects: a National Center for Digital Language Resources, a Land and Freshwater Environment Observing Network, a nuclear medicine research facility, a health and population monitoring point, a lab for natural science collections, shallow sea and coastal research infrastructure, genomics research distributive platform (including genetic research), bio-banking, marine and Antarctic research infrastructure, nano-micro manufacturing facilities, solar energy research facilities, material characteristics research facilities, and bio-geochemical research infrastructure platform. The first seven projects will be completed in fiscal year (FY) 2016/2017 and the latter six will be completed in FY 2020/2021.

1.3.2 Establish State-Owned Enterprises and R&D Centers to Vigorously Promote the Development of the Pharmaceutical, Bio-Economy and Other High-Tech Industry

South Africa would invest 2.7 billion rand in promoting the development of innovative industries, especially the pharmaceutical and biological industries. In the 2016 Presidential State of the Union address, President Zuma announced the establishment of a state-owned company Ketlaphela for the production of active pharmaceutical ingredients required for AIDS-resistant and other drugs, which aimed to improve South Africa's pharmaceutical industry capacity. The government would continue to invest 5.2 million rand for the next three years. The expected social benefits would include: creating jobs, technology transfer, reducing imbalance in technology revenues in the pharmaceutical industry, and providing quality medicines for combating cancer, tuberculosis and AIDS in South Africa.

In the field of bio-economy, after the South African Department of Science and Technology released the Bio-economy Strategy in January 2014, bio-economy has made a significant contribution not only to its GDP growth but also to South Africa's efforts to develop a green economy, to ensure food safety, to improve international competitiveness and to create jobs. In May 2016, South Africa established the first bio-manufacturing R&D center under the Science Industry Council to support small and medium-sized and micro enterprises in manufacturing and marketing biological products. SMEs and micro enterprises with advanced biotechnology would be able to use the center's advanced laboratory facilities for business incubation. It is expected that in the next five years, South Africa will create about 250 million rand of benefits per year in cosmetics, nutritional supplements and other bio-tech industries with local features. The anti-mosquito candles developed in the South African plant essential oil business incubator has started to make profits.

1.3.3 Launch a Number of Projects and Programs to Develop the Advanced Manufacturing Industry

Advanced manufacturing represented by 3D printing is a key area supported by South Africa. Main projects include: ① Technical localization projects. The South African Department of Science and Technology invested 105 million rand in the FY 2015/2016 to support the development of 147 national manufacturing enterprises, of which 32 companies were involved in large-scale localization projects. In the FY 2016/2017, the Department of Science and Technology would allocate 33 million rand to continue to support the above companies. ② Technical station projects. 18 affiliated technical stations were established near universities and research institutions to provide technology, skills and product production, export and other extensive technical support to over 2000 SMEs. ③ The Aeroswift Program. Known as the "Next Generation Additive Manufacturing Machine Program", the Aeroswift Program was developed by Aerosud ITC, a South African private company, and the National Laser Center of the South African Science and Industry Council, which created the world's largest powder additive production machine. In the FY 2016/2017, the Department of Science and Technology planned to continue to support the program and develop the Aeroswift commercialization strategy and development roadmap to create revenue sources for local manufacturing in the aerospace, automotive, medical and dental industries, and to promote it globally step by step. ④ Continue to invest in the "Fluoride Growth Program". The South African government invested 45 million rand in the Pelchem company implementing the program in 2016. The company has developed a number of commercial products. In addition, the South African Department of Science and Technology intended to focus on developing disruptive innovation technology in titanium powder production. The pilot factory would start running in October 2016.

1.3.4 Develop the National Strategy of Astronomy and Lead the Development of African Space Policy and Strategy

Since successfully becoming one of the two co-host countries of the National Science Project SKA (the square kilometer array radio telescope project) in 2012, South Africa has been strategically leading the eight African SKA members to collaborate on development, making Africa the forefront of the world in the field of astronomy, so as to promote economic restructuring and human resources development in Africa. To this end, South Africa introduced the *National Strategy on Multi-wavelength Astronomy* to further promote the development of African astronomy, especially the SKA mega-science project. The strategy puts forward the astronomy strategic objectives and key areas of development, laying out the plan for relevant frontier cross-sector key funding projects. The South African National Research Foundation is currently working on a future implementation plan.

Owing to the leading role of South Africa, the African Union (AU) took a substantive step in the field of space research. In February 2016, the 26th AU summit was held in Addis Ababa, capital of Ethiopia. The meeting formally adopted the *Africa Space Policy and Strategy*. The space program aims at establishing relevant research institutes, improving research and development capabilities, and developing space derivative services for the benefit of the African people. Major applications include: earth observation; satellite communications, navigation and positioning; disease outbreak prevention and management; natural resources and environment management; natural disaster management; weather forecasting; climate change modeling and monitoring; agriculture and food safety; regional peace and conflict.

1.3.5 Launch the Marine Economic Forum and Promote the Development of Marine Economy

South Africa launched its marine economic strategy in July 2014, with the overall strategic objective of contributing 177 billion rand to GDP in 2033, about 3.3 times that of 2010 and creating one million jobs. The marine economic strategy will fully tap marine potential, with focus on four key areas: marine transport and manufacturing, offshore oil and gas development, aquaculture, marine conservation and marine governance. In 2016/2017 the Department of Science and Technology plans to allocate 20 million rand to finance marine economic programs to support innovative value-added activities in aquaculture. In February 2016, the Department of Science and Technology launched the South African Ocean Research and Development Forum (SAMREF), symbolizing the offshore oil and gas development moving from the strategic level to the implementation stage. The forum will explore the potential development opportunities for marine resources and work with the South African Offshore Oil Association (OPASA) to promote research and cooperation between government research institutes and private enterprises.

1.4 Favourable Innovation and Entrepreneurship Policies

1.4.1 Establish the Technology Innovation Agency to Manage National Technology Innovation and Transfer

In November 2008, South Africa signed the *Technology Innovation Agency Act*, approving the South African Department of Science and Technology to set up the Technology Innovation Agency as a specialized technology innovation management agency in charge of overall management, coordination, and promotion of national technological innovation activities. The main functions of the Technology Innovation Agency include: strengthening knowledge and economic ties, stimulating technology-based services and product development, encouraging technology-oriented enterprise innovation and technology transfer, laying the technical foundation for economic and social development, providing intellectual property support platform, encouraging investment, foreign direct investment and fund investment, so as to promote the development of innovative talents. The main objectives of the Technology Innovation Agency include: vigorously promoting technological development, developing technological innovation enterprises, consolidating the foundations of high-tech industries, ensuring that scientific knowledge generated in South Africa is transformed into products or real productive forces at home to promote the production of high value-added manufacturing enterprises, so that the country can leap forward to the modern knowledge economy.

1.4.2 Set up Relevant Funds and S&T Programs to Support R&D Innovation and Technology Transfer

Lack of funding is an important obstacle to the technological innovation and technology transfer of South African research institutes and small and medium-sized enterprises. The South African government adopts the relevant science and technology programs and funds to provide grants, matching funding, equity investment and other ways to materialize multi-channel support for research institutes and enterprises on R&D innovation and technology transfer.

1.4.3 Improve the Innovation and Technology Service System, and Set up a Platform for Incubation and Technology Transfer

In recent years, South Africa has continuously established and improved the whole process service system for innovation. Besides setting up the above-mentioned funds and science and technology programs, South Africa has been continuously establishing and improving the enterprise incubation and technological innovation transfer platform, which sets a platform for technology transfer for research institutes and enterprises.

1.4.4 Emphasize on the Management of Intellectual Property and Support the Transfer of Technology Through Policies and Regulations

For a long time, South Africa's intellectual property output remained relatively low, and awareness of IP protection was not strong. To reverse this situation, South Africa introduced *the Policy Framework on Intellectual Property of Public Funded Research and Development* and the *Intellectual Property Law of Public Funded Research and Development* respectively in 2007 and 2008, so that intellectual property rights of public funded research and development are protected, managed and well utilized. In this way, it can strengthen the management of intellectual property rights, and promote the local transformation of scientific research. At the same time, South Africa establishes intellectual property management system to improve performance of public research institutes, to promote the commercialization of intellectual property, to stimulate economic and social development, to create a favorable environment for technological innovation and technology transfer, and to protect state-funded R&D projects of intellectual property rights.

1.5 R&D Expenditure, Research Institutes, and Talents

1.5.1 Fiscal Investment in Science, Technology and Innovation

Since its founding in 1994, the South African government has attached great importance to the development of science and technology. It published the *White Paper on Science and Technology: Preparing for the 21st Century* in 1996, the *National R&D Strategy* in 2002, and the *Ten-Year Innovation Plan* in 2010. All put stress on the need to strengthen R&D investment, especially the 2002 *National R&D Strategy* which clearly states the GERD should achieve 1% GDP in the next three years. Although the target has yet to be met due to GDP growth rate exceeding the R&D investment growth (see Fig. 3), since 2001, South Africa research expenditure has experienced continuous growth except fluctuations in 2008–2010. Total expenditure in FY 2001/2002 is R74.88 billion and FY 2013/2014 is more

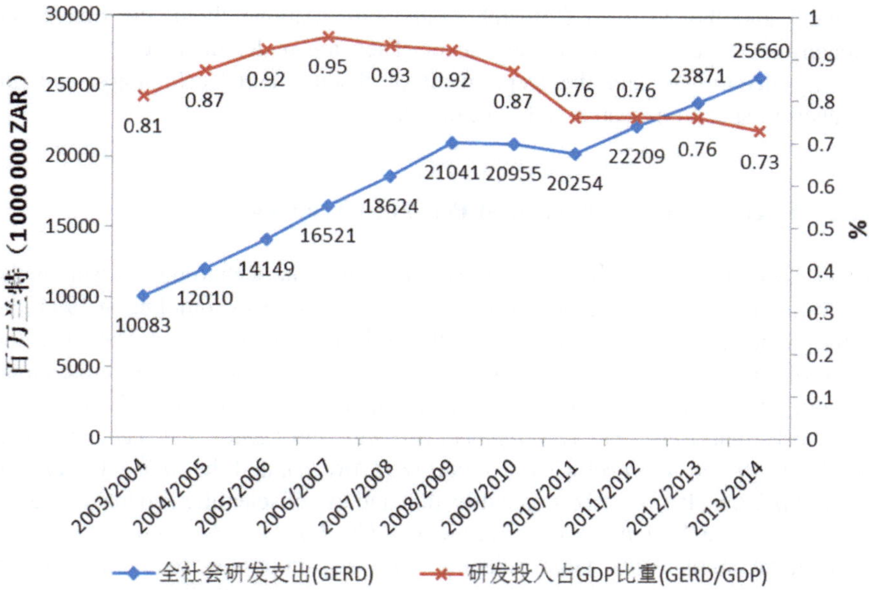

Fig. 3 Change of research expenditure in South Africa (2003–2014) *Source* OECD data base

than 3 times than that. In particular, since FY 2007/2008, the government has overtaken the business sector to be the main source of research investment. In FY 2011/2012, the government accounted for 43.1% of the GERD and business sector accounted for 39%.

In recent years, South African Minister of Science and Technology Mrs. Pandor has repeatedly said that STI has become the engine for South Africa to achieve the re-industrialization, informatization and modernization. The South African government will continue to use STI to promote economic restructuring. The latest statistics show that in FY 2016/17, the South African government invested R7.482 billion, same as the previous fiscal year, accounting for about 1/4 of the total R&D investment. Department of Science and Technology said it would work with the Department of Finance, and strive to achieve 1.5% GDP in 2019.

1.5.2 Overview of Research Institutes

The South African research system is relatively complete, mainly consists of higher education institutes (23 universities), national public scientific research council (8), other government research institutes (35), commercial research institutes (45) and non-governmental research organizations (about 80) and so on. As the country's most important scientific research force, South Africa's eight national scientific councils are: Agricultural Research Council, Council for Scientific and Industrial Research, Council for Geoscience, Medical Research Council, Institute of Mining and Metallurgy, South African Bureau of Standards, Human Sciences Research

Council, and the National Research Foundation. Among them, the first seven councils are essentially on national level. They are under dual guidance of the industry department and the Department of Science and Technology, and are engaged in specific research and development work.

1.5.3 Scientific and Technological Personnel Training

South Africa focuses on the development of science and technology as well as the cultivation of scientific and technological personnel. After its founding in 1994, the South African government recognized that science and technology talent is the basis of national innovation system and the key to realize knowledge-based economy. It has promulgated series of talent policies such as the *Skills Development Act*, the Human *Resources Development Strategy*, the *Center of Excellence Plan*, the *South African Research Chairs Initiative (SARChI)*, and the *Industry S&T and Talent Initiative*. It emphasizes on cultivation and use of scientific and technological personnel in the *National R&D strategy* and *The Ten-Year Innovation Plan*. The government implements large scientific projects, improves scientific research facilities, and adopts more attractive talent policy to retain their own science and technology elite talent and to attract world-class scientific and technological personnel, so that the scientific literacy and skills of the whole society is improved. Through unremitting efforts, the development of scientific and technological talents in South Africa has made great progress: first, the outflow of science and technology talent has slowed down, and the trend is reversed; second, the number of scientific and technological engineering talent has increased.

However, the shortage of young scientific and technological talents in South Africa has become the most important bottleneck restricting the development of science and technology in South Africa. The cultivation of young scientists has become the highest priority of science and technology expenditure in South Africa. Taking SKA's as an example, it is estimated that by 2018, South Africa's demand gap for in-depth analysis and big data will reach 23,000–31,000. According to the current training speed, about 10 years of training time will be needed. Therefore young scientists have become the most urgent issue of scientific and technological development in South Africa. In the FY 2015/2016, the Department of Science and Technology, through the Technology Innovation Agency, has set up a Seed Fund, Innovative Skills Development and Global Clean tech Innovation Program. More than 1000 young scientists have been funded for R&D and innovation activities. FY 2016/2017 initiatives include the holding of the first Youth Conference on Knowledge Economy to provide a platform for young scientists to learn how to create a business; continuous support for the mLab Southern Africa project R&D and innovation activities to provide incubators for young ICT entrepreneurs; SKA project to cultivate talents in the fields of science, technology and engineering; and joint training with international research institutes, local technology companies and other graduate students; along with the Da Vinci Institute and other vocational training institutes to carry out an internship plan to send non-employed graduate students to Top 100 technology companies to practice.

1.6 STI Development in Key Areas

1.6.1 Astronomy

a. In June 2016, the first 16 of the total 64 MeerKAT antennas of the world's largest radio telescope array-SKA was officially put into use. The first shot has already found 1300 galaxies in a "small corner" of the universe, far more than expected. Out of the 1300 galaxies, only 70 are known galaxies. As of May 2017, "MeerKAT" construction has made gratifying progress. 36 antennas have been successfully delivered, 57 base stations completed and 46 antenna lift constructed.
b. In February 2016, only hours after a small supernova was discovered, the Southern African Large Telescope (SALT) obtained its spectrum in the earliest manner, and then immediately cooperated with the United States Dartmouth University on analysis. In May, SALT once again achieved progress in astronomical research. It detected the first pulsating white dwarf, drawing international attention. SALT, located in Northern Cape, South Africa, is the largest single telescope in the southern hemisphere. Its hexagonal primary mirror array span reaches 11 m, and composes 91 one meter wide hexagonal mirror.
c. In May 2016, Kevin Govender, an astronomical scientist at Cape Town University, was awarded the International Astronomical Union (IAU) 2016 Medal, the first international astronomical award for South Africa, demonstrating South Africa's remarkable achievements in astronomy in recent years.

1.6.2 Information and Communication Technology

The South African High Performance Computing Center in Cape Town invented the first supercomputer in Africa, named Lengau, which means Cheetah in the local language, Tswana. It marks South Africa's entry into the world in high-performance computing level. The petascale system will provide services for African scientists to study SKA, climate change, energy and minerals. It is 8 times faster than the existing super computer computing.

1.6.3 New Energy

In March 2016, Hydrogen South Africa (HySA) Systems Competence Centre, affiliated to DST, and the South American Impala Platinum jointly developed

3 tons of prototype hydrogen fuel cell forklift prototype and its supporting gas station. Impala Platinum plans to convert all of its 33 forklifts to fuel cell drive. In addition, the Department of Science and Technology also sponsors to use hydrogen fuel cells for TB vaccine refrigeration backup power in Randburg District, Johannesburg.

1.6.4 New Material

South Africa leads Africa in the field of 3D printing technology, and in 2016 progress has been made in R&D and application. Hans Fouche developed a large Cheetah3D printer, which has been recognized by the 3D printing industry, and successfully passed the I Maker Lab authoritative test. South African local companies have also begun to use 3D printing technology in production. 3D printing and digital design has helped South African footwear industry to reduce the concept product time from 10–14 weeks to just three days.

1.6.5 Medicine

Virologist Maria Papatnama Sopoulos and pathologist Professor Penny Moore from South Africa Jinshan University made major breakthrough in the field of HIV vaccine, which has shown effect in the rabbit body. They are currently planning to experiment on monkeys and even the human body.

1.7 International Scientific and Technological Cooperation

The South African government believes that international scientific and technological cooperation relates to develop key areas, carry out cooperative research, and attract international R&D funds, talent, technology and other innovative elements, thus contributing to scientific and technological progress and economic development in South Africa and even Africa as a whole. South Africa is becoming one of the hotspots of international technology investment. After years of accumulation, more than 10% of South Africa's GERD comes from abroad. To date, South Africa has signed cooperation agreements or established and cooperative relations with more than 60 countries, regions and international organizations, and has jointly carried out hundreds of international scientific and technological cooperation projects, which mainly related to biotechnology, hydrogen economy, climate change, new materials, information and communication technology, agricultural research, health, nanotechnology, Antarctic research, society and social sciences, geophysics, oceanography, laser technology and applications. At present, the EU has become the most important scientific and technological partner in South Africa, and South Africa is now EU's fifth most important partner of FP7. It successfully participated in the EU FP7's 64 cooperation projects, with cooperative researches up to more than 230, and direct research funding totaling 25 million euros.

According to the *Ten-Year Innovation Plan*, South Africa will vigorously develop international scientific and technological cooperation, which serves as an important part of building a knowledge economy. South Africa not only intends to become an international research and development hub of key areas, but also strives to become the preferred investment destination in Africa. The FDI for R&D is expected to account for 1.5% GERD in 2018.

According to the latest OECD study, South Africa ranked first in terms of co-authoring papers with international authors in all BRIC countries. 15% R&D fund of South Africa comes from international cooperation projects, indicating that South Africa enjoys strong international scientific and technological cooperation support. The cooperation has the following characteristics:

First, South Africa increases cooperation with multinational companies. Recently, large multinational companies such as Pfizer, Nestle and Hitachi have worked with the South African Ministry of Science and Technology to invest in research, innovation and human capital development. The Bill & Melinda Gates Foundation is the most important strategic partner and charity organization in South Africa and invests heavily in scientific and technological innovation in the area of poverty reduction. In July 2016, Boeing worked with South African Airways for the first time to use aviation biofuels for African aviation commercial flights. Over the past three years, Boeing and South African Airways have been engaged in research and development of cost effective biofuels in line with market demand. It is expected to make breakthroughs in reducing bio-fuel manufacturing costs, as well as improving supply chains and expanding market supply in the future. In August 2016, the international giant GE invested 500 million rand in Johannesburg, South Africa to set up the first African innovation center, to show its advanced medical technology and to strengthen bilateral cooperation. It is reported that the innovation center will target the African infrastructure and medical market, and will also invest in local railway infrastructure projects with TRANSNET. In order to promote the development of ICT industry, South Africa is also widely working with foreign large-scale ICT giants, such as IBM, SAP, Nokia, Cisco, etc., which all have set up branches or investment projects in South Africa.

Second, South Africa attracts international high-end scientific and technological personnel through research chairs initiative. In 2006, South Africa launched the South African Research Chairs Initiative (SARChI), which sets up a total of 210 university research chairs in key research areas, and gives funding and policy priority to attract international top scientists. In this way, the country's science and technology competitiveness is upgraded while retaining domestic senior academic talent. Over the past 10 years, South Africa has been widely cooperating with Germany, Sweden, Britain, Italy, Norway, Ethiopia, Nigeria, Kenya and other countries to introduce top talent. So far the program has recruited 194 seats, 74% of which are from South African domestic universities and 26% from abroad. 74% are white, and 26% are colored (including Africans, Asian, etc.). At present the government spends 404 million rand to fund the program annually. The program is characterized by cooperation with the private sector, with government and industry funding ratio being 1:2. South African Ministry of Science and Technology decided to expand the scale of plan. It intends to attract more talent from abroad to boost international cooperation: it worked with Switzerland in the field of global environmental health in 2015, with the United Kingdom in the field of food safety and political science in 2016. In 2017 it plans to cooperate with Germany in nano science and advanced materials to set up research chief. South Africa has also focused on increasing the training of young talents in international scientific and technological cooperation. Recently, the Global Knowledge Partnership is being planned, that is, to train overseas doctoral students through international cooperation. Training agreements have started to be signed with relevant countries.

Third, make the Science Forum South Africa (SFSA) a world-renowned conference to widen international cooperation. Since 2015, South Africa holds Science Forum every December in the capital city of Pretoria. Extensive invitations go to governments around the world, enterprises, academia and civil society, providing a good platform for scientific and technological cooperation between Africa and other countries. The theme of the second Forum in 2016 is "Igniting Dialogues on Science". 1600 delegates from more than 40 countries were gathered to discuss topics such as scientific and technological consult, open access, African agricultural economy and social transformation, space science, knowledge economy, biotechnology, green economy and other issues. During the forum, a science and technology exhibition was also held. The forum has become an international platform for STI cooperation in South Africa and Africa as a whole. Africa has always had an urgent need for international STI cooperation. The aim of the forum is to promote international awareness and grasp such demand, and to establish a cooperative relationship that meets the common aspirations of both sides. SFSA, as a conference with African characteristics, will attract more international R&D cooperation in the future.

2 STI Cooperation with China

2.1 Bilateral Relations Between China and South Africa

In 1994, the establishment of the new South Africa marks the end of the racial segregation and rule of the white. China and South Africa established diplomatic relations on January 1, 1998. In 2000, the two countries established partnership relations; in 2004, a strategic partnership of equality, mutual benefit and common development was formed; in 2010, bilateral relations were upgraded to a comprehensive strategic partnership. During ten years, the bilateral relations achieved triple leap, a model of friendly cooperation between China and Africa and developing countries as a whole.

In March 2013, Chinese President Xi Jinping paid a state visit to South Africa, and the two sides issued a joint communique, marking the comprehensive strategic partnership between China and South Africa has entered a new historic stage. In December 2014, President Zuma paid a state visit to China, and the two sides signed the *Five-Ten Year Cooperation Strategic Plan 2015–2024*, injecting a new impetus for further development of the China-South Africa relations. Exchange activities of the Year of China in South Africa and Year of South Africa in China were held in 2014 and 2015. In December 2015, the China-Africa Cooperation Forum Johannesburg Summit was successfully held in Johannesburg, South Africa's economic center. The leaders of both countries witnessed the signing of a number of MOUs, including science and technology park project, pushing bilateral cooperation to a new level. In April 2017, the first meeting of the China-South Africa High-Level People-to-People Exchange Mechanism (PPEM) was successfully held in the South Africa. Chinese Vice Premier Liu Yandong attended the launching ceremony of China-South Africa science park during the visit to South Africa.

China and South Africa are both global emerging economies, and members of the BRIC countries. The two countries vary much from each other in terms of production factor endowments, the level of productivity and stage of economic development. Bilateral trade has a good economic basis, and therefore both sides enjoy more complementarities rather than competition.[1] At present, China is South Africa's largest trading partner, and South Africa is China's largest trading partner in Africa. Bilateral economic and trade cooperation has developed rapidly, and achieved fruitful results. According to statistics from the Ministry of Commerce, bilateral trade volume in 2015 was US$46.05 billion, of which China exported US $15.87 billion and imported US$30.38 billion. China has a clear competitive advantage for industrial products in South Africa, and South Africa has a competitive advantage for its primary products.[2] China mainly exports electric appliances and electronic products, textile and metal while imports mineral products.

According to information from the Ministry of Foreign Affairs issued in July 2016: more than 10 Chinese universities established relations of cooperation with South African universities. As of the end of 2015, China has about 7100 people studying in South Africa while received a total of 199 South African scholarship students. Nine South African institutes have set up Confucius Institute or Confucius Classroom, which is welcomed by the South African community generally.

2.2 STI Cooperation Mechanism Between China and South Africa

In order to promote scientific and technological cooperation and development, the two governments signed the Agreement on Science and Technology Cooperation

[1]Tian et al. (2014).

[2]Song (2017).

between China and South Africa in March 1999 and established the China-South Africa Joint Committee on Intergovernmental Cooperation in Science and Technology. Under the agreement, in order to strengthen scientific and technological exchanges and cooperation, the two countries' science and technology departments have set up a "China-South Africa Joint Research Program" to establish a funding system to support staff exchanges in project cooperation to assist research institutes and enterprises in key areas to carry out joint research. Each year the funding goes to no more than 15 projects. So far six meetings of the Joint Committee on Science and Technology Cooperation were held and seven rounds of 74 joint cooperation projects were co-financed. As of FY 2014/2015, the two sides have initiated the recruitment of the eighth round of the Joint Committee project and launched the first batch of projects in the water resource sector.

In December 2015, Chinese President Xi Jinping paid a state visit to South Africa. Both departments of Science and Technology signed a memorandum of understanding on the cooperation of Science Park under the witness of the leaders of the two countries. The main contents include: promoting cooperation in incubation of enterprises, high-skilled personnel development and information and communication technology research; initiating joint action plan for the implementation of Science Park projects to provide support. The main ways of cooperation include capacity building in planning Science Park, conducting technology transfer, commercialization of research, innovation management and enterprise start-up; promoting joint research and development projects in the areas of common interest in technology transfer and innovation cooperation to improve development and technology transfer of innovative products and services; enhancing financial innovation from microcredit to venture capital, as well as personnel exchange, information exchange, and innovative exhibitions.

In addition to the departments of Science and Technology, science and technology cooperation is also very active between the two countries' departments of health, agriculture, forestry, fisheries, water conservancy, environmental protection, mineral, energy, information and communication, transportation and other departments. A number of cooperation memorandums of understanding are signed, establishing a close tie between two sides. In the mean time, many South African universities, research institutes, scientific councils and their Chinese counterpart also formed wide exchanges and achieved substantive cooperation. For example, the Forestry and Agricultural Biotechnology Institute (FABI) of University of Pretoria and National Key Fungi Laboratory from the Chinese Academy of Sciences, Zhanjiang Eucalyptus Institute established a long-term research and doctoral training cooperation mechanism.

2.3 Status Quo of STI Cooperation Between China and South Africa

2.3.1 Key Areas

In recent years, as the political and trade relations between China and South Africa reaches a new high, the STI cooperation has also been elevated to a new stage. At present, the cooperation and exchange between China and South Africa have covered various fields related to science and technology. Pragmatic and in-depth cooperation has been carried out especially in biotechnology, new materials and advanced manufacturing technology, information technology and systems, environmental protection, mining technology, resource exploration, space technology, traditional knowledge systems and other areas, which promotes economic development and improvement of the people's livelihood.

2.3.2 Research and Development Projects

From 2007–2015, China carried out 30 projects in South Africa with total funding of 63.03 million yuan. In the BRIC countries, this scale is only second to Russia, both from the quantity and amount, which shows that China has maintained a good STI relation with South Africa. In the same period, China undertook 27 projects with other African countries which received 107 million yuan of funding, of which three projects were with Egypt, funding 15.7 million yuan; two are with Kenya and Algeria, funding 11.53 million yuan and 9.94 million yuan. It can be seen that the STI cooperation between China and South Africa, represented by joint research projects, is leading Sino-African STI cooperation.

2.3.3 Joint Exhibitions

From 2004–2012, China participated in the South African International STI Exhibition organized by South African Department of Science and Technology, and used this platform to showcase China's leading technology, practical technology and products for mutually beneficial cooperation between China and South Africa and Africa in science and technology and the business community. In October 2016, Chinese and South African Departments of Science and Technology co-sponsored the first China-South Africa Hi-Tech Exhibition. Sixty exhibitors from eight cities in China and more than 200 participants taking 170 projects including the life sciences, Chinese medicine, eco-environment protection, new energy, electronic information, advanced equipment manufacturing and other fields. More than 50 South African enterprises and institutes participated, fully demonstrating the latest achievements in bilateral STI. The exhibition also held STI cooperation seminars, three symposiums and cooperation docking activities in traditional medicine, advanced manufacturing, new energy. During the event, the Chinese exhibitors and South African professional visitors carried out 660 B2B special talks, reaching 158 initial cooperation intentions, involving technical cooperation, building joint R&D centers, trade agents, S&T personnel exchanges and other cooperation modes.

2.3.4 Model of STI Cooperation Between China and South Africa

The cooperation between China and South Africa in the field of space is an example of the complementary advantages of both sides. In September 2007, China announced CBERS-02 data was shared with Africa. In December 2008, the CBERS-02 South African ground station was successfully built and the data from satellites were successfully distributed in South Africa, and 13 countries in southern Africa could use the data for free. In December 2010, China, Brazil and South Africa jointly signed the *Memorandum of Understanding on the Direct Reception and Distribution of CBERS-03 Data in South Africa*, which agreed that South Africa could directly receive, use free of charge and distribute data from CBERS-03 to southern African countries.

Although high-tech is significant, it is more urgent to develop practical technology that can improve skills and productivity and increase jobs. African countries are still at a low level of development, and their primary task is to eradicate poverty and improve people's living standards. In this regard, China's technology and experience are very popular among African countries. In the vast rural areas of South Africa, most young people move to cities to seek jobs. Women, the elderly and children stay at home, and most families still live in poverty. In 2005, experts from Research Center of Juncao Technology, China Fujian Agriculture and Forestry University visited a village in KwaZulu-Natal, South Africa, teaching the local people hand by hand how to plant mushrooms. The village became the first Juncao cooperation project base. Now the village has nearly 100 women planting mushrooms, and the income increases by at least 1000 rand per month per household. KwaZulu-Natal province has nearly 100 grass test sites, and thousands of local farmers have benefited from it. But they are not alone. The Chinese experts also trained technicians from Lesotho, Rwanda, Kenya, Zambia, Cameroon, Zimbabwe, Malawi, Mozambique, Botswana and Uganda. These real S&T cooperation projects not only showcase the level of science and technology in China, but also quietly sowed the seeds of friendship between China and South Africa and Africa as a whole. South Africa also speaks highly of the science and technology cooperation with China.

References

Song, H. (2017). A study on the competitiveness of goods trade between China and South Africa. *Industrial & Science Tribune, 16*(1), 92–93.

Tian, Z., Liu, C., & Zhang, Y. (2014). Analysis on competition and complementarity of bilateral trade between China and South Africa. *Research on Development, 4,* 154–160.

Part III
Thematic Reports

Chapter 9
Study on Digital Technology in BRICS

Maoxing Huang, Tang Jie and Xinhuan Huang

During the 2016 BRICS Summit in Goa, India, strengthening exchanges and cooperation in digital economy among the BRICS countries was first proposed and the ICT Development Agenda and Action Plan were passed. Later, China, which assumed the rotating presidency of 2017 BRICS Summit, initiated that the priority be put on the establishment of a digital experience sharing mechanism and the development of digital economy. China proposed the formulation of supporting policies, the investment on digital economy and the promotion of experience sharing so that people can share benefits from the development of digital economy, and new impetus can be given to the economic development in BRICS countries[1] The digital economy in BRICS countries is still at the initial stage, enjoying great development potential and prospects. Different from the infrastructure of industrial economies, the informatization development in the BRICS ranks among the upper level in the world. Thus the BRICS enjoy a rare comparative advantage to develop digital economy. As the digital economy is increasingly gaining attention, the basis for the exchanges and cooperation among the member countries would be the regulation of its connotation and characteristics, the analysis of the brand-new economic laws, the prediction of development trend of technology, and the comparison of development strategies and existing problems in the bloc.

[1]Xin (2017).

M. Huang (✉)
School of Economics, Fujian Normal University, Fuzhou, China
e-mail: hmx2001007@163.com

T. Jie · X. Huang
China Science and Technology Exchange Center, Beijing, China

X. Zhao et al. (eds.), *BRICS Innovative Competitiveness Report 2017*,
Research Series on the Chinese Dream and China's Development Path,
https://doi.org/10.1007/978-981-10-8078-4_9

1 The Connotation and Characteristics of Digital Economy

The digital economy is also called the Internet economy, the Information Economy or the New Economy. At its initial stage, digital economy is mainly driven by the development of information technology and the internet, thus the definitions. The concept was put forward in 1995, yet no consensus was reached on its connotation and extension. The G20 2016 Summit made a comprehensive definition so far, that is, the digital economy refers to a broad range of economic activities that include using digitized information and knowledge as the key factor for production, modern information networks as an important activity space, and the effective use of information and communication technology (ICT) as an important driver of productivity growth and economical structural optimization.[2]

The concept extension of digital economy has also expanded in the past 20 years, according to American economist Mesenbourg (2001), and three main components can be identified[3]: the basic component being digital transaction infrastructure, including hardware, software, telecoms, networks, human capital, etc.; the next component is partial digital transactions, such as online sales and the digital management process within enterprises; and the last is e-commerce, which is the whole commercial activity from purchase, production to sales and distribution. Following comes social networking and Internet searching, incorporated into this system as the fourth component, which is a purely digital "virtual goods" transaction form. The fourth level of the digital economy, therefore, was also known as "virtual economy", which can be established on the basis of the third component or exist independently, and it can also promote the development of the first three levels. For example, social networking has greatly promoted the application and popularization of networks and communication technologies, and searching and crowd-sourcing has brought prosperity to digital transactions and e-commerce.

2 Development Trend of Digital Economy

Digital economy not only creates the opportunity to change the global economic downturn, and digitization is not only a kind of technology in itself, they will also bring new thought expansion, change the existing consumption and business models, and create a new and sustain able growth momentum. With the development of information and communication technology and its integration with traditional industries, digital technology will create comprehensive and fundamental

[2]China Daily: "G20 Digital Economy Development and Cooperation Initiative delivered at 2016 Hangzhou Summit renews impetus to global economy", Chinadaily.com.cn, Sept.4th, 2016. http://china.chinadaily.com.cn/2016-09/28/content_26926631.htm.

[3]Mesenbourg (2001).

changes to the foundation, environment, input, output and sustainability of economic innovation and development.

2.1 *Digital Technology Development Has Changed the Foundation of Economic Innovation*

Due to sustained investment in the past 20 years as well as the construction of digital infrastructures (broadband networks, e-commerce, and the financial system as representatives), the system pushes more and more second and third level new digital industries to be developed, such as Internet retailing, digital finance and sharing economy. The emerging digital industry is expected to maintain rapid growth for a long term, and its contribution to GDP growth will continue to be prominent, more capital investment will be attracted from the governments and private sectors. The development of these industries provides the market foundation to the further development of digital technology, which will regenerate its development, and put forward higher requirements for digital infrastructure.

The foundation for innovation in the era of digital economy will be the digital devices characterized by signal plus chip and digital networks. The capital and manpower needed for the construction of such infrastructure would be considerable, but the following achievements in innovation, growth, employment and the scale of talent training will also be profound and sustainable. Unlike the early stages of development, the future digital infrastructure construction will be increasingly driven by demands. From national security and smart city, to the enterprises' cloud-computing, big-data analysis, the supply chain management, and then to individual consumers' attention, social activity, interests and shares, these various levels of digital demands have given rise to a wide variety of software while the operations of the sein turn put forward higher requirements for digital hardware.

2.2 *Redefinition of Innovation in Digital Technology Application*

According to forecasts, the daily integration with digital devices per person in the world will be 4800 times in 2025, and the data creation volume will reach 163 ZB, ten times that of 2016, of which more than 25% is real time data, and the Internet of Things (IOT) data accounts for 95%.[4] Big data cannot always produce direct innovation and value; therefore, data analysis will become the main motivation for innovation. For example, with the help of macro-data monitoring, governments can

[4]IDC, "Data Age 2025", IDC White Paper: Framingham, MA, USA, 2017, http://www.seagate.com/www-content/our-story/trends/files/Seagate-WP-DataAge2025-March-2017.pdf.

quickly analyze economic forms, predict the commodity supply, manage disaster situations, and analyze demands for professional talents and medical facilities.

Digital content will become the main carrier of innovation. The traditional digital content industry was one of the first to realize industrial upgrading by way of digital technologies. One of the world's earliest digital regulations, the *Digital Economy Act 2010* introduced by the British government mainly aimed at the innovative protection of digital content and the regulation of the industry. With the aid of leading interactive technologies such as digital design, touch control and VR or AR, content industries like music, movies and games achieved rapid development. The supply end of digital content has also achieved crowd-sourcing using digital networks—contents of creativity and design coming from consumers worldwide.

The innovation subject has been changed. During the industrial economy era, the governments and multinational enterprises are the main bodies of global innovation activities. But digital networking changed the organization form of the global supply chain from the perspective of space and time, making those micro-enterprises the multinational service providers who produce and sell all kinds of virtual products, services, idea sin non-border digital networks, or participate in the "micro-work", "micro-payment", or "micro-shipment" in the global supply chain. Micro-enterprises or start-ups also can achieve horizontal division of labor worldwide with the aid of digital network, which helps to form a horizontal transnational production system based on modularization, mass customization and outsourcing production and those enterprises can provide technical services to firmly target customer groups through controlling standards and creating new standards.

2.3 Mobile Network Created New Modes and Formats

In recent years, with the popularity of mobile devices and the maturity of mobile network construction, a virtual world full of new modes and formats was born. Mobile internet is not a simple extension of the traditional internet. If people can't realize the brand-new management ideas within it, even the winners in the traditional way may not keep up with the new development trend in the digital age. First of all, users' mobile devices help enterprises identify the identity and data of the consumers. Through identification, multidimensional customer data will increase rapidly, which would create opportunities for providing personalized, consistent and accurate services. Second, the fragmentation of time represents huge consumer demands. This is a kind of demand that traditional Internet fails to meet, which is rapidly developed and fleeting with time. In order to seize such demand perfect services and products that can attract attentions from first sight is needed. Then, the portability of mobile devices offers a chance to track consumers' locations, product and service formats based on dynamic locations can thus be cultivated, the integration of online and offline modes is also this way realized. O2O model is required for the capture of all fleeting demands. Finally, the sharing practice from anytime and anywhere brings huge advertising and communication advantages, which also generates the sharing economy—people not only can share information, but also resources.

Mobile Internet products and services can't be independent anymore. Instead, they should set out from the different usage scenarios of users, focusing on users' 24 h demands and constantly satisfy them. Otherwise, users' attention span will soon be seized by competitors. As a result, the competition among Internet enterprises has gradually evolved into the competition between platforms. Super business platforms across multiple industries will be established by means of alliance, merger and integration among more and more enterprises, so as to meet the all-round needs of consumers. At present, among the world's largest 100 companies, the main income for 60 of them comes from the Internet business platform model. These enterprises have formed interaction networks with their alliances and users to the greatest degree by establishing free and open strategies. As the business ecosystem created by platforms and their alliances can provide users with a full range of products, services and experience, they are able to dig the maximum values from their permanent online users.

2.4 The Content of Digital Technology Innovative Education

With the penetration of digital network and digital devices into human life and work, digital literacy will gradually become the basic capability for every individual. The connotation of digital literacy includes the ability to capture, understand and integrate digital information, the skills such as web searching, hypertext reading, criticism and integration of digital information.[5] Due to the dual value of digital literacy to digital consumption and production, new requirements are raised for the reform of content and ways in the national education system.

Developed countries have already embedded the content of digital literacy into their education systems in various degrees, and have formed three kinds of digital literacy education modes. The first is the nation-led mode, represented by the United States, emphasizing on using national policies to guide the digital literacy education investment and related infrastructure construction; the second is the society-led mode, represented by the European Union, which emphasizes on the triple-party interaction of government, educational institutions and social powers, the promotion of citizens' digital literacy is mainly borne by libraries and other social institutions; and the third is the citizen-led mode, represented by Japan, which emphasizes on the strengthening of digital literacy content in elementary education, and form a mode where citizens can continuously improve their digital literacy through practice along with the development of science and technology, the

[5]Gilster and Glister (1997).

premise of which is that the whole society has a high level of elementary digital literacy education and digital practice resources. Among the three modes, the nation-led is indispensable at the initial stage. Governments expand investment in digital literacy education organizations and educators through policies, and enhance the overall power of the social digital literacy education. Then it will gradually transform into the society-led mode, and promote digital literacy education popularization. The ultimate goal is, under the premise that the overall social digital literacy and public digital infrastructure have achieved relatively higher level, to finally realize citizen-led digital literacy education mode.

3 Status Analysis on the Innovative Development of Digital Economy in BRICS

3.1 The Development of Digital Economy in BRICS

As an authoritative definition for the connotation and denotation of digital economy is still absent, there is no unified measuring standard about its scale and its contribution to GDP growth. At present, different research institutions conducted estimation according to some similar statistical methods and dimensions. For example, the Boston Consulting Group in the US published *G20 Countries Internet Economy* in 2012. According to its statistics and prediction, the average share of BRICS Internet economy in GDP was 3.1% in 2010, and the ratio was expected to rise up to 4.0% in 2016. The Internet economy in China occupied 5.5% of GDP, which is the highest among the BRICS and was expected to rise up to 6.9% by 2016, the proportion and growth rate are far higher than the average level of the BRICS. By 2016, the Internet economy in India is expected to gain a growth of 23% compared with that of 2010, which would be ahead of the other BRICS countries. In 2010, The Internet economy in Russia accounted for 1.5% of the GDP, the lowest in the BRICS. But it had a rapid growth and is expected to reach 2.7% in 2016, achieving a growth rate of 18.3%. The proportion of Internet economy in GDP and its growth rate in Brazil and South Africa are lagging behind in the BRICS. China Info 100 Research Team conducted an estimation using the new statistical methods and dimensions of digital economy, that China's digital economy accounts for 30.1% of the GDP in 2016, the growth rate of 16.6%, much faster than that of the United States (6.8%), Japan (5.5%) and the UK (5.4%).

Here are some comparisons about the indexes that reflect the development of digital economy. First of all, "fixed broadband subscribers per 100 people" and "network bandwidth" can describe the construction coverage and level of the national broadband network. As shown in Figs. 1 and 2, the fixed broadband coverage in China, Russia and Brazil is higher, but China's broadband construction level is relatively low. South Africa is different from China. Although the coverage level is low, its network bandwidth is much higher than the other BRICS countries,

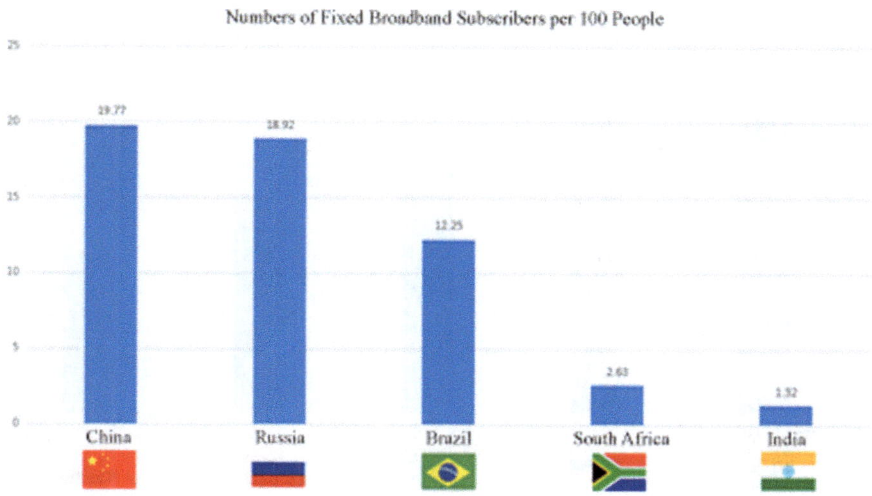

Fig. 1 Numbers of fixed broadband subscribers per 100 people in BRICS. *Source* Data published by World Bank in 2015

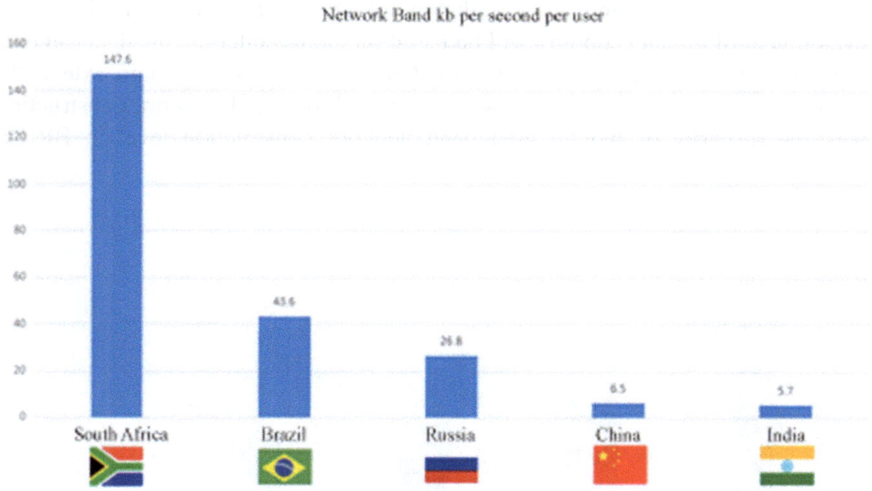

Fig. 2 Network bandwidth in BRICS. *Source* International Telecommunication Union, ITU World Telecommunication/ICT Indicators June 2016 (June 2016 edition)

which shows that there is great disparity in the network construction level in South Africa. The situation in China is relatively equal. India lags far behind other BRICS countries in both its coverage and construction level.

Second, numbers of mobile phone subscribers per 100 people and those of mobile broadband subscribers per 100 people can further illustrate the construction

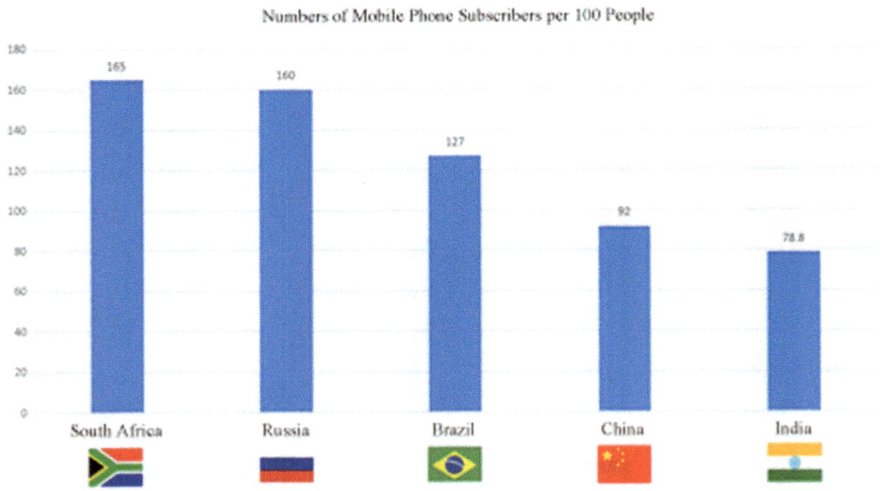

Fig. 3 Numbers of mobile phone subscribers per 100 people in BRICS. *Source* Data published by World Bank in 2015

level of mobile broadband and the digital literacy of the people, as the future development of digital economy will be based on mobile Internet, the data can also reflect national development potential of digital economy to a certain extent. As shown in Figs. 3 and 4, compared with the fixed broadband network construction level, the disparity of mobile broadband network construction level in BRICS

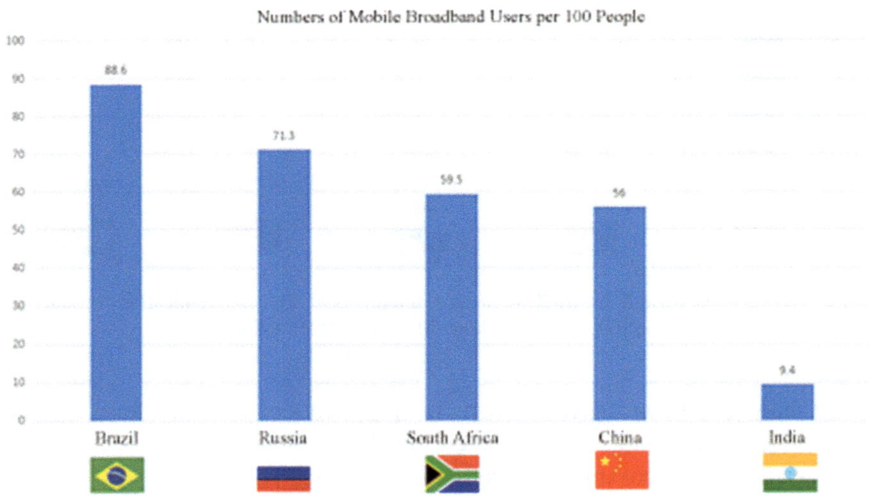

Fig. 4 Numbers of mobile broadband users per 100 people in BRICS. *Source* International Telecommunication Union, ITU World Telecommunication/ICT Indicators June 2016 (June 2016 edition)

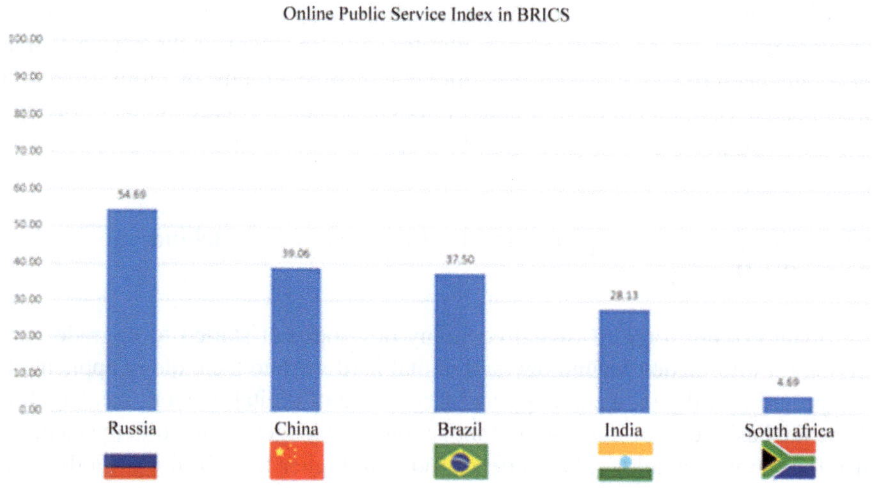

Fig. 5 Online public service index in BRICS. *Source* Report on the Group of Twenty (G20) National Innovation Competitiveness Development (2016–2017)

countries (India excluded) is quite small, which demonstrates all BRICS countries has invested greatly in the construction of mobile network, and they all have good development potentials. The mobile network construction in South Africa is similar to its fixed network which presents polarization.

Finally, "online public services index" reflects the social digitization process of the BRICS countries. As shown in Fig. 5, Russia's advantage is obvious. The government has put huge investment in public service digitization. The disparity of online public services index between China, Brazil and India is not significant, they all positioned in the middle and lower level in the world. The digitization of public services in these countries is still in its initial stage and has large room for improvement. Government can invest more to further promote the development of the digital economy. Compared with other countries, South Africa is lagging far behind, the reason of which is greatly associated with the maldistribution of the broadband network construction.

3.2 *Problems of Digital Economy Development in BRICS*

From the development level of digital economy in the BRICS, we can see that all governments have shifted to the development of digital economy as their main approach for economic transformation and innovative development due to its strong momentum. According to *The Global Information Technology Report 2016* published by World Economy Forum, the ranking of networked Readiness Index

(NRI)[6] shows that the BRICS countries are climbing up in the past ten years, but the development in recent two years is quite disappointing. Part of the reason is that, there are deep-rooted gap between different income groups in terms of digital broadband access, digital literacy and services acceptance, a large number of people are unable to participate in the rapid development of digital economy.

3.2.1 The Strategic Layout of Digital Economy Is Left Behind the Development Stage

The main driving force of digital economy development started to upgrade from relying on information technology and digital hard ware to technology application. The progress of digital technology and large-scale of digital network has begun to shape the basic form of innovation. Innovation is no longer confined to the category of information technology, but depends more on higher level business models and industrial innovation in digital economy represented by special software, online retailing, digital finance and sharing economy. These innovations can not only promote the efficiency of traditional businesses, but also work in a way of nearly zero-cost. They no longer rely on the investment on R&D, but on the innovation of ideas, such as the digitization of procedure, distributed manufacturing, free services, and sharing economy in finance, tourism, education, real estate, transportation and other sectors. The digital strategies in the BRICS countries now still remain in promoting technological development, strengthening the manufacture of digital devices and the popularization of Internet, while the understanding of the connotation has not kept up with the speed of industrial development. Digital strategies formulated by these governments can no longer cover the third and fourth level of digital economy, thus cannot create greater social and economic influence by introducing and popularizing information technology.

3.2.2 Financial Deficits in Digital Infrastructure Construction

The innovation and development of digital economy are all based on digital data. Thus digital infrastructure is a key factor to decide the development and potential of digital economy in any country, and it will also decide the influence and application of digital technology in various social sectors. However, there are still huge financial deficits for large-scale digitization, even in the UK and the US, the highest level of digital economic development countries in the world. EU's report in 2016 recorded the long-term financial deficit for broadband upgrading as a main problem facing the development of digital economy. According to the statistics by World Bank in 2015, that the population ratio of people who cannot access to the Internet accounts for

[6]A comprehensive index put forward by World Economy Forum, which represents the influence of ICT development to society and economy.

more than 58.7%, that is, about 1.81 billion people are still unable to receive Internet services. If further construction can bring more high-speed mobile network with business and service innovation, more capital are well needed, which means there are still huge financial deficit in infrastructure construction of digital economy in the BRICS. According to *The Global Information Technology Report 2016* published by World Economy Forum, though the majority of statistical indicators have showed a rising trend, digital infrastructure indexes are of least obvious and the infrastructure in BRICS countries has been worse since the year of 2012.

3.2.3 Co-Existence of Visible and Invisible Digital Divide

Whether in the overall development of digital economy or the infrastructure construction, there is imbalance in many aspects in the BRICS countries. In general, China and India enjoy a faster development. Digital economy plays an increasingly important role in national economy and people's life. While the impact of digital economy on national economy in Brazil, South Africa and Russia is very limited. Compared with China and India in terms of total volume of digital economy and the development level of digital industries, there is an obvious "digital divide". But Russia has good prospects for development, as the digital infrastructure and the foundation of public digital literacy are quite good, and its growth in digital economy is equivalent to that of China and India. But the economy in South Africa and Brazil depends more on resources and traditional industries, the digital economy has not yet received enough attention. In addition, different countries hold various views about data standardization, the digital standards and market regulations are not uniform, which leads to invisible "digital divide" in data storage, wireless communications, security maintenance and the construction of Internet of things. This goes against the integration of respectively technical advantages of BRICS countries as well as the promotion of higher level innovation in digital technologies. The BRICS countries gradually realized that in order to develop digital economy, countries need to open up and share digital technologies and markets, but their development strategies remain on how to strengthen the construction of domestic digital infrastructure, which fail to view how to achieve infrastructure and platform connectivity from a much higher perspective.

3.2.4 Seriously Inadequacy in Digital Economy Governance and Supervision

The rapid development of digital economy and digital network created a new virtual world. From the very first day, governance and security problems of digital economy such as digital copyright, network security and personal privacy accompanied its birth. Attentions and disputes on the digital security grow almost at the same rate of data growth. On the one hand, digital economy is stimulating a new driving force for the innovative development of the economy and society, and it is of vital importance to create a good governance environment for the sustainable development of digital economy. On the other hand, the development speed of digital technology and the growth of scope are too fast that some means of governance and regulations soon become invalid. And the cross-border and cross-industry characteristics of digital network make it unable to be independently managed and supervised by any country. The shortage of construction funds resulted in the relatively backward construction of network security in the BRICS, and criminal activities such as Internet gambling, money laundering, and drug trafficking continue to increase, there even exist some illegal activities such as the hacker attack against government networks, and the steal of all kinds of commercial and state secrets. These irregularities disturbed the order of the digital economy development, becoming an unstable factor to the innovation foundation, thus urging for timely management and planning. Governments need to build up a flexible governance framework for sustainable development, which should have the shape of technology and social foresight, and is able to adapt to the changing circumstances.

3.2.5 Opportunities Missed Due to the Shortage of Digital Literacy Education

The overall development of digital economy puts forward higher requirements for the digital skills and literacy of all ordinary citizens. First of all, in the era of digital economy, workers need to possess both digital and professional skills. Researchers pointed out that the concept of production over the next 10 years will be totally changed, the main industrial manufacturing machines will be the main labor force, so millions of workers who only have basic professional skills will lose their jobs. New jobs created by digital economy will invariably require for digital skills, while some popular jobs will be redefined by data management and analysis skills. Second, in the era of digital economy, citizen's digital literacy will also be the consumer basis for market prosperity. Digital literacy level directly determines the size of the digital consumption demand, on the other hand, it is of vital importance to protect the security of digital network, maintain the stability of digital market and participate in the governance of digital society.

The European Union's latest survey showed that over the past decade, the employment growth of information communication technicians has exceeded 4%, while the rate for graduates of information communication technology decreased by 40%.[7] Compared with the developed countries, the gap between BRICS citizens in their quality and literacy is larger, which results in development contradictions in two aspects: first, the governments try to increase the level of public digital services through digital input so as to improve the efficiency of government and society, but the number of service users are limited by their citizens' quality and literacy;

[7]European Commission,"Europe's Digital Progress Report", 2016, https://ec.europa.eu/digital-single-market/en/news/europes-digital-progress-report-2016.

second, the gap between the growth of demands from some people with high quality and literacy for the application of digital technology and network and the commitments and input made by governments is enlarging. The population base for digital consumption in the BRICS countries is enormous, but the shortage of digital literacy education makes the governments and enterprises lose the market and dividend of the increasing number of digital population.

4 Key Areas of Innovative Development of BRICS Digital Economy

At present, digital economy has become the new driving forces behind global economic development and growth. BRICS nations should promote pragmatic cooperation in key areas including innovation of information communication enabling technology, ICT infrastructure construction, expanding and strengthening digital economic development, boosting e-government development and enhancing cooperation on cyber security governance, so as to inject new dynamism into BRICS economic development.

4.1 Promoting Innovation and Infrastructure Construction of Information Communication Enabling Technology

BRICS countries should support qualified ICT companies to conduct cooperation and innovation in such 5 major enabling technologies as broadband, data center, cloud computing, big data and the Internet of Things speed up the construction of ICT infrastructure, in particular, increase broadband network coverage, and improve the network broadband quality, so as to achieve connectivity of their ICT infrastructure. BRICS should enable the new generation of information technology with the Internet as its core to be integrated into every aspect of economic and social life, reshape the national innovation system, the industrial ecology, the competitive landscape, enterprise organizations and individual production lifestyle, and become an important engine driving BRICS economic transformation and upgrading. In order to realize sustainable economic development, BRICS countries should attach more importance to the investment and application of such four major enabling technologies as data center, cloud computing, big data and the Internet of things. Driven by the demand for cloud computing, big data and the Internet of things, industrial cloud platform and community participation can help extend the digital supply and distribution chain, which may be 1000 times of its current size in the future. The global data center space will have grown from the current 480–600 million square meters by 2020.

According to the current trend of funding sources, there is a huge capital gap in investment in the 5 major enabling technologies. To this end, BRICS nations should actively make innovations on investment and financing models, develop the financing channels for these technologies, and accelerate innovation and infrastructure construction of information communication enabling technology. First, BRICS should engage in a wider range of a higher level of innovative cooperation and cross-border infrastructure investment cooperation, so as to build an open, inclusive, balanced and beneficial-to-all cooperation framework, put in place a system of market access negative listing, in particular, promote a high level of opening of the financial service industry to the both domestic markets and outside world, push for the foreign capital management to be transformed from a pre-operation approval process into an operational and post-operation supervision model, so as to create an international rule of law based business environment, and sign bilateral or multilateral protection agreements and agreements on avoidance of double taxation on the 5 major enabling technologies and infrastructure investment, so as to protect interests and rights of investors. Second, BRICS should vigorously create new ways of financing, and make an integrated application of partnerships such as Build-Operate-Transfer (BOT) and Transfer-Operate-Transfer (TOT) derived from the Public-Private-Partnership (PPP), so as to attract investment of private or social capital into ICT infrastructure construction. Third, BRICS should enhance cooperation in the financial sector to provide financial institutions with a pragmatic cooperation platform, jointly study a long-term planning of cross-regional financial cooperation, set up cross-regional and multilateral financial institutions, and encourage mutual investment in ICT innovation within the BRICS block. The BRICS New Development Bank should actively collect financing for its ICT infrastructure projects and make an investment priority plan of enabling technological innovations for each member state.

4.2 Expanding and Strengthening Digital Economic Development Potential

With the rapid development of digital economy, the ICT integrates comprehensively into the manufacturing industry, service sector, business startups and innovations. As a result, new business models and patterns emerge constantly and production and service models of all industries are reshaped. BRICS should seize this opportunity to expand and strengthen the digital economic development potential, vigorously promote the in-depth integration between the new generation of ICT and manufacturing, so as to achieve a smart, digital and network-based and service-oriented production. First, smart manufacturing should be vigorously advanced. BRICS should encourage qualified manufacturing enterprises to develop smart control system, put into use high-end CNC machine tools and industrial robots, improve the integration of major complete sets of equipment and production line system, and carry out a trial of developing smart workshops, smart factories, and smart enterprise. Second, BRICS should develop the Internet-based personalized customization, crowd sourcing design, cloud manufacturing and other new manufacturing models, systematically integrate customer demand with information about enterprise research and development, production, sales and service, so as to come up with an overall solution based on customer demand. Third, BRICS should guide manufacturing enterprises to apply the new generation of ICT to develop new business models such as online customization, and online and offline integrated development, and create new organization forms such as specialized division of labor, virtual operations, and collaborative manufacturing. Fourth, BRICS should push for manufacturing enterprises to be service-oriented based on the ICT. Manufacturing enterprises should rely on the manufacturing business to actively develop information technology and other production services, actively develop service business related to enterprises' products, and develop themselves from products sellers into suppliers of complete sets of solutions and then transform into service suppliers.

BRICS nations should develop new modes and new patterns of the service sector based on the ICT. First, through separating value-added applications from basic platforms, traditional service industries should derive value-added service formats based on data services from existing basic business formats. Second, effective division of labor should be promoted between the service sector's big production (basic platform) and "small" production (value-added applications). As a result, through the world-class platform, the service sector will be promoted from the overall small production state to a world-class level of socialized mass production.

BRICS countries should make full use of the ICT to push for the public to start businesses and make innovations. First, the business startups and innovations should be based on the network platform. BRIC countries can make full use of platform interaction, sharing and integration of their various types of innovative elements and business startups resources to improve the efficiency of resource allocation, promote business startups and innovations by the public on a wider range and at a higher level and create a new engine and grow a new economy. In addition, small, medium and micro-sized enterprises can use cross-border e-commerce platform to build new distribution channels in domestic markets and actively expand overseas markets so as to enjoy a broader growth space. Second, the business startups and innovations based on the Internet crowd sourcing should be made to promote cross-disciplinary integration and innovation of product technology and push for cost reduction and quality and efficiency promotion of enterprises. Competent individuals and qualified enterprises should be guided, through the Internet crowd sourcing platform, to achieve the creation, updates and aggregation of knowledge so as to create a new model for aggregation and sharing of wisdom of crowds. Qualified large and medium-sized manufacturing enterprises should be pushed to gather cross-regional standardized production capacity through the Internet crowd sourcing platform to meet the manufacturing needs of large-scale standardized product orders.[8]

4.3 Promoting E-Government Development

According to the *United Nations e-Government Study Report 2016* issued by the Department of Economic and Social Affairs of the United Nations, China's e-government development index was 0.6071, ranking No. 63, Russia was 0.7215, ranking No. 35, Brazil 0.6377, ranking No. 51, South Africa 0.5546, ranking No. 76, and India 0.4638, ranking No. 107. It can be seen that development of e-government among BRICS nations is not balanced.

BRIC countries should focus on the following key areas to further promote the development of e-government. First, BRIC governments should continue to make broadband networks more accessible and available, so that people can obtain more information about education, social welfare, health, finance, employment, labor and the environment. Second, BRICS should continue to deepen e-government applications based on the mobile Internet, make use of mobile application programs and social media to provide people with equal services to form a wide-ranging service pattern featuring multiple levels and wide coverage, and in particular, to enable the mobile service to cover more people in the poorest areas and the areas with low population density and to advance implementation of sustainable development initiatives and new models of the delivery of derivative services. Third, BRICS should further improve the level of e-participation and actively disclose government information. Through the broad e-participation and active disclosure of government information, BRICS should continue to improve decision-making efficiency and quality of administrative departments, ensure supervision covers everywhere so that corruption loses its breeding space, and facilitate the public to re-use and re-develop data so as to inspire people to make innovations, generate new services and create new opportunities for economic development. Fourth, BRICS should promote continued and organic integration between the new generation of ICT and government governance and give full play to the potential of big data, the Internet of things, cloud computing and other new generation of ICT to be applied in e-government. On the one hand, BRICS should integrate government information resources, build a government supporting platform, explore demands for social services, make governments' public services personalized and government decision-making intelligent, and to enhance warning and emergency response capabilities of the government crisis management. On the other hand, BRICS

[8]*Guidelines to Accelerate the Building of Supporting Platform for Business Startup and Innovations by the Public* of The State Council of the P.R.C, on the website of the central government of the P.R.C. on Sept. 26th, 2015 http://www.gov.cn/zhengce/content/2015-09/26/content_10183.htm.

should develop intelligent livelihood services by assisting governments in providing new solutions to typical problems in areas of focus such as medical care, education, and transportation and continuously improve community safety, health services, smart home and intelligent logistics.

4.4 Strengthening Network Security Management Cooperation

In the era of digital economy, cyberspace is regarded as the "fifth space" after the land, sea, air and space. Network security governance has become an unprecedented concern of the international community. BRICS countries must first establish a network security cooperation governance concept. On the basis of safeguarding their national network sovereignty, BRICS nations should reach a consensus for common development of network security and a unified agenda for coping with cyber threats, build the basic framework of network security cooperation governance, deepen cooperation and exchanges in network security governance, and promote mutual tolerance, mutual understanding, mutual trust and mutual learning. Second, BRICS should establish multi-level cyber security governance consultation mechanism, develop responsible cyber security governance rules, norms and principles, advance the formulation of treaty on cyber security cooperation governance within the BRICS block, so as to safeguard national cyber security and people's rights to the free use of the network in accordance with law. At the same time, BRICS should maintain an open attitude, and actively participate in the international cyber security governance process, push for changes in the global Internet security governance system, and jointly build a community of a shared future of cyberspace. Third, BRICS nations should propose basic cybercrime types for each member state to develop further according to their respective national conditions. In accordance with local culture and legislation, each country can handle various cybercrimes in different ways, form a multi-lingual governance dictionary in English, Russian, Chinese and other languages to describe national laws and local cultures and to associate these laws and cultures with standardized cybercrime types.[9] Finally, BRICS must be committed to promoting cooperation and innovations on cyber security technology and privacy protection technology in the new generation of enabling technology environment. Cyber security technology mainly includes virus protection technology, intrusion detection technology, security scanning technology, authentication and signature technology, VPN technology, application security technology etc. Privacy protection technology includes static anonymous technology, dynamic anonymous technology, anonymous parallelization, data watermarking technology, encryption storage

[9]Zuo Shengdan. BRICS Cyber Economy and Cyber Security Seminar: Cope with Challenges on Cyber Security with Cooperation. chinanews.com. May 20, 2017. http://www.chinanews.com/cj/2017/05-20/8229589.shtml.

technology, risk adaptive access control, large data audit technology, data traceability technology, role and property based access control technology etc.

5 Strategies to Improve Innovative Development of Digital Economy of BRICS Countries

5.1 Strengthening Strategic Coordination of Digital Economy Development of BRICS Countries

BRICS countries are also members of Group 20. Thus, BRICS countries should map out their digital economic development strategy under the guidance of G20 Digital Economy Development and Cooperation Initiative approved in G20 Hangzhou Summit so as to improve the strategic coordination of digital economy development. Firstly, BRICS countries should stick to the ideas of working together, mutual trust and mutual benefit, improve openness and cooperation in strategies of digital economy development and take the initiative to build more platforms for strategic cooperation so as to deepen the inclusive development of digital economies. They should have a clear position of the development of digital economy; strengthen the link and synergy of developmental strategies so as to form a differentiated and coordinated development. Secondly, integrate the advantages and complementary resources to develop digital economies and create more areas of common interests, growth areas of cooperation and new highlights of win-win results. By integrating the advantages and complementary resources, more bilateral and multilateral digital cooperative projects can be materialized and some demonstration projects can be built. Improve the match able development of digital industries in different countries and optimize the layout of industrial chains. Maintain the coordinated development of physical economy and virtual economy, stimulate the new momentum of economic development and achieve economic growth in stability. Thirdly, further improving the international rules and standards of technical innovation of digital technology, cross-border data flow, cross-border E-commerce and E-government, strengthening the coordination between BRICS countries of network infrastructure and platform interconnectivity planning, perfecting the international tax policies that matches the digital economy development, jointly coping with the tax base erosion and profit shifting and creating a good environment for the coordinated development of digital economy. The fourth is to summarize the successful practices of coordinated digital economy development, shape reproducible guidelines, policies, procedures and innovation, and share good experience of the BRICS countries in the cooperative research and development of new digital technology, creating new business model and new industries so as to promote the coordinated development of global digital economy.

5.2 *Developing Inclusive Digital Economy*

In order to overcome the negative impact of the digital divide on the development of the digital economy in BRICS countries, BRICS countries should constantly enhance the inclusiveness of digital economic development. Firstly, countries with better conditions for development and good application results of ICT should help and guide the relatively backward countries to promote informatization strategy, use a new generation of ICT to transform traditional industries and give rise to new industries and promote industrial transformation and upgrading. The New Development Bank of the BRICS countries should set up special funds to promote the construction of information infrastructure in the relatively backward countries of ICT and especially, to accelerate the construction of Internet construction in backward regions and rural areas; to promote the innovation of information communication technology in the relatively backward countries of ICT, especially innovations in mobile Internet, Internet of things, cloud computing, big data and other enabling technologies; launch innovative and entrepreneurial activities based on the Internet platforms and crowd funding to inject new vitality into the economic development of countries with comparatively weak ICT. Second, BRICS countries should further develop multi-form and multi-language online products and services including English, Russian and Chinese so as to have an easier access to the sharing of Internet services for people from different groups and classes in BRICS countries. They should enable the vulnerable groups to master and improve the basic abilities and skills in the use of ICT by strengthening basic education in digital media, information and digital literacy, to learn via the Internet and improve work skills and entrepreneurial ability. Third, they should use digital inclusiveness and digital technology as the key elements to promote inclusiveness of the digital economy, use the digital technology to develop network education, remote medical treatment and promote targeted poverty alleviation so as to promote the elimination of poverty and the improvement of people's livelihood.

5.3 *Emphasizing the Cultivation of Digital Talents*

Digital talents are the first resource for the BRICS countries to develop data economy. To this end, the BRICS should make relevant calculation of digital talent demands according to their own digital development strategy, and make a prediction and overall plan of the supply and demand of digital talents in the future according to the existing digital talent reserves, the trend of development and the situation of digital talent output by the institutions of higher learning. The BRICS should focus on the cultivation of top leading digital talent and filling the gap of talent team building, set up top digital talent cultivation project, increase funds for the digital talents under the age of 35, established incentive and training mechanism on the basis of the evaluation to promote a further growth of their ability and career development. National education departments should set up a committee for digital experts on different levels of education institutions and provide tailored guidance on planning digital talent development, setting target of personnel training, digital talent training methods, digital talent plan approval, and maintain the quality of digital talent in education institutions on different levels. Colleges and universities should strive to create a high-end research institutes in the field of STEM, deepen a new generation of information and communication technologies such as the big data analysis, Industrial Internet, the Internet of Things and information technology security, combine STEM with other disciplines such as economics, law, political science, sociology and management, support innovative entrepreneurial projects from colleges and universities so as to promote the transformation of information communication technology into economic and social fruits. The BRICS should increase the requirement of information technology of vocational education to meet the changing needs of existing career and new business; popularize programming education, advocate the mentor-mentee system and prepare talented person for the digital economy. In the field of vocational education, they should create flexible and personalized digital learning experience, strengthen the digital training for employees from small and medium-sized enterprises, improve the assessment and certification of staff training and enable the staff to master the necessary skills of information technology and Internet use.

References

Gilster, P., & Glister, P. (1997). *Digital literacy*. New York: Wiley Computer Publisher.

Jiang, Q. (2015). Future of internet plus and China's economy. *People's Forum (Academic Frontier), 10,* 52–63.

McKinsey & Company. (2015). *Raising your digital quotient*. Shanghai: Jiaotong University Press.

Mesenbourg, T. L. (2001). *Measuring electronic business: Definitions, underlying concepts, and measurement plans*. US Bureau of the Census: Maryland.

Negroponte, N. (2017). *Being digital* (Y. Hu, & H. Fan, Trans.). Publishing House of Electronics Industry: China.

Rifkin, J. (2014). *The zero marginal cost society: The internet of things, the collaborative commons, and the eclipse of capitalism* (Experts from CCID, Trans.). China: CITIC Press.

Tapscott, D. (2016). *The digital economy* (C. Bi, Trans.). China: China Machine Press.

Wang, Y. (2016). Status quo and development trend of global e-government, enlightenment in China: An interpretation of united unions e-government survey 2016. *E-Government*, *9*, 62–69.

Xin, L. (2017, April 6). BRICS business council held mid-term meeting in India. *People's Posts and Telecommunications News*, 001.

Chapter 10
Study on Technologies for Financial Inclusion in BRICS

Xijun Zhao, Boyang Wei and Mengzhu Shao

Financial inclusion, initially proposed by the UN in 2005, stresses utilizing complete financial infrastructure to expand financial services to underdeveloped areas and low-income groups with affordable cost. Compared with traditional finance, Financial Inclusion attaches more attention to the equitable access of the whole society to financial services. Vast requirements such as financial management, investment and consumption of the high income group can be satisfied through financial services. Now, low-income group is entitled to financial services because this is an equal financial service system aimed at serving the broad masses. At the G20 Hangzhou Summit in 2016, developing financial inclusion had been an important topic attracting vast attention from member states of the G20.

Based on microcredit and microfinance, the concept of financial inclusion experienced a rather long period of formation and transformation based on their constant development and perfection process. International development of Inclusion Finance mainly experienced the following four stages. In the 19th century, financial cooperative, a formal saving and credit institution aiming at rural and urban poverty population initially emerged in Europe, is the earliest development stage of finance inclusion. In the 1970s, Bangladesh and Brazil took the lead in microcredit business. Other countries followed suit in succession, utilizing microcredit as an important method to serve micro enterprises and families underprivileged family in order to eradicate poverty. In the 1990s, Microfinance showed up, providing all-around financial services such as credit, saving and insurance for low-income groups. In the 21st century, many have suggested that financial institutions that serve the poor should be integrated into the national mainstream financial system. The construction of Financial Inclusion has drawn attention from more and more countries.

X. Zhao (✉) · B. Wei · M. Shao
China Science and Technology Exchange Center, Beijing, China

X. Zhao et al. (eds.), *BRICS Innovative Competitiveness Report 2017*,
Research Series on the Chinese Dream and China's Development Path,
https://doi.org/10.1007/978-981-10-8078-4_10

With the development of global society and economy, the concept of inclusive growth and sustainable development has also injected new connotation into financial inclusion. The core of financial inclusion is to enhance the availability of financial services, not only to promote sustained economic growth, but also help alleviate poverty, improve the position of vulnerable groups, and improve social and economic inequality. Therefore, compared with developed countries, the development of Financial Inclusion in the developing countries with underdeveloped society and economy and incomplete financial market is even more important. Emerging market countries represented by BRICS countries have explored and practiced earlier in system design, business model and technological innovation of Financial Inclusion, and achieved remarkable results. Research on the development process of the technology innovation of Financial Inclusion in BRICS countries will be beneficial to our understanding to the Financial Inclusion development strategy and its implementation, so as to promote the financial system to be further deepened and more inclusive.

1 Innovative Development of Financial Inclusion Technology in Brazil

1.1 Overview

More than 50% Brazil's major financial institutions including banks, credit cooperatives and SCMEPPs are collectively located in the southeastern Brazil, while the northern region accounted for only less than 10%. It is clear that Brazil's wealth and financial resource are asymmetrically distributed. Since the 1990s, in order to enable the vulnerable groups such as small and micro enterprises and low-income groups subject to financial services and financial products, the Brazilian central bank mainly launched two stages of financial inclusion policies. The first stage was to diagnose and cooperate. The Brazilian central bank firstly started financial inclusion business through interdepartmental cooperation. For example, the Brazilian central bank conducted communication and discussion with all departments under the help of financial inclusion forums and issued Financial Inclusion Report (FIR) to provide support for research on financial inclusion. At the same time, regarding the Financial Inclusion system design, the Brazilian central bank and the Ministry of Finance, Department of Social Development and other departments actively cooperated and formed a comprehensive, multi-sector Financial Inclusion system. The second stage is to promote international cooperation. Brazil has been actively involved in the discussion of the Financial Inclusion Experts Group (FIEG) and actively cooperated with international institutions such as the Alliance for Financial Inclusion (AFI) and Consultative Group to Assist the Poor (CGAP) to provide sufficient external resources for the development of financial inclusion in the country.

1.2 Typical Model and Products of Financial Inclusion in Brazil

1.2.1 Correspondent Banking System

Correspondent banking system is an important innovation in Brazil's efforts of promoting the development of financial inclusion. Brazil has rather uneven population distribution, wealth levels, and financial penetration degrees in its vast territory and land. The number of branches per million in northern Brazil is 0.75, 45% lower than the national average number (1.36), while in the southern region, the number is 28% above the national average. There is about one financial institution per 1000 km^2 in the north and central parts of the country, and about 12 financial institutions per 1000 km^2 in the southeastern region. Correspondent bank system came into being to enable the residents in the remote and sparsely populated areas have equal access to financial services.

Banking Correspondent in Brazil refers to the third party agents on behalf of banks to provide customers with services of saving, credit, payment, transfer, insurance and other "banking services outside networks" in the region scarce for bank branches. These agents can be post offices or large retailers, or individual and small shops. This business model began in the 1970s, with the Brazilian central bank starting to allow banks to hire other companies as agents to recover collaterals, and bear the legal responsibility of agents. In 1979, the law allowed agents to conduct business on behalf of financial institutions to submit loan applications, analyze credit status, collect debt and manage of credit transactions. In 1999, the agents further expanded its financial services scope on behalf of financial institutions, including opening accounts, deposits and withdrawals, transactions and payments, granting governmental subsidy and pensions, acting the receipt and payment order of savings account, purchasing and redeeming investment funds. In 2000, the law broadened the scope for the establishment of agents to further improve the autonomy of financial institutions. Since then, in the agency model, the service network of financial institution has expanded rapidly with vigorous development of agent business, and greatly improved the penetration of Brazil's financial services. Through cooperation with entities such as retail stores, lottery shops, pharmacies and post offices, financial institutions expanded and broadened financial services utilizing these entities outlets. In this model, financial institutions have worked with business entities to complement the gaps in financial services in the underdeveloped areas and remote regions, successfully providing financial services to remote regions and low-income groups, expanding the coverage and availability of Brazilian financial services, at the same time reducing the cost of banking to expand the scope of financial services. In addition, the banking sector utilizing business entities outlets makes it possible to comprehend customer's requirements in a more comprehensive way, so as to provide them with targeted financial services. As of 2013, Brazil has formed a huge network of correspondent banks with the agent outlets up to 400,000, covering all of Brazil's urban areas.

1.2.2 Promotion of Simplified Accounts

Promoting a simplified account is another action of innovation in carrying out financial inclusion in Brazil. In 2004, regulator sectors of the country proposed to promote the account classification system, and allow individuals to open simplified account. With low opening conditions and simple application process, no need of proving income or initial margin, customers can open accounts merely with their personal ID card or the proof of house accumulation funds, which encourages and helps remote areas and low-income groups to open bank accounts and access to financial services. As of 2010, the number of users of simplified accounts has exceeded 107 million. The implementation of the simplified account further has enhanced the participation of residents in the fundamental business such as bank deposits and loans. By the end of 2014, the number of loan recipients from commercial banks in Brazil was 405.03 per thousand, and the number of which deposited in commercial banks per thousand was as high as 627.31. Furthermore, Brazilian regulators have also increased the availability of the special groups such as low-income group, farmers and women to financial credit services by policies such as reducing microcredit conditions and streamlining credit processes, so as to increase the penetration degree of financial credit services (Fig. 1).

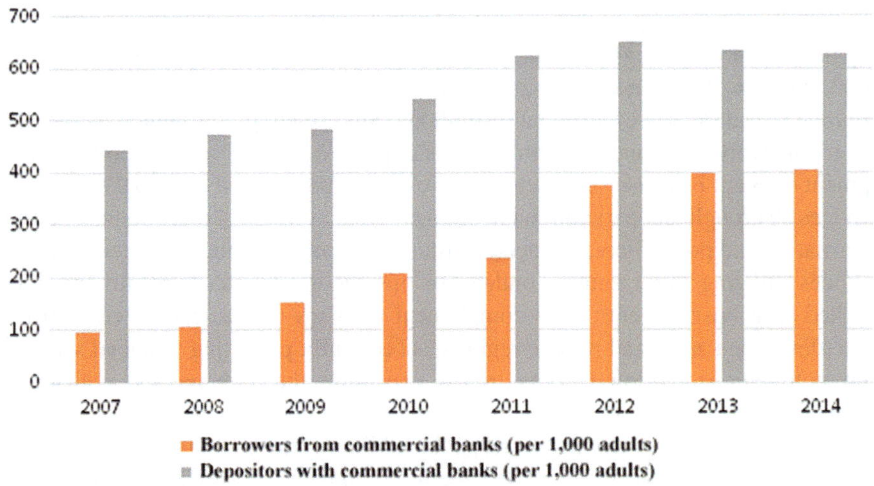

Fig. 1 Availability of deposit and loan services in Brazil from 2007–2014. *Source* Financial Assess Survey IMF

2 Innovative Development of Financial Inclusion Technology in India

2.1 *Overview*

As per the Report on Indian Financial Inclusion Development in 2014, India has an average of about 7.5 bank branches per 100,000 population in 2012, yet the poverty-stricken area has only 3.5 bank branches per 100,000 population. In terms of financial business structure, India's savings and credit services availability also shows a clear inequality. About 50% of Indian residents have savings accounts, while only one in seven Indian residents have access to bank credit. From a statistical perspective, increasing the penetration degree of financial credit comprises an important part of Indian Financial Inclusion development. In recent years, through the efforts of the Reserve Bank of India and the regulatory authorities, financial penetration degree in India has been significantly improved. However data from the Financial Access Survey conducted by the IMF shows that compared with developed countries and some major developing countries, financial exclusion degree in India still maintains at a low level, and Financial Inclusion degree in India needs to be further improved (Table 1).

Generally, India's implementation of financial inclusion strategy can be divided into three stages. The first stage: 1950–1960, the cooperative financial institutions provided loans for agriculture. The most important cooperative financial institutions

Table 1 Data of financial inclusion in different countries (2014)

Country	Numbers of ATM	Numbers of bank branches	Numbers of ATM	Numbers of bank branches	Bank deposits	Bank loans
	Per 1000 km		Per 0.1 million		% of GDP	
India	54.90	39.69	17.80	12.87	63.71	50.30
China	65.49	9.59	54.44	7.97	137.40	92.39
Brazil	21.64	3.97	114.79	21.05	35.33	41.11
Mexico	22.49	6.84	48.74	14.75	22.12	19.40
Philippines	52.64	19.56	23.36	8.68	45.85	24.89
South Africa	20.81	3.44	66.25	10.95	40.57	65.47
Russia	13.60	2.72	185.16	36.99	39.44	47.47
Germany	248.82	29.36	123.06	14.52	28.01	21.36
France	106.91	37.60	108.03	37.99	35.46	38.55
UK	285.21	54.95	129.76	25.19	129.41	124.36
USA	N/A	9.09	N/A	32.22	58.09	43.77
Japan	386.96	102.86	127.45	33.88	134.11	96.28

Note UK data were issued in 2013
Data Source Financial assess survey IMF

are primary agricultural credit association, which is formed by raising funds from farmers and in return, to provide loans to farmers. The county level primary agricultural credit association can also be combined with each other to jointly form a regional center bank responsible for providing loans for the members of Agricultural Credit Association. 1970–1980 have witnessed the second stage. Regional Rural Banks and the National Bank for Agriculture and Rural Development were gradually established in India. Rural banks started involving in rural financial market as an official financial institution, and Indian the National Bank for Agriculture and Rural Development provided support to rural financial institutions as policy related bank. The third stage began from 1990 to this day as the implementation of Self-Help Groups (SHGs)-Banking Project has delivered convenient financial services to low-income groups. India's innovative self-help group guarantee model has greatly contributed to the development of microcredit businesses.

2.2 *Typical Model and Products of Financial Inclusion in Brazil*

2.2.1 Priority Sector Loans

During the early stage of India's promotion of Financial Inclusion, the most prominent feature is implementing state intervention. In a broad sense, priority sectors include sector of agriculture, handicrafts, micro-enterprises, education, real estate, import, export and trade. Loan policy for priority sectors requires domestic commercial banks to lend to the priority sectors more than 40% of the total required loans, and foreign banks needs to lend at least 31%. As of March of 2014, about 14.1 million Rupees loans were released to the priority sectors. This mandatory credit policy entitles the vulnerable sectors to the special rights of financial service. Financial institutions provided low-income groups such as farmer household with financial services and credit supports in accordance with policies, which to a certain extent facilitated the penetration and inclusion of financial services and became the early attempt of India's financial inclusion. But this strong intervention policy distorts the efficiency of resource allocation. Subsequently, Reserve Bank of India and the Financial Sector Legislative Reforms Commission gradually utilized a method of combining mandatory regulations with marketization to benignly guide the financial institutions to tilt toward the vulnerable sectors.

2.2.2 Self-help Groups-Banking Projects

Self-Help Groups were originally made up of 15–20 members from low-income families, most members are women, who deposit savings to the group and provide loan services to needy members. Launched by National Bank for Agriculture and

Rural Development, SHGs-Banking Project is to provide policy support for SHGs and encourage low-income groups such as farmers and women to collect petty cash for the group to carry out microcredit within the institution. Under this framework, members of the group deposit regularly and open account in the commercial banks or cooperative bank of vicinity in the name of the group as a group fund. After internal evaluation, needy members will have access to the group fund with self-decided interest rate and repayment period by the group.

The relationship between the SHGs and banking institutions can be divided into three categories. Firstly, model of non-governmental organization and governmental agency-self-help groups-banks, in which SHGs are established by non-governmental organizations or governmental agencies. Operation of SHGs gradually becomes mature, and after each indicator meeting the request of the bank, a loan will be released by the bank to the SHGs, and the loan repayment will be jointly born by the group members. Besides, the model is with rather low cost; the loan interest rate is set by the bank; and the loan is repaid by the group, the risk of default is whereby greatly reduced. Therefore it has obtained great support from financial and governmental institutions of India. Second model is known as the bank-SHGs model. In this model, SHG is established by banking institutions, and the banking institutions are responsible for training the group members so that the SHGs have more possibilities to meet the bank's lending standards and whereby to obtain loans. In the last model, banks-SHGs-non-governmental and other institution, SHG is established through bank financing, and non-governmental institutions and other institutions merely function as financial intermediaries.

As an intermediary, SHGs bridges the gap between financial institution and low-income groups. Compared with single factor credit model, this model is able to better address the problems of difficult financing for vulnerable groups such as farmer household due to low income and lack of mortgage and guaranty. The model of group style, to a large extent, alleviates the uneven information between banks and the borrowers, whereby to lower the risk of bank credit and effectively mobilize the initiative of banks to provide financial services for low-income groups.

2.2.3 Business Correspondent System

Only retired bank and government employees, micro-financing institutions, non-governmental organizations and other civil society organizations are allowed to act as business correspondent for the Reserve Bank of India. In addition, individual shop owners, small scale companies for saving plans and insurance as well as retired teachers have been gradually included in the scope of business correspondent, further strengthening the connection between banks and the needy and low-income groups, lowering the cost for banks in providing financial services to the rural and remote areas, as well as creating favorable conditions for the expansion of financial services.

2.2.4 Application of Information Technology

Combining information technology with Financial Inclusion business, the Reserve Bank of India utilizes advantages and convenience of information technology to provide technology assistance for Financial Inclusion. For example, the Reserve Bank of India encourages the state governments to adopt e-welfare payment systems for wage payments and social security expenditures, greatly increasing the speed and efficiency of payments and liquidation. Commercial banks can be free from space constrains by using smart card to provide banking services for the residents of remote areas, which broadens the coverage of financial services. The retail business can also play its role in providing withdraw and payment services for customers through the bank network terminals, so that financial activities can be closely connected with resident's daily life to ensure the maximum satisfaction to the demand of residents for financial services.

3 Innovative Development of Financial Inclusion Technology in South Africa

3.1 Overview

South Africa's economy has the classic binary characteristics of urban and rural areas and black and white people. The prosperity of the first world economy is a sharp contrast to the destitution of the third world economy. Since South Africa vigorously pushed forward international transactions in 1994, first world economic and financial sectors have been developed rapidly. Inflows of foreign capital and enterprises have injected new vitality into South Africa's first world economy. However, due to long-term exclusion and discrimination, the development of the third world economy with black and other colored people as its main powers has always been hindered severely. Those vulnerable groups have always been on the brink of South Africa's economy, and their access to financial services is fairly limited. At the beginning of the 21st century, only half residents of South Africa could get banking service. Over 90% of the groups that experienced financial exclusion were black and other colored people. In rural areas, only about 20% of the population have bank accounts, while nearly 60% of the urban population do. Financial exclusion in South Africa is extremely severe because of culture, race and other factors. Therefore, advancing financial inclusion is of great significance to the inclusive development strategy of South Africa.

3.2 *Classic Models and Products of Financial Inclusion in South Africa*

3.2.1 Microfinance

In order to provide financial services to more black and poor people, South Africa implements differentiated institutional arrangements. For example, the South African authorities have implemented an immunity act that allows poor potential customers without official identification or household register certificate to skip verification and directly open bank accounts, and thus encouraging them to accept financial services. Moreover, South Africa also lowers the maximum limit of microfinance (in South Africa, microfinance refers to cash loan that is less than or equal to 10,000 ZAR, which is about 1515.15 US dollars), restricts microfinance's highest amount of interest and etc., thereby improving the feasibility of microfinance business in low-income groups. A series of policy and system designs have provided strong guarantee for the rapid development of microfinance in South Africa.

Distinctive microfinance credit reporting system is South Africa's important pillar to facilitate the development of microfinance as well as an important content of the country's innovation model of financial inclusion. South Africa's microfinance market is dominated by banks and microfinance companies. About 5–8% of the total amount of consumption loans is from microfinance market, making it become an important component of South African residences' consumption funds sources. In 2002, Microfinance Regulatory Council established National Loans Register, and as the database of the country's loan information, it is in charge of collecting the transaction information of microfinance institutions and the credit information of clients. In 2005, Microfinance Regulatory Council was admitted into South Africa's National Credit Regulator, becoming a part of consumer credit management. Sophisticated institutional guarantee and information system have attracted traditional financial institutions to get involved in microfinance business, which further promoted the development of microfinance markets. Relatively mature information sharing mechanisms have been formed among national credit reporting agencies, banks, non-banking credit providers and loan clients, and thus pushing forward the orderly and healthy development of microfinance market becomes an important force for South Africa to further advance financial inclusion.

3.2.2 Black Economic Empowerment

The primary problem South Africa needs to solve to achieve inclusive development is the exclusion of black economy and financial activities. In 1990s, the South African government and commercial institutions jointly launched Black Economic Empowerment, aiming to address the extreme imbalance of South Africa's social and economic development. The transaction capability and financial participation of black community were dramatically improved through pushing forward the mixed ownership reform of South African companies, helping black people become a part of boards and participate in companies' management and decision-making, encouraging black people and black enterprises to participate in financial activities and other measures. Meanwhile, financial sectors of the country's National Economic Development and Labor Council commit to encouraging banking system to provide financial services for diversified client groups. Over the past 20 years or so, there has been as many as 500 billion ZAR of black capital, and the gap of wealth between the black and the white was drastically narrowed. With the accumulation of wealth, more and more black people begin to involve and participate in financial activities of various kinds, which tremendously elevated finance's position in the economy and life of black community.

3.2.3 Mobile Payment

Due to relatively backward infrastructure construction of South Africa, it costs the country's financial institutions a lot to provide financial services for residents in remote areas, which inhibits their enthusiasm in developing new client groups. On the other side, it also costs the residents plenty of fares and time to get to banking outlets, which reduces the residents' impetus to participate in financial activities. Therefore, solving the problem of financial institutions' high business expansion costs becomes the key for South Africa to expanding the popularization and inclusiveness of financial services. With the development of IT and the Internet technology, mobile communication technology becomes an important breakthrough to lower operating costs of banks. South Africa's rapidly developed mobile payment technology has become a new channel for rural and low-income groups to obtain basic financial services. Dominated by banks and MTN Group (the largest telecommunication operator in South Africa), mobile payment, which utilizes communication technology to achieve remote payments, has drastically lowered the threshold for residents to acquire financial services from banks and the transaction cost, giving low-income people, women, peasant households and other vulnerable groups in remote areas the chance to participate in financial activities. Moreover, transfer, living payment, withdrawal, micro saving and other financial businesses have been made possible by communication technology, enriching the contents of financial services.

4 Innovative Development of Financial Inclusion Technology in Russia

4.1 Overview

Russia is a large country with vast territory, but its population distribution is extremely uneven. Due to the influence of weather, terrain and other factors, most population in Russia are concentrated in western developed areas, causing significant differences in economic structure, infrastructure, financial development and other aspects among various regions of Russia. Russia's financial system is relatively mature on the whole. The country has formed a diversified financial system which is dominated by the Central Bank of Russia and a relatively developed financial market. However, the distribution of Russian financial resources is extremely unbalanced, with over 80% of them concentrating around the two big cities of Moscow and St. Petersburg, forming a pattern featuring "strong center and weak surroundings". To make the financial resources of core cities radiate to peripheral cities and rural areas is one of the important contents of Russia's financial inclusion strategy. At present, Russia has made preliminary achievements from the exploration of the inclusive development of finance, with remarkable improvement in the range of financial services and the participation of residents in financial activities.

4.2 Classic Models and Products of Financial Inclusion in Russia

4.2.1 Microfinance

Microfinance plays an important role in the development process of Russia's financial inclusion. By the end of the first quarter of 2013, the total number of microfinance institutions in Russia has reached 2939, with as many as 1574 newly increased ones in 2012 alone. Different from South Africa's microfinance market which relies on perfect credit reporting system, Russia's microfinance market flourishes because of system design. Institutions providing microfinance services in Russia include consumer cooperatives, fund companies, private microfinance institutions and so on. Every kind of microfinance institution can carry out business under the legitimate forms of organization and is guaranteed by the basic legal framework. The loose market environment, specific institutional guarantee and the improvement of industrial self-discipline and clients' credit consciousness provide microfinance institutions with diverse capital sources and make the market develop steadily. The service range of microfinance institutions has also been expanded, covering not only loan, but also deposit, consultation, leasing and guarantee. Thanks to the development of microfinance market, participation of low-income groups and small and medium-sized enterprises in the finance spectrum has been promoted constantly. Moreover, Russia also tries to conduct microfinance business through Automatic Teller Machine (ATM), which will further facilitate the development of microfinance business. It is worth noting that in recent years, turmoil in the international crude oil market and markets of other staple commodities as well as the frequent fluctuations of ruble exchange rate have made traditional bank credit in Russia tighter and tighter and more and more customers turn to microfinance institutions, and thus the demand of microfinance market keeps surging.

4.2.2 Financial Consumer Education

To carry out financial inclusion, it is not only necessary to make innovation in business models and products, but also necessary to expand financial demand. Enhancing consumers' financial education and improving consumers' understanding of financial systems are basic requirements of advancing the development of financial inclusion. Russian residents lack the awareness of personal finance and endowment insurance, as well as trust to financial sectors. Therefore, some consumers are unwilling to accept financial services. In order to promote the financial education for students and low-income groups with potential for financial services, the Russian government launched a five-year nationwide financial literacy program, whose duration was from June 2011 to June 2016. The program mainly included educating students and low-income groups on financial knowledge, strengthening the legislative protection to consumers, conducting pilot study and regular monitoring and evaluation and etc., aiming to improve Russia's financial ecological environment, increase the efficiency of the public's financial behaviors, and enhance the protection to financial consumers, thereby encouraging residents to actively accept financial services and actively participate in financial activities, so as to offer soft infrastructure to the inclusive development of finance.

5 Innovative Development of Financial Inclusion in China

5.1 Overview

The economic and financial development in China's urban and rural areas is severely out of balance because the country has long been affected and restricted by dual structure. Due to greater risk and lacking mortgage assets, small and micro businesses are always excluded from traditional financing institutions. On this occasion, the Chinese government officially took the carrying out of financial inclusion as the basis state policy in 2013. The development of financial inclusion in China can be divided into four phases, first, during the micro-credit period in the 1990s, China mainly conducted poverty alleviation projects complying with the model of country bank in Bangladesh, making many people wrongly understand the financial inclusion into poverty alleviation. Second, from 2000 to 2005, also as the period of micron finance, some rural financial institutions have started providing diversified financial services including loan, payment and insurance to peasants and low-income groups. Third, China has entered the stage of financial inclusion from 2005 to 2010, and attracted domestic financial institutions, large-scale, small and medium size commercial banks to join. It enjoyed wider service objects, more diversified and networked service mode. Fourth, China came into the stage of internet financial inclusion in 2010. With the rapid development of the internet and popularizing rate of mobile phones, the internet financial inclusion has become an important carrier of realizing financial inclusion. Featuring low cost, low threshold and broad coverage, with intelligent services, the internet finance has effectively made up shortages of financing institutions' clients including dispersal, low benefit and high cost.

Overall, the financial inclusion in China has two major participants. One is the financial institution entities including commercial banks, small loan companies, village banks and so forth. The other is the internet-finance. Its rapid growth makes financial inclusion enjoy more abundant contents, more extensive and widespread serve objects, and play a significant role in promoting the development of financial inclusion.

5.2 *Classic Model and Products of Financial Inclusion in China*

5.2.1 Community Bank of CMBC

In 2013, while continuing to deepen small and micro finance, the China Minsheng Banking Corporation Ltd. (CMBC) proposed to speed up the advancement of financial strategies in housing estates, and jointly officially launched Minsheng housing estates financial service stores with government, real estate agencies and property management companies. It was the first stock-holding bank in China to carry out layout and establishment of community bank. The community bank of CMBC has following characteristics; (1) Build marketing alliance with merchants within the scope of 1.5 km to jointly develop clients and improve cross-selling. For example, merchants give out food and hair coupons to clients, and if they successful recommended clients to open bank account at the CMBC, the bank would offer deal base to them, achieving combination of card, net, point and circle. It not only improves cross-selling, but excavates financing demand of neighboring owners, making for expanding the bank's number of clients. (2) Integrate features of housing estates to push out series Intellectual-home products, build platform of core products for housing estates finance and provide owners with multiple financial services. (3) Cooperate with government and property management companies in handling all kinds of convenience businesses. In financial strategies of the CMBC, diversified convenience businesses have become a key manner to enhance connection with clients and improve service quality. The community banks could help one-stop payment including fees in property, parking, water and electricity as well as telephone through cooperation with government and property management companies, and application of mobile payment technology, so as to more comprehensive serve owners. Through starting community banks, the CMBC has expanded its financial services into community and resident level and combined financial services with residents' daily life, which will bring benefits of expanding contents and scope of its financial services and further cultivate new client groups, thus improving its coverage and availability.

5.2.2 Practice of Financial Inclusion in Rural Area of Jilin Province Rural Credit Bank

Jilin Province Rural Credit Bank has made a bold trial in exploring the development path of financial inclusion, and achieved great innovation in expanding service area, improving means of service and enriching debit and credit products as well as collaterals, which won a wide base of clients and improved coverage and influence of financing services in rural areas. The exemption of handling charge in deposit, withdrawal and remittance and all standing businesses of peasant household loan in Jilin proposed by Jilin Province Rural Credit Bank have provided conditions for peasants to participate in financial activities. In terms of means and convenience of financial services, Jilin Province Rural Credit Bank has a completely broad coverage in physical branches, peasant-helping service points and flowing service vehicles, and complete mobile terminal systems such as ATM and POS machines, forming a service system integrating self-service banking, mobile banking and WeChat banking. At the same time, it has also built agricultural loan, personal loan, micro loan and other characteristic business centers to satisfy diversified demands of clients. In terms of financial products and collaterals, Jilin Province Rural Credit Bank supports featured agriculture and green industry in line with local features and has achieving prominent outcomes by taking the lead in promoting guarantees of direct subsidies for grain and right to derive benefits from land as well as the three rights mortgage.

5.2.3 Mobile and Internet Finances Help Financial Inclusion

Emerging technologies such as internet, mobile internet and big data advance the rapid development of internet finance. Taking full advantage of internet technology is an important feature of China's financial inclusion development at the present stage. One the one hand, the features in trans-time-and-space and interconnectivity of internet finance have expanded service channel and manner of traditional financial institutions, reduced service costs and effectively improved cover degree and convenience of financial services. For instance, with the help of big data technology, Shanghai Pudong Development Bank brought out new modes of financial services such as e-commerce transaction and online lending in 2013, which realized the comprehensive online operation in such areas as online management, online data, online approval and online loan, and effectively satisfied small and micro e-commerce with urgent needs during process of financing. Chongqing Rural Commercial Bank officially launched an APP named Jiangyu

Mobile Finance in 2013, enabling clients independently check accounts, pay living costs, set mutual transfers in "due on demand" and "regular intervals", purchase finance products, check credit card bill, initiate collect and other personal financial services. On the other hand, directly rely on Internet technology as the carrier, and facilitate innovative financial services and produces including mobile payment and third-party payment, P2P internet loan and crowd funding as well as internet finance products help greatly enhance penetration of financial areas. Network and mobile payment have provided residents with more convenient channels in transfer accounts and payment, clients will not be limited by branches and counters of financial institutions. In addition, those vulnerable groups, who cannot get loan from traditional financial channels could obtain "small-amount, short-time and flexible borrowing and returning" loan with the help of internet financial platform to satisfy their financing demands. Internet finance products enjoy simple operation and high profitability, which can effectively solve redundant mobility of users. In 2016, frequency internet payment of non-bank institutions payment (third-party payment) has exceeded the summation of e-payment payment of all country's banks. Trading volume of P2P network debit and credit platform and financing amount of crowd funding platform have also demonstrated blown out growth. All these indicate that with the development of the internet, financial services will extend and cover in larger areas in cities, towns and rural areas (Fig. 2).

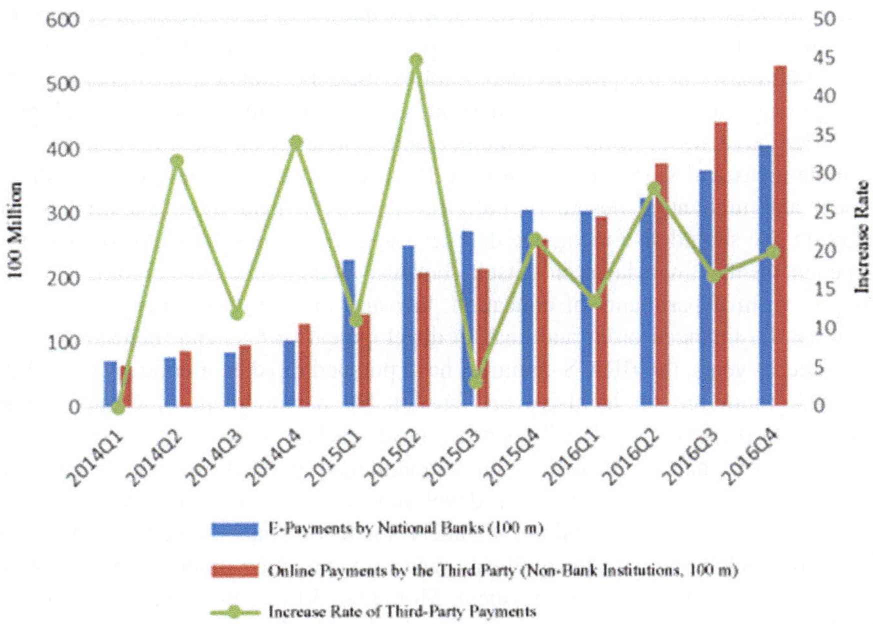

Fig. 2 Comparison of number of payments by non-financial institutions and financial institutions. *Source* Data Collected by the Author

6 Summary and Prospects

The BRICS has carried out the beneficial exploration and practice in strategic policy and system design, business mode of financial institutions, application of information technology and consumer financial education and so on, but the various countries has taken different development model. China, India and Brazil emphasize on providing financial services to more poverty stricken population and plan to lead the poor and low-income groups to participate in the financial activities via more means of commercialization, marketization on the basis of widespread financial and non-financial institutions and developed scientific and technological level. In South Africa, the black are still at a disadvantage compared to the white due to discrimination against the black race. Therefore, South Africa considers race, culture and other factors as important factors in the process of exploring the model of financial development. The development of the Russian financial system is relatively good, and the microfinance business is stable and mature. It is more focused on promoting the construction of the software facilities of inclusive finance such as education and personnel training and so on.

In addition, there is a big gap between developing countries and developed countries in inclusive finance. Due to differences in economic level and financial infrastructure, focuses are different in constructing the inclusive financial system. Developed countries have high economic level and good social welfare. Diversified and multi-level financial institutions can meet the needs of all kinds of customers. Therefore, the government pays more attention to regulating and perfecting the system and laws in the process of promoting inclusive finance, thus guaranteeing the coverage and expansion of financial services. On the contrary, developing countries have lower level of economy and deficiency in the financial system. Remote areas and vulnerable groups are lack of financial services. Thus, business model and innovative design of products should be paid more attention to. In addition, we should also recognize that development of financial inclusion not only depends on the financial sector itself, but also needs effective coordination and support from departments of education, law and finance. Only this way can we promote the financial order and healthy development of financial inclusion.

In recent years, the BRICS countries have pursued rapid economic growth while focusing more on the inclusive and sustainable development of economy and finance. At present, the BRICS countries have reached a consensus on the implementation of financial inclusion and strengthened international exchanges and cooperation to actively explore the development path of the financial inclusion. According to the data from Global Financial Inclusion Database of World Bank in 2014, the BRICS have made significant results in spreading financial services such as personal savings, credit and payment. More than 50% of the population who are aged 15 or older have a financial account. Except for India, the proportion of net usage for payments or transactions have also reached more than 10%, with China and South Africa developing faster in Internet payments or transactions. Moreover, the BRICS countries have also seen significant improvements in remittances,

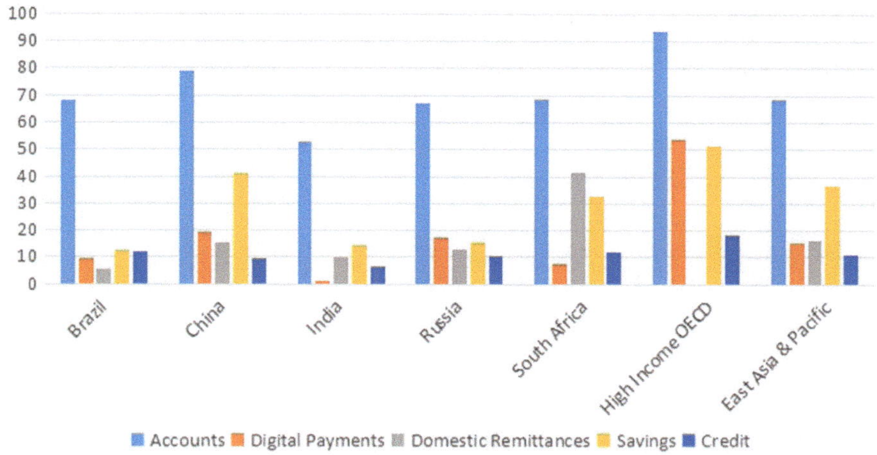

Fig. 3 Indicators of financial inclusion in some countries and regions in 2014. *Note* Accounts for the proportion of people aged 15 and older who have a financial institution account; Digital payments is measured by the proportion of people aged 15 and older who use the internet for payments and transactions; Domestic remittance represents the proportion of the population of the domestic remittance business in the population aged 15 and over in the past year; Savings accounts for the proportion of people aged 15 and older who have been saving in financial institutions over the past year; Credit says the proportion of people over the age of 15 over the past year has been in financial institutions. *Data Source* Global Financial Inclusion Database, WB

savings and credit services. However, compared with high-income countries, the financial penetration and popularity of BRICS countries still have a lot of room to improve, and there is still a long way to go in the development of inclusive finance (Fig. 3).

Due to the large population size and rapid economic development, the BRICS countries have become increasingly important in international financial and economic activities. The development of the financial sector in the BRICS countries will also have a far-reaching impact on the development of global inclusive finance. In the future, digital technology will become an important driving force for the development of inclusive finance. At the 2016 G20 Hangzhou Summit, "focusing on digital inclusive finance" and "the construction of financial data collection and indicators" became important topics of this meeting. China submitted three important documents to the G20 Summit, including *G20 High-Level Principles for Digital Financial Inclusion*, *G20 Financial Inclusion Indicators* and the *G20 Action Plan on SME Financing*, which will provide important guidance for the development of global financial inclusion. Digital financial inclusion will integrate the Internet technology with modern finance, using digital means to provide low cost, high efficiency of financial services for a broader area, a wider range of groups, to further improve the financial service's popularity and availability and to be an inevitable trend of each country in developing financial inclusion.

Chapter 11
Study on Energy Technology in BRICS

Shirong Zhang, Pei Zhang and Mingyuan Gao

China leads the BRICS countries in the field of energy science and technology. Russia, Brazil, India and South Africa have their own distinctive national conditions and advantages in energy and technology. The development of the energy and technology in BRICS countries in general, especially in the frontiers, however, still lags behind that of developed countries.

1 Research on Energy Science and Technology in China

In the course of the development of China's energy utilization, it can be found that there are many variables at different stages that affect the evolution of the energy security system, and the variable that is the slowest in change and has the most far-reaching effect is science and technology. At present, in the field of energy, the important role of this slow variable, science and technological innovation is becoming increasingly prominent and has gradually become the "order parameter" of China's energy security system. Therefore, scientific and technological progress is the driving force behind China's energy security problem. The advent of the high oil price era has created conditions for research, development and promotion of energy and new technology. The fourth technological revolution, represented by the new energy technology will definitely come. China will no longer rely on the traditional fossil energy through the development of new technologies and new energy.

China's coal-based energy supply pattern and its increasing dependence on oil imports have necessitated its independent innovation in this regard, especially considering that the technological research & development and commercial promotion of the new energy is expected to fundamentally solve China's energy

S. Zhang (✉) · P. Zhang · M. Gao
China Science and Technology Exchange Center, Beijing, China

X. Zhao et al. (eds.), *BRICS Innovative Competitiveness Report 2017*,
Research Series on the Chinese Dream and China's Development Path,
https://doi.org/10.1007/978-981-10-8078-4_11

problems in the long term. Only through becoming the pioneer of the fourth technological revolution, can China have the opportunity to break through the energy bottlenecks and have its own say in this field.

1.1 China Has Made Many Achievements in Energy Science and Technology

Since the 1980s, the oil and gas exploration and development technology has been an important part in China's national science and technology research program. Thanks to years of efforts, China has bridged its gap in oil & gas technology with that of foreign countries and is now among the top countries in this field. In line with the demand for oil and gas of the national economic development, China has launched the national science and technology research program, the exploration and development research on large and medium-sized gas field. With the provision of a stable source of gas for the West-East Gas Transmission Project as the main task, the project has made many innovative achievements in the natural gas geology theory and beneficiation law, achieved breakthroughs in the key technologies of natural gas exploration and development, basically established oil and gas exploration evaluation system with independent intellectual property rights and suitable for basins of various geological characteristics and a high level of technical equipment series, which has provided a strong theoretical guidance, technical support and equipment support for China to explore large and medium-sized gas fields.

All China's 15 large gas fields are discovered under the relevant theoretical and technical support following the implementation of the national scientific and technological research. The discovery of Sulige Gas Field is also part of the achievements of scientific and technological program. As China's first gas field of world-class reserves, Sulige Gas Field is an important gas source for the smooth implementation of the West-East Gas Transmission Project and the construction of Shaanxi-Beijing Second Pipeline Project. It is of great strategic significance and economic value.[1]

The application of scientific and technological research achievements has played an important role in stabilizing China's oil production, promoting the development of China's natural gas industry and related industries, ensuring the safety of China's oil and gas resources and promoting the sustained and stable development of the national economy. The scientific and technological research achievements also provide a strong technical support for the discovery of China's offshore oil and gas fields and the stable production for Daqing oil fields. The field of oil exploitation has embraced good news one after another following the application of new

[1]Source: Science and technology for China's energy security, http://www.china.com.cn/chinese/TEC-c/158596.htm and www.people.com.cn.

technology. On May 4, 2007, China National Petroleum Corporation (CNPC) announced that it has discovered in the Bohai Bay area a large oil field of one billion tons of reserve—Jidong Nanpu oil field, which is an exciting discovery in China's oil exploration history over the last 40 years. It is of great significance for the implementation of the CPC Central Committee and the State Council's strategic approach on the oil industry, which is "stabilizing the eastern part and developing the western part", for the stable growth and sustainable development of China's crude oil production and the enhancement of China's energy supply security. The oil and gas exploration in the Jidong beach area began in 1998, and self-exploration and cooperative exploration were carried out in the following 14 years, with no breakthroughs. Since 2002, CNPC began to adjust its strategies through strengthening the organization and leadership of exploration, innovating the management system, changing the exploration ideas, strengthening the comprehensive geological research and fine three-dimensional seismic exploration and supporting the application of large displacement inclined shaft and horizontal well drilling technology. In so doing it has overcome many geological exploration and construction problems. In September 2004, a major breakthrough was made in Laobaonan No. 1 well exploration, with industrial oil of a daily production capacity of 700 m^3 discovered in the Ordovician petroleum system. On this basis, through nearly two and a half years of overall exploration, a total of four oil formations and the basic implementation of three grade oil and gas geological reserves (equivalent of 1.02 billion tons) were discovered in Jidong Nanpu Oilfield. Among them, there is a proven reserve of 405.7 million tons, controlled reserve of 298.34 million tons, forecast reserve of 202.17 million tons, natural gas (dissolved gas) geological reserve of 140.1 billion cubic meters (oil equivalent of 111.63 million tons)[2].

In fact, the national science and technology research program provides a strong scientific and technological support for strategic security of China's oil and gas resources and forms backbone for China's energy security system[3].

On May 18, 2017, Jiang Daming, the Minister of Land and Resources announced at the "Blue Whale No. 1" drilling platform in the South China Sea that China succeeded in testing its first sea gas hydrate (also known as flammable ice), which meant that staff in Chinese science and technology field officially opened the door to "flammable ice age", the deposits of which is equivalent to that of up to 100 billion tons of oil. This is a historic breakthrough in China's energy development, which has an important and far-reaching impact on the promotion of energy production and consumption revolution. South China Sea is the main area where flammable ice in China sits. The deposits of the national flammable ice are equivalent to 100 billion tons of oil, of which 80 billion tons lies in the South China Sea.

[2]China discovers 1-billion-tone large oilfield, People's Daily, Edition 1, May 4, 2007.

[3]Science and technology for China's energy security, source: http://www.china.com.cn/chinese/TEC-c/158596.htm and www.people.com.cn.

Nuclear energy designates the future direction of energy. As a safe and environmentally friendly nuclear power technology, China's "Hualong No. 1" has gone global. This is an indication that China's nuclear power technology has been increasingly recognized by the world. With China's own intellectual property rights, Hualong No. 1 nuclear power technology has got a total of 743 patents and 104 software copyrights, covering various sectors like the design technology, patent design software, fuel technology, operating and maintenance technology. China holds over 85% of these patents and copyrights.

It is not only the only way to handle the increasingly serious energy and environmental problems for China to strengthen the development and utilization of renewable energy, but also the only way to achieve sustainable development. Biomass energy is clean and renewable energy. China is also actively promoting the development of biomass energy and is putting in place relevant incentive policies. "Renewable Energy Law" was introduced on January 1, 2006 and later some deputies to the National People's Congress proposed to introduce more practical, feasible and operational specific rules. In November 2006, the Ministry of Finance, the National Development and Reform Commission, the Ministry of Agriculture, the State Administration of Taxation and the State Forestry Administration jointly issued "Implementation Opinions on the Development of Bio-energy and Bio-chemical Tax Support Policy", which clarified the significance of developing bio-energy and bio-chemistry in replacing the fossil energy, promoting farmers' income and improving the ecological environment. "Opinions" also said that China had promoted the use of ethanol as fuel in some areas during the "Tenth Five-Year" period, which had achieved good social and ecological benefits. With the rise of international oil prices, there is urgent need to accelerate the implementation of oil substitution strategy and development of bio-energy and bio-chemical industry in a positive and orderly manner. In the next phase, the emphasis will be put on the development of bio-oil alternatives like bio-fuel ethanol, bio-diesel and bio-chemical products and on the guidance of the development of other bio-energy products. As China's bio-energy and bio-chemical industry is still in its infancy, the enactment and implementation of relevant fiscal and taxation support policies will provide a strong guarantee for the healthy development of bio-energy and bio-chemical industry.

As of June 2017, China's installed capacity of wind power totaled 149 GW (1 GW = 1 million kilowatts) and the installed capacity of photovoltaic power totaled 77GW, both of which topped the world; the value of China's semiconductor lighting industry is over RMB 420 billion, becoming the world's largest product development and production base and application market; China had saved electricity of about 100 billion kilowatt hours annually; China is a global leader in clean energy. It is not only the world's largest clean energy market, but also the largest clean energy production base. By the end of 2016, the output of China's new energy vehicles had topped 500,000 and the ownership exceeds 1 million, each accounting for 50% in the world.

Besides, in terms of energy transmission, China's UHV transmission technology is also leading the world. UHV power grid refers to the 1000 kV AC or 800 kV DC power grid. 20 years ago, China built its first one-million-volt research line in Wuhan. In 2010, the world's first 1000 kV high voltage AC test project, with a total length of 640 km was put into commercial operation. Liu Zhenya, the former Chairman of the Board and CPC Secretary of State Grid Corporation remarked at the Paris Climate Summit in 2014 that UHV is a must in the construction of the global energy Internet. As a technology unique to China, UHV represents the development of all mankind. China is willing to share this advanced technology with the world.

1.2 STI Is Key to China's Future Energy Security

The large-scale industrial development of shale gas in USA marked an important reform in global energy sector and will exert significant impact on the world's energy structure and natural gas supply & demand situation. China will gradually catch up with USA in this regard.

The fundamental solution to China's energy problem is scientific and technological innovation, upon which the understanding of energy is dependent. The discovery of the Nanbao Oilfield in 2007 was a case in point. Objectively the large oil fields near Nanbao were there, but they had not been detected by the two-dimensional seismic technology until the latest 3D terrestrial imaging technology was applied. It is also true for thermonuclear fusion. Deuterium and tritium are there in the seawater, but only under corresponding scientific and technical conditions, can they be converted to energy that can be utilized by humans. In the future, with the development of science and technology, many things that are not mentioned in today's energy field will become new and alternative energy. The word "energy" will carry more meanings.

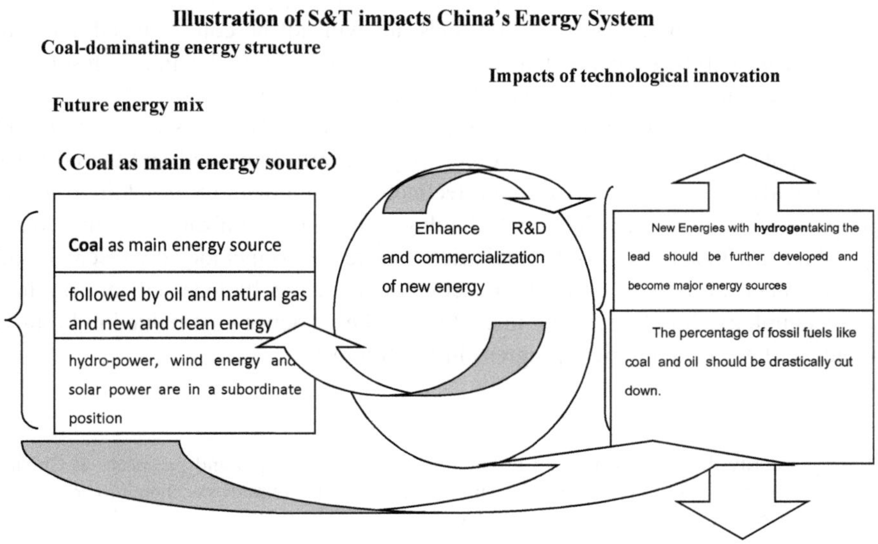

STI is the core means of ensuring stable supply and energy security in China. China must on the one hand focus on the use of clean energy, on the other on the upgrading of traditional energy. It is a long process to maintain the operation of China's energy security system and promote its continuous improvement through the "slow variable" of technological innovation. And that calls for continuous development of scientific and technological innovation, the close integration of society and science & technology and the abandonment of people's dependence on traditional ideas. It can exert decisive impact on the supply, application and environment security of energy.

Technological innovation can increase stable and reliable energy supply for the country through the adoption of new technologies, reduce environmental pollution and improve environmental quality through adopting new energy conservation technologies. At present, China must enhance energy and technological innovation capability, adjust and optimize energy structure and build a modern energy industry system in line with the requirements of accelerating the transformation of economic development in order to guarantee a reliable energy supply for economic and social development. For example, great importance should be attached to energy efficiency in the power industry in summer in order to strengthen the emergency response and ensure safe and stable operation. China should promote nuclear power construction as per the safety-first principle, develop water, wind, solar and biomass energy in an active and orderly manner and improve the development of new energy industries.

Ethanol and other types of bio-fuels are part of strategies to guarantee energy security, keep air clean and mitigate global warming. The biggest challenge to bio-fuel development in China is the concern about "food security" as agriculture is a strategic but a disadvantaged industry in China. The technology of producing cellulose ethanol using corn straws and other waste raw materials by Danish Novozymes company can "turn trash into treasure", which undoubtedly is the best way to solve the problem.

These essential scientific research and new technology development can ensure that gasoline will be replaced by bio-fuels, air will not be contaminated and that more energy will be saved, which will result in sustainable utilization of resources. In the long term, technological innovation and progress matters a lot for fuel ethanol to become a sustainable and market-oriented industry. The unit yield of corn and corn-to-ethanol conversion rate have been significantly improved and the cost has been greatly reduced in USA through technological innovation. In addition to the progress of industrial production technology and the optimization of the corresponding raw material crops, technological progress also includes the research and development on the new generation fuel ethanol such as biomass ethanol production technology. It is foreseeable that China will experience cost reduction brought about by these technological progress in the future.[4]

[4]Energy security through science and technology, source: Status quo and prospects of China's ethanol fuel industry—white paper of industrial research, http://www.cheminfo.gov.cn/UI/Information/Show.aspx?xh=123&tblName=focus.

The development of clean, efficient, safe and reliable nuclear fission energy is a strategic choice to solve the future energy supply and guarantee the sustainable development of China's economy and society. However, the issue of the utilization efficiency of nuclear fuel and the safe disposal of spent fuel must be addressed, which is a common challenge facing the international nuclear circle. In June 2017, the Institute of Modern Physics, Chinese Academy of Sciences proposed the new accelerator-driven advanced nuclear energy system, which can increase the utilization rate of uranium from "less than 1%" to "more than 95%" and shorten the radiation life from hundreds of thousands of years to about 500 years. These have laid the groundwork for exploring a more efficient and safer nuclear fuel cycle system and is expected to make nuclear fission a sustainable, safe and clean strategic energy in many years to come.

On January 13, 2017, China National Energy Administration released the "Energy Technology Innovation the Thirteenth Five Plan", which proposed that in the "thirteenth five-year" period more efforts will be put on making breakthroughs on key technologies, materials and equipment in order to make great improvements in energy self-innovation capabilities, the international competitiveness of the energy industry and initially form energy technology innovation system. The plan said that in the "Twelfth Five-Year Plan" period, China's energy technology innovation capability and the level of equipment localization have significantly improved. Some areas have reached the international advanced level, but there is still need to follow the pace of upgrading the energy industry, focus on breaking the key technical bottlenecks, and provide technical support for the comprehensive construction of China's safe, green, low-carbon, economic and sustainable modern energy industry system. The plan also said that in the "thirteenth five-year period", energy technology revolution should be promoted in accordance with relevant government rules and regulations and more focus will be put on the development of clean and efficient fossil energy technology, new energy power system technology, safe and advanced nuclear power technology, strategic energy technology and energy base materials technology.

2 Research on Energy Science and Technology in Russia

2.1 Status Quo

Russia, as the world's largest country in terms of area, has a wealth of energy. Its natural gas proven reserves amounts to 48 trillion cubic meters, accounting for 35% of the world and ranking the first; oil reserves of 10.9 billion tons, accounting for 13% of the world's proven reserves; coal reserves of 201.6 billion tons, ranking the second. Although Russia boasts rich energy reserves, its energy structure is relatively simple, oil and gas being the main energy. Its exports are highly dependent on oil and gas resources. As the energy crisis continues to escalate, some areas in

Russia also face great challenges. Meanwhile, with the rise in the cost of energy exploration and sanctions imposed on Russia, it is imperative that Russia continue to expand its energy technology scope and utilize rich energy types. Therefore, in the new energy technology, Russia began to seek innovation and has made some breakthroughs in recent years.

According to the report of a Russian science and technology website on June 10, 2016, Moscow Topiyev Institute of Petrochemical Chemistry of Russian Academy of Sciences initiated the technology of using the associated gas released from oil extraction to produce environmentally friendly non-frozen fuel. Popov, a representative from the Institute pointed out that the development of Arctic has posed an increasing demand for non-frozen fuel. The Arctic near Russia has rich hydrocarbon resources, which can be used as raw materials to produce non-frozen fuel. Compared with diesel, this kind of dimethyl ether fuel is more environmentally friendly, which can increase the engine efficiency by 2%. Above all, it will not freeze at ultra-low temperature, hence more effective in Arctic. It can be used in diesel engines, gas turbines and boilers.

In terms of wind energy potential, Russia ranks the first in the world. The development of wind energy is ongoing, although under-developed in the country. On June 13, 2016 Xinhua News Agency quoted the Moscow Institute of Iron and Steel as saying that scientists from the institute developed a hybrid power generation device, which can convert solar and wind energy into electricity and can be widely used in a variety of climatic conditions. It is of great significance to provide power for those areas with poor traffic conditions. It is said that several parameters of the device are superior to those of foreign products. Compared to the existing wind power generation device, wind power and solar power can increase its generation efficiency by 15–20%. In addition, due to its simple internal structure, it is easy to be repaired when there is a fault. When the light and wind are sufficient, its power generation efficiency can reach 300–500 watts, which is the equivalent of up to 4 MW h of the annual power generation. Its theoretical operating life is expected to be no less than 20 years.

In short, in the field of energy technology, Russia has not made big progress although its efforts have never faded, which not only enriched its energy types, but also provided impetus to the future development of energy in Russia.

2.2 Strengths and Future Directions of Energy Science and Technology in Russia

In the field of energy technology, although Russia has achieved something in solar, wind and other energy types, the development and utilization of nuclear energy is its advantage. Based on the foundation of the Soviet Union, Russia has made great achievements in the development and utilization of nuclear energy. In 2012, Russia's nuclear energy exports totaled USD 66.5 billion. Its installed capacity of nuclear power amounted to 24 GWe, which is expected to reach 60 GWe by 2030; in 2016 Russian nuclear energy company (Rosatome) not only completed the construction of nuclear reactors in China and India, but also beat competitors from France and Japan in June and reached agreement with Finnish nuclear power company of building another nuclear power plant. Based on the strong advantages of the Soviet, Russia continues to invest and develop new technologies, build new reactors, improve its nuclear capabilities and maintain its leading edge of nuclear energy in nuclear development, which not only enriches its energy types, but enhances its comprehensive national strength.

For the future development of energy technology, since the outbreak of Ukrainian crisis, Russia has adopted corresponding counter measures to the sanctions imposed against it. Under the harsh international energy strategic environment, Russia introduced "2035 Russia Energy Strategy Draft" (Hereinafter referred to as "2035 Energy Strategy") in February 2014, which said that Russia will reduce its dependence on the energy economy, adjust energy structure, enhance energy and technological innovation, especially the cultivation of talents and expansion of Asia-Pacific market and strengthen energy infrastructure.

As per the "2035 Energy Strategy", Russia will enhance energy and technological innovation and improve energy infrastructure. In terms of technological innovation, Russia has accelerated the cycle of "basic research, practical research, research and development, test product production, mature product processing". At the same time, it will gradually reduce the dependence on foreign energy technologies and equipment, allow only the introduction of certain key technologies and form key technology alliance with foreign leading companies within the scope of national policy. In terms of infrastructure construction, Russia will rely on the balanced development of energy and transport infrastructure to promote the establishment of a new model of energy space development and overcome the "bottleneck" of energy infrastructure development and focus on the construction and improvement of energy infrastructure in East Siberia and the Far East.

On the external side, Russia should open up Asia-Pacific market and achieve diversification of energy exports. "2035 Energy Strategy" believes that there is limited growth in the energy consumption of Europe and the Asia-Pacific region will embrace growth of energy consumption in the future. Therefore, it should increase exports, especially high-quality petroleum products to the Asia-Pacific market. The country should solve tensions with traditional energy consumers in Europe without compromising its national interests and strengthen cooperation with Asian consumers through dialogue, especially to sign gas supply contract with China. In the process, Russia should also give play to its advantages in nuclear energy, attract more foreign cooperation and cooperate on solar, wind and other new energies. As Russia's international talents are scarce, in the whole process it should focus on the cultivation and exchanges of international talents.

3 Research on Energy Science and Technology in Brazil

3.1 Overview

Following the outbreak of the oil crisis in the 1970s, Brazil formulated energy diversification and new energy research & development strategies, with the energy sector becoming a top priority. On the one hand, it intensified the exploration and exploitation of oil, especially the seabed oil; it made full use of abundant water resources and put more efforts on developing hydropower (for example, the construction of Itaipu Binacional and Tucurui Hydro-power Station in the mid-1970s); it developed agricultural energy resources in the country and implemented "national bio-diesel production and use program"; on the other hand, it had put the development of new alternative energy on the agenda, for example the formulation of "National Alcohol Project" in 1975, which carried out tests to extract alcohol from sugarcane, cassava and wood to use as fuel.

Since the 1990s, Brazil has been committed to the diversified development of energy and strive to make science and technology play a greater role in the field of energy. According to figures from the Brazilian authorities, Brazil's energy production structure in 2004 was as follows: 39.1% for oil, 14.4% for hydro-power, 13.8% for alcohol, 13.2% for wood and charcoal, 8.9% for natural gas, 6.7% for coal and 1.5% for nuclear energy.

Historically, Brazil had been an "oil-poor country". After the oil crisis in the 1970s, Brazil began to launch the energy diversification strategy in which it strove to apply science and technology in the field of energy and make it play a greater role. As there is less oil reserve in the land of Brazil, it began to pay attention to the development of offshore oil fields and increased investment on exploring and developing offshore oil fields. In terms of offshore oil field development, Brazil insisted on independent research and development and has made great achievements. In the following decades, Brazil gradually mastered the advanced deep-sea exploration technology and used advanced oil extraction technology to speed up the pace of offshore oil extraction. For example, it discovered oil fields in Camp stead Basin in Rio de Janeiro in 1974, the Marlim Oilfield in 1985 and the Baracyda & caratinga field in 1987.

From 1986 to 1991 in particular, the Brazilian Oil Company implemented "deep-water oilfield mining innovation and development technology" and increased research investment in the deep sea oil field, which ultimately resulted in low-budget, high-safety and high-output production. By the end of 2000, the vast majority of Brazilian oil production base were in the sea, with 13 large fixed offshore drills and 21 large floating drilling platforms. In 2005, the Brazilian State Oil Company discovered high quality crude oil in 1332-m deep underwater 160 km north of Rio de Janeiro. By 2006, Brazil's average daily output of oil reached 1.91 million barrels. The annual average growth of offshore oil production maintains at more than 10%, making it the third largest oil producer in Latin America.

3.2 Future Directions

On the basis of consolidating traditional energy, Brazil treated clean energy as its own energy development strategy and put forward energy development plan stretching as far as 2030. Brazil plans to use clean energy as the main energy for industrial and civilian use that can replace gasoline. Clean energy includes electricity, wind, nuclear, hydro and bio energy, among which bio-energy is the strategic focus.

Brazil has abundant water resources, which ranks the fourth in the world. Above all, most of rivers in Brazil are plateau rivers, laying foundation for it to utilize water for power generation. The potential of Brazil's Amazon River can be up to 105 million kilowatts, and now only a small part has been developed. The Itaipu Hydro-power Station co-developed by Brazil and Paraguay is the world's second largest hydro-power station just following Three Gorges Hydro-power Station in China. Therefore, in the long-term, the hydro-power stations in Brazil will play a greater role with the advent of high-tech era.

Brazil boasted a proven uranium reserve of 215,300 tons before 1979, at the time ranking the fifth in the world. On the technical side, Brazil is the only country in the southern hemisphere that has the capacity to independently develop uranium enrichment technique and that enjoys competitiveness in this area. Brazil has a huge nuclear energy potential which is of great help to ease the country's energy shortages. In the 1980s, Brazil achieved enriched uranium technology. "2004–2008 Brazilian nuclear energy development plan" mentioned that Brazil would achieve full self-sufficiency of nuclear fuel required by the domestic nuclear industry in three to five years.

Since the launching of the ethanol development program in the 1970s, Brazil has made great progress in this regard. In 1975, the Brazilian government promulgated the "National Alcohol Project", allowing the addition of a certain percentage of alcohol in gasoline to produce alcohol gasoline; by mid-1980s, the use of ethanol in Brazil reached its peak, with the number of alcohol vehicles accounting for 94.4% of the total vehicles; to the late 1980s and early 1990s, due to the fall in the international oil prices, the high cost of producing ethanol using sugar cane, a lot of ethanol production plants closed or switched to produce sugar. The fuel ethanol production was depressed in this period. But Brazil has not given up the plan. In 1993, the Brazilian government promulgated that 20–25% of anhydrous alcohol should be added to the petrol in all service stations. At the same time, Brazil has launched a "flexible fuel" car program. In 2002, Brazil's flexible fuel vehicle sales rebounded, which in turn greatly contributed to the production of ethanol. In October 2004, the world's first alcohol fuel aircraft was tested success in Sao Paulo. Now, Brazil is the only country in the world that does not supply pure gasoline and that achieves the most success in the development and utilization of ethanol fuel.

Besides ethanol fuel, Brazil also actively developed bio-diesel technology. According to the head of the Renewable Energy Program of the Brazilian Institute of Agriculture and Animal Husbandry, Brazil has sufficient raw material to produce bio-diesel, with 2 million hectares in the land of the northeastern region suitable for planting castor. The annual output of castor is expected to reach 2 million tons in the next few years, which can produce bio-diesel of 112 million liters and can create 100,000 jobs. Brazil already has all the conditions for becoming the world's largest bio-diesel producer.

In December 2004, the Brazilian government proposed the "National Bio-diesel Production and Use Program". From January 1, 2008, Brazil began to implement the mandatory requirement to add 2% of bio-diesel to the diesel fuel. On July 1, 2009, the proportion was increased to 3%; the plan of increasing the proportion to 5% was implemented in 2010 ahead of schedule. This has greatly increased the sales potential of bio-diesel in Brazil, with many large thermal power plants, railways, ferry buses, buses and truck companies beginning to use bio-diesel to reduce emissions. At present, Brazil has greatly enjoyed the benefits of technological development to energy. With the advance of global governance and increasing attention on climate change, the development of low-carbon and green economy has become a trend. The development of green energy will be inevitable. It will be the strategic focus for Brazil to conform to the changes in the historical trend and develop clean energy.

4 Research Energy Science and Technology in India

4.1 Status Quo

As our neighbor, India is similar to China in many aspects. With large area and a huge population second only to China, the country's energy is mainly based on coal and oil. The population, GDP and development speed has rendered India one of the world's leading energy consuming powers.

Over the past 10 years, the Indian economy has experienced rapid growth and growing demand for energy. At present, more than 73% of India's oil depends on imports, with the annual cost totaling $21 billion and accounting for about 4% of its GDP. According to a forecast by Indian authorities, 90% of oil and gas need to be imported by 2030. For India to solve the huge gap between the energy demand and supply, it is not an issue concerning the selection of which kind of energy and how to develop it in a big way, but an issue concerning how to use modern science and technology to provide impetus for the energy supply. The development of unconventional and renewable energy is the way forward.

India has been working to promote energy diversification to ensure energy security. One of the key initiatives is to search for alternative energy through encouraging large-scale development and utilization of renewable energy such as solar, wind energy and hydro-power. Next comes the development of nuclear power. India plans to increase the nuclear power generation from the current 3500–60,000 MW by 2030.

In August 2006, Indian Planning Commission organized experts to draft a 182-page "Energy Integrated Policy Report" (hereinafter referred to as the "Report"), which will be used as guidelines to develop energy policies for India's "Eleventh Five-Year Plan (2007–2012)". The basic objective of the report is to foster a low-cost and efficient energy market system in India and focus on expanding energy sources and improving energy efficiency. Some practices in the report which involve technology include the achievement of sufficient and stable supply of strategic energy and energy security, the development of non-conventional and renewable energy, diversification of energy sources; enhanced demand-side management; increased energy and technology investment and the adoption of energy technology route.

In recent years, India has been committed to the development of renewable energy in view of the high energy consumption and the wide application of technology in the energy sector. These renewable energies include fuel-wood cultivation, bio-gas digester construction, solar thermal energy, solar hot water, photovoltaic solar energy, bio-diesel and bio-alcohol; despite great potential, India's energy technology development still face many problems. According to the report, there will not be great changes in India's energy structure in the next 25 years. Coal and oil will continue to serve as the strategic energy, accounting for some 80%. The percentage of nuclear power and natural is expected to rise, with the nuclear power accounting for 1.53% of the country's total power generation. But whether the percentage of nuclear power will rise sharply or not will be closely related to the nuclear agreement between India and USA.

Hydro-power plays a complementary role in India's energy sector, the main task being bridging the supply gap during peak hours. Bio-energy (mainly bio-diesel) and other renewable energy will not become India's strategic energy in the future, but these energies can benefit decentralized users. India has a clear idea about the choice of technological route in energy research and development. The focus is to improve the efficiency of coal utilization technology, coal pit gasification solar power, bio-diesel and alcohol fuel. In the long run, renewable energy such as solar energy is also important for India to achieve energy independence.

4.2 Future Directions

Since the 2000s, India has begun to attach great importance to the positive role of science and technology in the field of energy, adjust the energy structure and strive to increase the percentage of solar energy, wind energy, bio-energy and other low-carbon renewable energy and nuclear energy, hydrogen and other clean and efficient energy in its energy composition in order to establish a green energy supply system that supports sustainable economic and social development.

India's top priority is to adopt energy technology route through formulating proper energy policies. India will encourage the development of energy technologies that are close to commercialization and have a clear time schedule. The report clarifies the new energy technology route to improve energy production and utilization efficiency and ultimately achieve energy security and independence. The report points out that the country should encourage the development of solar technology (solar thermal technology and photovoltaic solar technology), bio-fuel technology (bio-diesel, bio-ethanol, bio-material planting technology, charcoal gasification technology and community bio-gas digesters construction, etc.), comprehensive utilization of nuclear energy technology, hybrid fuel vehicle technology, high-energy battery technology and gas-water compound technology.

As the new energy industry is the focus of competition between countries and the strategic point of international competition in the future, India attaches great importance to the strategies of new energy development and has reaped great progress in this regard.

Firstly, India will intensify the development of solar energy and expand the solar photovoltaic industry. In view of the imbalanced domestic energy structure, the development of solar energy is a good choice. In the process, it should make full use of modern science and technology and make the development of solar photovoltaic technology and industry a top priority. India has rich solar power, with about 250–300 days of light per year and 4–7 kW-h of heat per square meter of land per day. Therefore, India has a natural advantage in solar energy utilization.

Secondly, the country will tap into the potential of wind power and strive to be a wind energy powerhouse. In recent years, India's wind energy utilization is moving forward. The Indian government has introduced a raft of initiatives to encourage its development, such as no tax for wind power sales in the first five years; the establishment of wind power business park; provision of various preferential services to private investment; to speed up the transfer of wind power technology through the establishment of wind power technology center; duty-free import of special parts and components for ten wind power equipment; special tax policies. In addition, it also developed special preferential policies and development strategies to support the use of wind energy in rural areas. The Indian government has put in place wind energy utilization and commercialization plan and introduced preferential policies for developing wind energy. India's wind power will embrace a bright future.

Thirdly, the country should focus on nuclear fusion power generation technology and engage in technical cooperation with other countries. From various parameters, nuclear energy is the only efficient, clean energy that can achieve large-scale stable supply and are relatively cost-effective. India is the world's second largest nuclear fusion research developing country second only to China. The Indian government believes that nuclear fusion will become a clean energy to meet its future surge in electricity demand and has approved plans to participate in the construction of advanced nuclear fusion reactors. It will provide USD 620 million in support to the International Thermonuclear Reactor (ITER) project Cooperation. Besides, it has signed nuclear energy for civil use agreement with USA, the world's nuclear technology power in order to gain access to nuclear fuel and technology.

Fourthly, it ought to fully develop bio-energy technology. The launching of such projects as bio-gas utilization, bio-oil development, bio-power generation has played a significant role in promoting the development of new energy in rural areas of India. In the past 10 years, India has installed about 2.5 million bio-gas production plants, of which the annual heat production is the equivalent to that of burning about ten million tons of wood. The plants can also produce about 50 million tons of condensate organic fertilizer per year. India's huge population and relatively backward infrastructure mean great potential in the energy and technology sector.

5 Research on Energy Science and Technology in South Africa

5.1 *Status Quo*

South Africa, located in the southern tip of the African continent, is a middle-income developing country, but also the most economically developed country in Africa. Compared with other African countries, South Africa has stronger energy industry foundation, and its technology is more advanced. South Africa's electricity sector is more developed; its power generation accounts for 2/3 of Africa's total; but about 92% of South Africa's electricity comes from thermal power. ESKOM, a state-owned South African power company, is the world's top ten power generation company and 11th largest electricity sales company. It has the world's largest dry-cooled power station, generating 95% of the electricity used in South Africa and 60% of the electricity used in Africa. The Koeberg nuclear power plant, the only nuclear power plant on the African continent, is built near Cape Town with a generating capacity of 1.8 million kilowatts. In addition, South Africa's SASOL is a world-leading company in commercializing synthetic fuels from coal and natural gas. The company produces about 1/4 of South Africa's total liquid fuel supply. In recent years, however, due to power production, management lags and other reasons, nation-wide power shortage is getting even more serious. South Africa is rich in coal resources and coal export constitutes a major source of foreign exchange revenue. However, due to domestic factors like policy instability and coal industry slump, thermal power generation is also facing a huge crisis. ESKOM has said that the company is facing a serious problem of coal supply strain, and needs new coal resources as reserves.

At the same time, facing the global environmental governance issues and under the new energy initiative, South Africa has begun to develop and utilize its own new energy. Thanks to the excellent location, South Africa's inland solar energy and offshore wind energy resources have the potential for development and value creation, especially solar energy; South Africa is one of the countries with the richest solar energy. In March 2009, the South African National Energy Management Committee (NERSA) developed a mandatory renewable energy feed-in tariff; in

2010, the draft Integrated Resource Plan (IRP) 2010–2030 was enacted. In 2011, the tender for new energy projects was launched; each year a new energy project tender is completed according to the capacity as stipulated in the draft. Currently, the implementation is sound.

Despite a series of new energy measures, the electricity market in South Africa is still facing supply shortage. There are three main reasons: (1) South Africa's domestic policies are unstable. During development process, time-consuming decision-making and lack of attention are common; (2) Although South Africa is the most developed country in Africa, it is still a developing country with insufficient economic growth. South Africa's infrastructure, such as roads and machinery are backward and aging and are not updated in time, which also greatly hinders the development of energy technology in South Africa; (3) From the perspective of South Africa's energy technology itself, South Africa's science and technology is still in its infancy in many ways. Science and technology underdevelopment and the shortage of professionals are the main reasons why South Africa's energy technology is underdeveloped; they are also why South Africa's energy condition is not properly improved.

5.2 Future Directions

South Africa, as the most developed country in Africa, has advantages over other African countries in developing energy technology. South Africa enjoys supports from domestic economy and advantages to attract foreign funds and technology. In the face of severe energy situation, thermal power relying on coal has clearly lost impetus; so on the future energy development path, South Africa should focus on developing and utilizing new energy.

Firstly, to develop and utilize new energy, South Africa must improve its infrastructure. The backwardness of infrastructure in South Africa has seriously hampered its new energy development; that's why it is necessary to improve infrastructure. In the process of infrastructure construction, South Africa can not only rely on its own human and material resources, but also can turn to other countries. China's "Belt and Road" initiative, seen by many South African scholars as the gospel of South Africa, will be very helpful to infrastructure construction in South Africa. At the same time, South Africa's own rich natural resources and human resources will attract more Chinese enterprises to invest in its construction, which will greatly speed up infrastructure construction in South Africa, so as to provide the basis for developing its energy technology.

Secondly, in the field of energy technology, South Africa must strengthen the development of qualified professionals. Professionals, especially internationalized professionals, are the most important factor in energy technology development. South Africa is lacking in this regard, and should therefore support the development of domestic professionals in relevant fields, encouraging them to study abroad. South Africa should also constantly introduce foreign professionals to work on energy technology development in South Africa. In recent years, in order to attract world-class talents, retain domestic talents, and lead the development of technological innovation, South Africa has launched the "centre of excellence", the South African Research Chairs Initiative (SARChI) and a series of personnel training programs.

Thirdly, South Africa should continue to develop new energy sources including the above-mentioned inland solar energy and offshore wind energy, give full play to its geographical advantages, and actively develop new energy to enrich its energy mix. In this way, South Africa can mitigate and even avoid the energy crisis. To this end, South Africa has taken a series of new energy development measures, among which the "new IRP plan" is of representative significance. Under the new plan, South Africa's total power generation capacity is expected to reach 89,532 MW by 2030, of which coal power generation accounts for 45.9%, renewable power generation 21%, nuclear power 12.7%, open cycle gas turbine (OCGT) 8.2%, combined cycle gas turbine (CCGT) 2.6%, hydro-electric power 5.3%, and pumped-storage hydropower 3.3%. If the plan is to be implemented successfully, it will serve the purpose of providing South Africa with electricity security to ensure rapid economic growth.

Finally, in addition to developing its naturally advantageous solar energy and hydro-energy etc., South Africa can also turn to nuclear energy. Nuclear energy is efficient and clean with steady large-scale supply and cost advantage, and should therefore be the focus of South Africa's future development. And South Africa is currently the only country with a nuclear power plant on the African continent. As early as 2014, South Africa shifted its energy policy focus to developing nuclear power and planned to establish fuel cycle capacity in its own country; it has signed an intergovernmental agreement on nuclear energy and industrial strategic partnership with Russia. In recent years, South Africa and China have also signed similar nuclear agreements. Since a comprehensive partnership was established between China and Africa in December 2015, China and South Africa have cooperated in wind energy and solar energy, and there is also a broad space for development in nuclear energy. In the recent tender of South African nuclear power plants, China, South Korea, Russia, France and Japan stood out. China has the world's largest nuclear power capacity, and cooperation between China and Africa will promote nuclear energy development in South Africa. The construction of nuclear power plants, and the development of nuclear energy technology, will certainly optimize the South African energy mix, and ease the power shortage crisis in South Africa. China's nuclear power technology export to South Africa is also just around the corner.

6 Summary

BRICS countries are all emerging developing countries, so in energy technology cooperation, an open, coordinated and sustainable global operating system must be built.

According to data from the International Energy Agency (IEA), in 2016, the world's new renewable energy met the majority of new demand for electricity; applications of battery-technology-based new energy storage technology increased by over 50%; energy technology innovation has played an increasingly prominent role in energy security and economic restructuring. Among them, the production and use of fuel ethanol have contributed to countries' implementation of alternative energy strategies to achieve the "20 in 10" target to reduce the dependence of imported oil and GHGs emissions.

In the 21st century, when the world is being beset by the struggle for oil and gas resources, the "new energy era", marked by the peaceful use of nuclear energy, has come quietly. A country's technology for peaceful use of nuclear energy is an important indicator of its comprehensive technological strength. In the late 1980s, the International Thermonuclear Experimental Reactor (ITER) project was born. In May 2006, the signing of the ITER project joint implementation agreement marked the entry of international nuclear fusion into the experimental reactor research stage, and China has become an important member of the project. Once the nuclear fusion achieves a technological breakthrough in a complete sense and is commercialized, the global energy crisis will be expected to be completely resolved. The peaceful-use-of-nuclear-energy industry plays an important role in each country's national economic development, national defense and people's livelihood.

From the "fuel wood era" to "coal age", "oil and gas times" and even "new energy era", the international status of the BRICS countries represented by China has undergone many changes. Experience and lessons can be drawn as follows: the backwardness of the technology with energy technology being a precursor will inevitably lead to the decline of national strength and even the loss of sovereignty; national sovereignty is the precondition for advancement of science and technology and its application in society; future competition among countries lies in the competition of comprehensive national strength with energy technology and new alternative energy being the core.

Chapter 12
Study on Agricultural Technology in BRICS

Xiangyu Guo, Dan Wang and Xinli Zhao

1 Introduction

BRICS countries are endowed with rich agricultural resources and boast a large agricultural population. They feature prominently in global agricultural area. From 2010 to 2016, their gross farm production increased from 1.4 trillion dollars to 2.3 trillion dollars with their shares in global total climbing from 47 to 57%. Their grain production accounted for over 40% of the global total. The four countries use less than 30% of the national land to raise 43% of the global population. Agricultural trade within BRICS countries has grown rapidly. China's import and export of farm produce to other BRICS countries have increased by 50 and 23%, exceeding the global average of 14%. In recent years, agricultural modernization drive in BRICS countries has sped up with elevated agricultural level. However, global warming, frequent natural disasters, land and water shortage and increased global population have put pressure on the supply and demand of farm produce. Facing all this, BRICS agriculture is still facing the risk of declined grain producing capacity and unguaranteed food security.

Agricultural technology is integrated into the three factors of productivity and makes the three more efficient in the sustainable development of productivity in rural areas. Therefore, agricultural technology innovation is an important factor in promoting modern agricultural development. To guarantee national food security, enhance agricultural efficiency, reduce pressure on environmental resources and promote sustainable agricultural development, countries around the world should make full use of the role technological innovation plays in agricultural modernization

X. Guo · X. Zhao (✉)
China Science and Technology Exchange Center, Beijing, China
e-mail: Zhaoxl@cestc.org.cn

D. Wang
Northeast Agricultural University, Harbin, China

X. Zhao et al. (eds.), *BRICS Innovative Competitiveness Report 2017*,
Research Series on the Chinese Dream and China's Development Path,
https://doi.org/10.1007/978-981-10-8078-4_12

drive. To meet the needs of the modernization drive and potential technological revolution, Northeast Agricultural University and China Science and Technology Exchange Center jointly carried out research in global agricultural innovative capacity. Based on related research findings, this research has followed closely cutting-edge research dynamics in technological economics, agricultural economics, management, econometrics and statistics. It has analyzed development level, changing features, internal factors and trend of agricultural technology innovation capacity of major countries from 2001 to 2016. The research report focuses on such development level of BRICS countries. Based on the national agricultural technology innovation system, evaluation indicators system and mathematics models set up by the research team, the report analyzes and evaluates the agricultural technology innovation capacity of BRICS countries from 2001 to 2016 in a comprehensive, in-depth and scientific way, which reveals the features and differences among countries, identifies their comparative advantages and weak points, tracks their development trajectory and ways for improvement and provides valuable theories and solutions for countries to enhance their agricultural technology innovation.

2 Evaluation of Agricultural Technology Innovation Capacity in BRICS Countries

To understand fully the status of BRICS countries in global agriculture, this report takes G20 as reference and it is based on the evaluation indicators system and mathematics models (for details please see Appendix A). It evaluates and analyzes the development of agricultural technology innovation capacity of BRICS countries from 2001 to 2016 and compares their differences. Grain supply of G20 accounts for 80% of the global total. G20 countries include strong agricultural countries like the U.S. which believes in science to enhance the quality of agricultural modernization and also developing countries which are accelerating its efforts for agricultural modernization.

2.1 Overall Report on Evaluation

2.1.1 Evaluation of Agricultural Technology Innovation Capacity

From Table 1, in overall ranking, from 2001 to 2016, Brazil and China have ranked higher, India has remained the same and Russia and South Africa have ranked lower. In 2016, China (10th) is the only one in the second tier. Brazil (11th), South Africa (13th) and Russia (14th) were in the third tier. India (17th) was in the fourth tier.

In terms of overall score, in 2001 BRICS average score was 30.12, 12.75 less than G20 average of 42.87. In 2016 BRICS average score was 33.02, 10.76 less

Table 1 2001–2016 evaluation of agricultural technology innovation capacity of BRICS

Item country	Overall change		2001		2006		2011		2016	
	Score	Ranking	Score	Ranking	Score	Ranking	Score	Ranking	Score	Ranking
Brazil	5.78	2	29.45	13	33.20	11	37.57	10	35.23	11
China	5.95	1	30.30	11	32.24	13	33.61	13	36.25	10
India	2.96	0	24.85	17	26.48	17	25.47	17	27.81	17
Russia	1.61	−2	30.02	12	28.20	15	25.38	18	31.63	14
South Africa	−1.85	−3	36.00	10	36.72	10	36.74	11	34.16	13
Highest	5.95	–	36.00	–	36.72	–	37.57	–	36.25	–
Lowest	−1.85	–	24.85	–	26.48	–	25.38	–	27.81	–
Average	2.89	–	30.12	–	31.37	–	31.75	–	33.02	–
G20 highest	6.19	–	77.04	–	73.99	–	72.20	–	73.36	–
G20 lowest	−4.52	–	20.14	–	21.94	–	23.04	–	25.41	–
G20 average	0.91	–	42.87	–	43.85	–	43.33	–	43.78	–

than G20 average of 43.78. In 16 years, the average gap between BRICS and G20 was reduced by 1.99. In terms of changes in overall score, China has increased the most with 5.95. Brazil has increased 5.78. India and Russia have increased by 2.96 and 1.61. South Africa has decreased by 1.85. In general, BRICS overall capacity has climbed by 2.89, overtaking G20's increase of 0.91.

2.1.2 Evaluation of Secondary Index of Agricultural Technology Innovation

(1) **Evaluation of Basic Capacity of Agricultural Technology Innovation**

From Table 2, in terms of overall ranking of basic capacity of agricultural technology innovation, from 2001 to 2016, Brazil, China and Russia have remained the same. India and South Africa have decreased by one place. In 2016, China (10th) was the only one in the second tier. India (12th), South Africa (13th), Russia (14th) and Brazil (15th) were in the third tier.

In terms of overall score, in 2001 BRICS average score was 20.33, 11.21 less than G20 average of 31.54. In 2016, BRICS average score was 25.44, 9.13 less than G20 average of 34.57. In 16 years, the average gap between BRICS and G20 was reduced by 2.08. In terms of changes in overall score, BRICS have all climbed up. China has increased the most by 12.44. Russia has increased by 5.74. Brazil, South Africa and India have increased by 3.52, 2.40 and 1.49. In general, BRICS overall basic capacity has climbed by 5.12, overtaking G20's increase of 3.03.

(2) **Evaluation of Environmental Capacity of Agricultural Technology Innovation**

From Table 3, in terms of overall ranking of environmental capacity of agricultural technology innovation, from 2001 to 2016 India and Russia have climbed higher. China has remained stable. South Africa and Brazil have dropped. In 2016, Russia (10th) was the only one in the second tier. South Africa (11th), China (12th) and India (13th) were in the third tier. Brazil (16th) was in the fourth tier.

In terms of overall score, in 2001 BRICS average score was 39.29, 10.69 less than G20 average of 49.98. In 2016, BRICS average score was 39.17, 11.13 less than G20 average of 50.30. In 16 years, the average gap between BRICS and G20 was increased by 0.44. In terms of changes in overall score, Russia has increased the most by 3.04. China has increased by 2.27. India has increased by 1.04. However, South Africa and Brazil have dropped by 2.11 and 4.82. In general, the score of BRICS has showed negative growth by 0.12 while that of G20 has increased by 0.32.

(3) **Evaluation of Production Capacity of Agricultural Technology Innovation**

From Table 4, in terms of overall ranking of production capacity of agricultural technology innovation, from 2001 to 2016 Brazil has climbed by five places. China and India have remained the same place. Russia and South Africa has dropped by 2

Table 2 2001–2016 evaluation of basic capacity of agricultural technology innovation of G20 countries

Item country	Overall change		2001		2006		2011		2016	
	Score	Ranking	Score	Ranking	Score	Ranking	Score	Ranking	Score	Ranking
Brazil	3.52	0	15.96	15	20.18	15	23.57	12	19.47	15
China	12.44	0	25.35	10	28.78	10	33.38	10	37.78	10
India	1.49	−1	22.61	11	24.22	12	22.23	14	24.10	12
Russia	5.74	0	16.87	14	20.55	14	21.49	16	22.60	14
South Africa	2.40	−1	20.86	12	24.87	11	22.62	13	23.27	13
Highest	12.44	–	25.35	–	28.78	–	33.38	–	37.78	–
Lowest	1.49	–	15.96	–	20.18	–	21.49	–	19.47	–
Average	5.12	–	20.33	–	23.72	–	24.66	–	25.44	–
G20 highest	12.44	–	76.04	–	75.26	–	72.95	–	73.74	–
G20 lowest	−3.90	–	9.30	–	10.27	–	12.08	–	14.03	–
G20 average	3.03	–	31.54	–	34.82	–	34.8	–	34.57	–

Table 3 2001–2016 evaluation of environmental capacity of agricultural technology innovation

Item country	Overall change		2001		2006		2011		2016	
	Score	Ranking	Score	Ranking	Score	Ranking	Score	Ranking	Score	Ranking
Brazil	−4.82	−3	36.89	13	36.93	13	38.64	14	32.07	16
China	2.27	0	37.85	12	37.69	12	39.54	13	40.12	12
India	1.04	2	34.75	15	34.82	14	30.47	15	35.79	13
Russia	3.04	1	41.84	11	39.54	11	39.70	12	44.88	10
South Africa	−2.11	−1	45.11	10	43.83	10	46.01	10	43.00	11
Highest	3.04	–	45.11	–	43.83	–	46.01	–	44.88	–
Lowest	−4.82	–	34.75	–	34.82	–	30.47	–	32.07	–
Average	−0.12	–	39.29	–	38.56	–	38.87	–	39.17	–
G20 highest	7.51	–	82.94	–	81.63	–	78.01	–	83.08	–
G20 lowest	−4.82	–	25.49	–	25.74	–	26.02	–	21.85	–
G20 average	0.32	–	49.98	–	49.91	–	50.25	–	50.30	–

Table 4 2001–2016 evaluation of production capacity of agricultural technology innovation in G20 countries

Item countries	Overall change		2001		2006		2011		2016	
	Score	Ranking	Score	Ranking	Score	Ranking	Score	Ranking	Score	Ranking
Brazil	18.65	5	35.50	13	42.49	11	50.50	9	54.15	8
China	3.14	0	27.70	16	30.25	15	27.91	15	30.83	16
India	6.36	0	17.19	19	20.38	19	23.69	17	23.55	19
Russia	−3.95	−2	31.34	15	24.51	16	14.95	19	27.39	17
South Africa	−5.84	−3	42.03	11	41.44	13	41.58	12	36.20	14
Highest	18.65	–	42.03	–	42.49	–	50.50	–	54.15	–
Lowest	−5.84	–	17.19	–	20.38	–	14.95	–	23.55	–
Average	3.67	–	30.75	–	31.81	–	31.73	–	34.42	–
G20 highest	18.65	–	82.8	–	84.71	–	68.74	–	66.11	–
G20 lowest	−18.62	–	17.19	–	20.38	–	14.95	–	23.55	–
G20 average	−0.62	–	47.09	–	46.83	–	44.94	–	46.47	–

and 3 places. In 2016, Brazil (8th) was the only one in the second tier. South Africa (14th) was in the third tier. China (16th), Russia (17th) and India (19th) were in the fourth tier.

In terms of overall score, in 2001 BRICS average score was 30.75, 16.34 less than G20 average. In 2016 BRICS average score was 34.42, 12.05 less than G20 average. In 16 years, average gap between BRICS and G20 was reduced by 4.29. In terms of change in overall score, Brazil has increased the most by 18.65. India and China have climbed by 6.36 and 3.14. South Africa and Russia have dropped by 5.84 and 3.95. The overall score of BRICS has showed a trend of increasing by 3.67 while that of G20 has fallen by 0.62.

In terms of ranking and its change in secondary index, ranking of BRICS countries remained in the middle and latter part. The only exception is Brazil. It ranked 13th in 2001 and 8th in 2016. The other BRICS countries have remained basically the same in 2001–2016. In general, ranking of BRICS overall capacity remained in the latter part with no obvious changes.

Note: (1) The countries in the report are placed in alphabetical order; (2) To show a country's level of agricultural technological capacity compared to G20 countries, 1st to 5th place is defined as first tier, 6th to 10th place as second tier, 11th to 15th as third tier and 16th to 19th as fourth tier. (3) Readers can visit the website of Northeast Agricultural University http://www.neau.edu.cn/info/1216/29169.htm to see detailed evaluation of agricultural technology innovation of G20 countries.

2.2 *Report on Evaluation Result*

2.2.1 Brazil's National Agricultural S&T Innovation Capability (NASTIC): Assessment and Analysis

This part analyzes in detail score changes of national agricultural technology innovation capacity of Brazil and its ranking in G20 countries in the 16 years from 2001 to 2016.

The ranking and score change of Brazil's national agricultural technology innovation capacity in G20 countries are shown in Fig. 1.

(1) In terms of overall ranking, in 2016 Brazil's national agricultural technology innovation capacity ranked 11th in G20 and in 2001 it ranked 13th. In general, the ranking has increased steadily during the evaluation period.
(2) In terms of score, in 2016 score for Brazil's agricultural technology innovation capacity was 35.23, 38.13 lower than the G20's highest and 8.58 lower than the average. Compared with 2001, it has increased by 5.78, 9.46 less than the score gap with the highest score in 2001 and 4.80 less than the score gap with the average score of G20 in 2001.

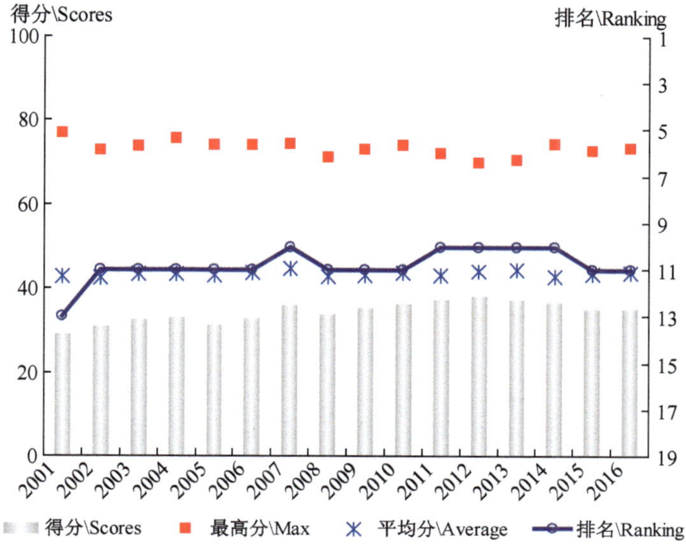

Fig. 1 2001–2016 score and ranking change of Brazil's national agricultural technology innovation capacity

(3) In terms of ranking change of secondary index, in the three indicators, one (environmental capacity for technological innovation) has dropped and one (production capacity for technological innovation) has climbed up (Fig. 2; Table 5).

(4) In terms of ranking of quaternary index, in 2016 Brazil had 7 strength indicators: ranking of agricultural discipline of its universities and research institutions, foreign direct investment in agriculture, density of pesticide use, bio-fuel production output, number of papers relating to agricultural technology, index number of agricultural production and net export of agricultural products; 9 advantage indicators: R&D level of agricultural enterprises, public education input, liberal and democratic atmosphere of scientific innovation, reasonable agricultural policies, support for agricultural credit, agricultural water use, citizens' agricultural patents, number of new plant varieties, annual agricultural growth rate; 20 middle indicators: GDP level, fiscal revenue level, proportion of irrigated area, extent of agricultural mechanization, popularity of information technology, number of agricultural researchers, farmers' educational level, health status, agricultural R&D input, cluster development status, international agricultural research cooperation, political stability, support for agricultural innovative technology, difficulty of starting business, market maturity, equality of rural education, density of fertilizer use, productivity of agricultural land, productivity of agricultural labor, Engel coefficient in rural area; 8 disadvantage indicators:

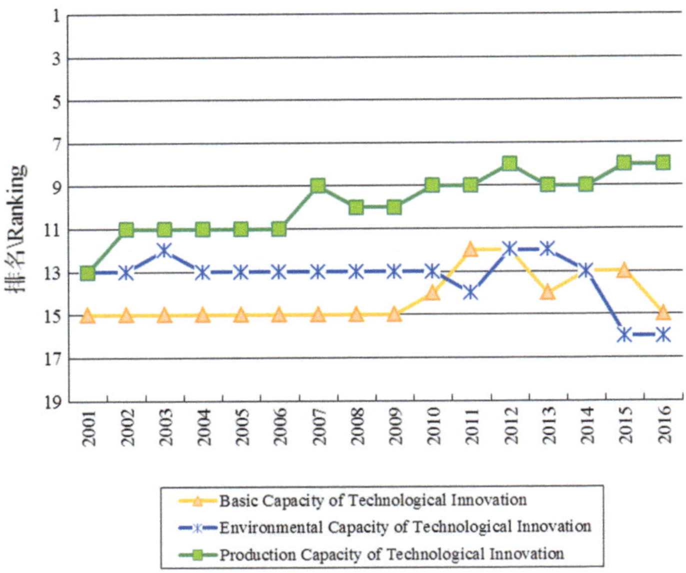

Fig. 2 2001–2016 ranking change of secondary index of Brazil's national agricultural technology innovation capacity

transportation infrastructure, industry-university-research cooperation, general public's curiosity, IPR protection, quality of education system, accessibility of technical training, use of agricultural energy and emission of agricultural greenhouse gas.

In terms of changes in index ranking, 15 of the 17 quaternary indicators have climbed up: ranking of agricultural discipline of its universities and research institutions, foreign direct investment in agriculture, liberal and democratic atmosphere of scientific innovation, difficulty of starting business, quality of education system, agricultural water use, bio-fuel production output, number of papers relating to agricultural technology, citizens' agricultural patents, number of new plant varieties, index number of agricultural production, productivity of agricultural land, productivity of agricultural labor, annual agricultural growth rate, net export of agricultural products; 12 indicators remained in the same place: GDP level, fiscal revenue level, R&D level of agricultural enterprises, number of agricultural researchers, health status, international agricultural research cooperation, support for agricultural credit, equality of rural education, density of fertilizer use, density of pesticide use, use of agricultural energy, Engel coefficient in rural area; 17 indicators have dropped in ranking: proportion of irrigated area, extent of agricultural mechanization, popularity of information technology, transportation infrastructure, farmers' educational level, public education input, agricultural R&D input, industry-university-research cooperation, cluster development status, general public's curiosity, political stability, IPR protection, reasonable agricultural policies,

Table 5 Score and ranking of Brazil's national agricultural technology innovation capacity

Item year	Basic capacity of technological innovation		Environmental capacity of technological innovation		Production capacity of technological innovation		Technological innovation capacity	
	Score	Ranking	Score	Ranking	Score	Ranking	Score	Ranking
2001	15.96	15	36.89	13	35.50	13	29.45	13
2002	16.44	15	36.86	13	40.85	11	31.38	11
2003	17.07	15	37.64	12	43.94	11	32.88	11
2004	16.93	15	36.12	13	47.44	11	33.50	11
2005	17.22	15	36.96	13	40.83	11	31.67	11
2006	20.18	15	36.93	13	42.49	11	33.20	11
2007	21.71	15	36.61	13	50.83	9	36.39	10
2008	19.05	15	35.27	13	48.24	10	34.19	11
2009	21.77	15	36.85	13	47.94	10	35.52	11
2010	22.20	14	37.87	13	49.48	9	36.52	11
2011	23.57	12	38.64	14	50.50	9	37.57	10
2012	23.73	12	39.85	12	52.22	8	38.60	10
2013	23.12	14	39.55	12	50.32	9	37.66	10
2014	23.51	13	35.33	13	51.84	9	36.89	10
2015	22.82	13	31.67	16	51.84	8	35.44	11
2016	19.47	15	32.07	16	54.15	8	35.23	11
Score change	3.52		−4.82		18.65		5.78	
Ranking change		0		−3		5		2
Advantage/disadvantage		Middle		Disadvantage		Advantage		Middle

support for agricultural innovative technology, market maturity, accessibility of technical training and emission of agricultural greenhouse gas.

2.2.2 China's National Agricultural S&T Innovation Capability (NASTIC): Assessment and Analysis

This part analyzes in detail score changes of national agricultural technology innovation capacity of China and its ranking in G20 countries in the 16 years from 2001 to 2016.

The ranking and score change of China's national agricultural technology innovation capacity in G20 countries are shown in Fig. 3.

(1) In terms of overall ranking, in 2016 China's national agricultural technology innovation capacity ranked 10th in G20 and in 2001 it ranked 11th. In general, the ranking has dropped and then climbed up during the evaluation period.

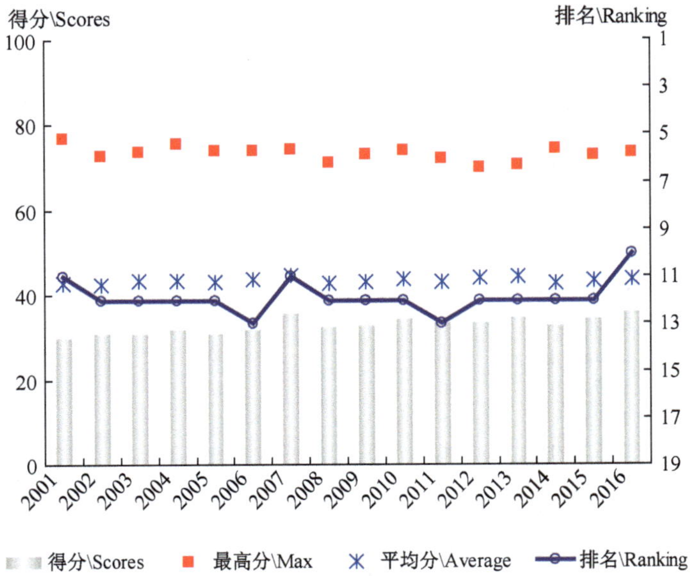

Fig. 3 2001–2016 score and ranking change of China's national agricultural technology innovation capacity

(2) In terms of score, in 2016 score for China's agricultural technology innovation capacity was 36.25, 37.12 less than the highest score of G20 and 7.56 less than the average score. Compared with 2001, score for China's agricultural technology innovation capacity has climbed by 5.95, 9.63 less than the score gap with the highest score in 2001 and 4.97 less than the score gap with the average score of G20 in 2001.
(3) In terms of ranking change of secondary index, three secondary indicators have remained stable (Fig. 4; Table 6).
(4) In terms of ranking of quaternary index, in 2016 China had 8 strength indicators: ranking of agricultural discipline of its universities and research institutions, foreign direct investment in agriculture, international agricultural research cooperation, general public's curiosity, reasonable agricultural policies, support for agricultural credit, index number of agricultural production, annual agricultural growth rate; 13 advantage indicators: GDP level, fiscal revenue level, irrigated area proportion, transportation infrastructure, R&D level of agricultural enterprises, public education input, industry-university-research cooperation, cluster development status, liberal and democratic atmosphere for scientific innovation, quality of education system, agricultural energy use, bio-fuel production, productivity of agricultural land; 15 middle indicators: extent of agricultural mechanization, number of agricultural researchers, farmers' educational level, health status, agricultural R&D input, political stability, IPR protection, support for agricultural technology innovation, difficulty of starting business,

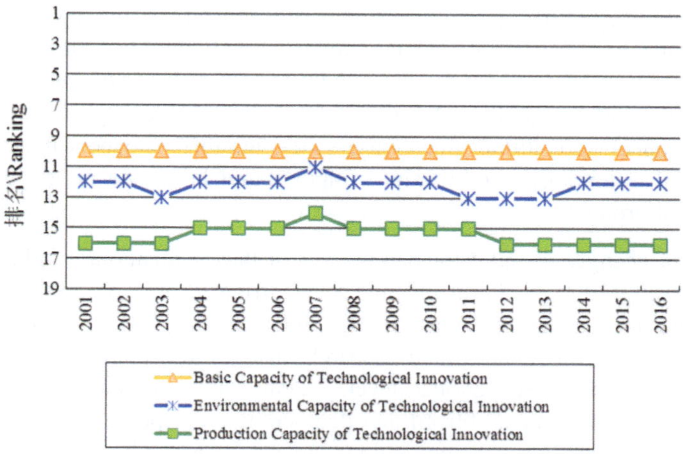

Fig. 4 2001–2016 ranking change of secondary index of China's national agricultural technology innovation capacity

Table 6 Score and ranking of China's national agricultural technology innovation capacity

Item year	Basic capacity of technological innovation		Environmental capacity of technological innovation		Production capacity of technological innovation		Technological innovation capacity	
	Score	Ranking	Score	Ranking	Score	Ranking	Score	Ranking
2001	25.35	10	37.85	12	27.70	16	30.30	11
2002	26.32	10	37.59	12	29.48	16	31.13	12
2003	26.97	10	37.08	13	29.92	16	31.32	12
2004	27.66	10	38.10	12	30.69	15	32.15	12
2005	28.69	10	38.48	12	26.57	15	31.25	12
2006	28.78	10	37.69	12	30.25	15	32.24	13
2007	30.65	10	41.90	11	35.56	14	36.04	11
2008	27.74	10	40.46	12	29.97	15	32.72	12
2009	31.34	10	40.90	12	26.93	15	33.06	12
2010	32.24	10	39.39	12	32.77	15	34.80	12
2011	33.38	10	39.54	13	27.91	15	33.61	13
2012	33.51	10	39.37	13	28.60	16	33.83	12
2013	34.00	10	39.27	13	31.97	16	35.08	12
2014	35.93	10	38.84	12	24.63	16	33.13	12
2015	37.23	10	38.16	12	28.53	16	34.64	12
2016	37.78	10	40.12	12	30.83	16	36.25	10
Score change	12.44		2.27		3.14		5.95	
Ranking change		0		0		0		1
Advantage/ disadvantage		Advantage		Middle		Disadvantage		Advantage

equality of rural education, accessibility of technical training, agricultural water use, citizens' agricultural patents, number of new plant varieties, Engel coefficient in rural area; 8 disadvantage indicators: popularity of information technology, market maturity, density of fertilizer use, density of pesticide use, emission of agricultural greenhouse gases, number of papers related to agricultural technology, productivity of agricultural labor and net export of agricultural products.

In terms of changes in indicator ranking, 17 of the 24 quaternary indicators have climbed up: GDP level, fiscal revenue level, irrigated area proportion, extent of agricultural mechanization, transportation infrastructure, ranking of agricultural discipline of its universities and research institutions, number of agricultural researchers, farmers' educational level, health status, agricultural R&D input, cluster development status, liberal and democratic atmosphere for scientific innovation, political stability, IPR protection, support for agricultural credit, quality of education system, equality of rural education, accessibility of technical training, agricultural energy use, number of new plant varieties, index number of agricultural production, productivity of agricultural land, productivity of agricultural labor, annual agricultural growth rate; 10 indicators have remained the same: popularity of information technology, R&D level of agricultural enterprises, public education input, foreign direct investment in agriculture, international agricultural research cooperation, general public's curiosity, agricultural water use, emission of agricultural greenhouse gases, citizens' agricultural patents, Engel coefficient in rural area; 10 indicators have dropped in ranking: industry-university-research cooperation, reasonable agricultural policies, support for agricultural scientific innovation, difficulty of starting business, market maturity, density of fertilizer use, density of pesticide use, bio-fuel production, number of papers related to agricultural technology and net export of agricultural products.

2.2.3 India's National Agricultural S&T Innovation Capability (NASTIC): Assessment and Analysis

This part analyzes in detail score changes of national agricultural technology innovation capacity of India and its ranking in G20 countries in the 16 years from 2001 to 2016.

The ranking and score changes of India's national agricultural technology innovation capacity in G20 countries are shown in Fig. 5.

(1) In terms of overall ranking, in 2016 India's national agricultural technology innovation capacity ranked 17th in G20, which was the same as in 2001. In general, it has remained stable during the evaluation period.
(2) In terms of score, in 2016 score for India's agricultural technology innovation capacity was 27.81, 45.55 less than the highest score and 16.00 less than the average score in G20. Compared with 2001, score for India's agricultural

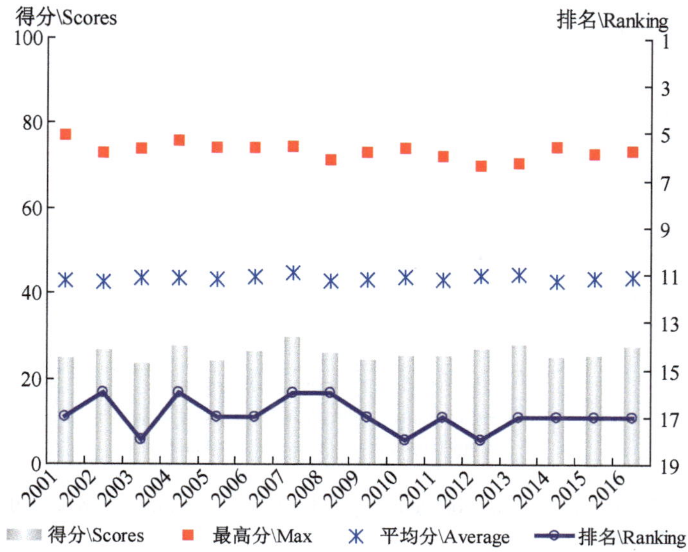

Fig. 5 2001–2016 score and ranking change of India's national agricultural technology innovation capacity

technology innovation capacity has climbed by 2.96, 6.64 less than the score gap with the highest score and 1.98 less than the score gap with the average score in G20.

(3) In terms of ranking change of secondary index, one of the three indicators (basic capacity of technological innovation) has dropped and one (environmental capacity of technological innovation) has climbed up (Fig. 6; Table 7).

(4) In terms of ranking of quaternary index, in 2016 India had 5 strength indicators: irrigated area proportion, international agricultural research cooperation, density of fertilizer use, bio-fuel production, index number of agricultural production; 12 advantage indicators: extent of agricultural mechanization, transportation infrastructure, ranking of agricultural discipline of its universities and research institutions, R&D level of agricultural enterprises, public education input, industry-university-research cooperation, cluster development status, IPR protection, reasonable agricultural policies, support for agricultural credit, quality of education system, net export of agricultural products; 6 middle indicators: agricultural foreign direct investment, liberal and democratic atmosphere for scientific innovation, general public's curiosity, accessibility of technical training, density of fertilizer use, annual agricultural growth rate; 21 disadvantage indicators: GDP level, fiscal revenue level, popularity of information technology, number of agricultural researchers, farmers' education level, health status, agricultural R&D input, political stability, support for agricultural scientific innovation, difficulty of starting business, market maturity, equality of rural education, agricultural water use, agricultural energy use, emission of agricultural greenhouse gas, number of papers related to

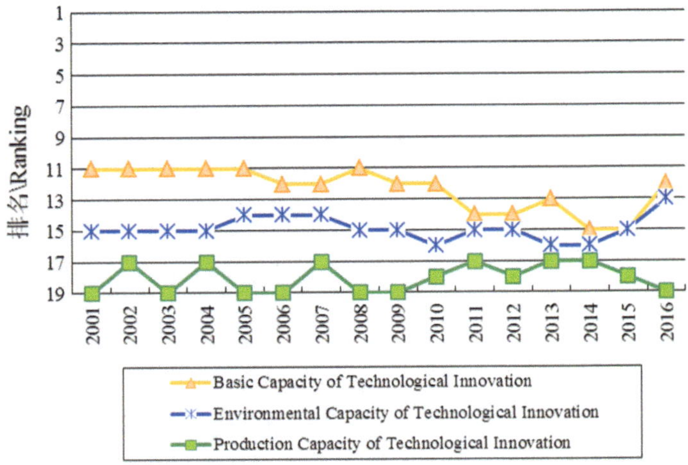

Fig. 6 2001–2016 ranking change of secondary index of India's national agricultural technology innovation capacity

Table 7 Score and ranking of India's national agricultural technology innovation capacity

Item year	Basic capacity of technological innovation		Environmental capacity of technological innovation		Production capacity of technological innovation		Technological innovation capacity	
	Score	Ranking	Score	Ranking	Score	Ranking	Score	Ranking
2001	22.61	11	34.75	15	17.19	19	24.85	17
2002	22.85	11	34.95	15	22.79	17	26.87	16
2003	24.28	11	34.15	15	12.77	19	23.73	18
2004	24.13	11	34.21	15	25.10	17	27.82	16
2005	23.43	11	34.81	14	14.51	19	24.25	17
2006	24.22	12	34.82	14	20.38	19	26.48	17
2007	25.36	12	34.12	14	30.54	17	30.01	16
2008	23.04	11	31.49	15	24.05	19	26.20	16
2009	22.78	12	32.38	15	18.42	19	24.52	17
2010	22.95	12	30.06	16	23.55	18	25.52	18
2011	22.23	14	30.47	15	23.69	17	25.47	17
2012	22.70	14	32.96	15	25.29	18	26.98	18
2013	23.61	13	33.31	16	26.96	17	27.96	17
2014	21.75	15	30.46	16	23.05	17	25.09	17
2015	21.45	15	31.71	15	23.45	18	25.54	17
2016	24.10	12	35.79	13	23.55	19	27.81	17
Score change	1.49		1.04		6.36		2.96	
Ranking change		−1		2		0		0
Advantage/disadvantage		Middle		Middle		Disadvantage		Disadvantage

agricultural technology, citizens' agricultural patents, number of new plant varieties, productivity of agricultural land, productivity of agricultural labor, Engel coefficient in rural area.

In terms of changes in indicator ranking, 12 of the 17 quaternary indicators have climbed up: GDP level, irrigated area proportion, extent of agricultural mechanization, transportation infrastructure, industry-university-research cooperation, general public's curiosity, reasonable agricultural policies, support for agricultural credit, equality of rural education, index number of agricultural production, annual agricultural growth rate, net export for agricultural products; 21 indicators have remained in ranking: fiscal revenue level, popularity of information technology, R&D level of agricultural enterprises, number of agricultural researchers, farmers' education level, health status, cluster development status, international agricultural research cooperation, liberal and democratic atmosphere for scientific innovation, political stability, IPR protection, support for agricultural scientific innovation, market maturity, density of fertilizer use, agricultural water use, agricultural energy use, emission of agricultural greenhouse gases, bio-fuel production, number of new plant varieties, productivity of agricultural land, Engel coefficient of rural area; 11 indicators have dropped in ranking: ranking of agricultural discipline of its universities and research institutions, public education input, agricultural R&D input, agricultural foreign direct investment, difficulty of starting business, quality of education system, accessibility of technical training, density of fertilizer use, number of papers related to agricultural technology, citizen's agricultural patents and productivity of agricultural labor.

2.2.4 Russia's National Agricultural S&T Innovation Capability (NASTIC): Assessment and Analysis

This report will analyze Russia's 2001–2016 NASTIC scores and Russia's ranking among the G20.

Russia's NASTIC scores and ranking among the G20 are as illustrated in Fig. 7.

(1) Overall ranking: In 2016, Russia ranked the 14th among the G20, 2 places down as compared with in 2001. 2011 witnessed the gravest drop. The ranking shows a downward trend with temporary fluctuation.
(2) Score: In 2016, Russia scored 31.63 points in NASTIC, 41.73 points lower than the G20 highest and 12.18 points lower than the G20 average. As compared with in 2001, Russia's score increased by 1.61 points; the gap with the G20 highest of the year narrowed by 5.29 points and with the G20 average, by 0.63 point.
(3) Ranking in terms of secondary indicators: Among 3 secondary indicators, Russia went down in one indicator (national agricultural science and technology innovation output capability, NASTIOC) and went up in one indicator (national agricultural science and technology innovation environment capability, NASTIEC) (Fig. 8; Table 8).

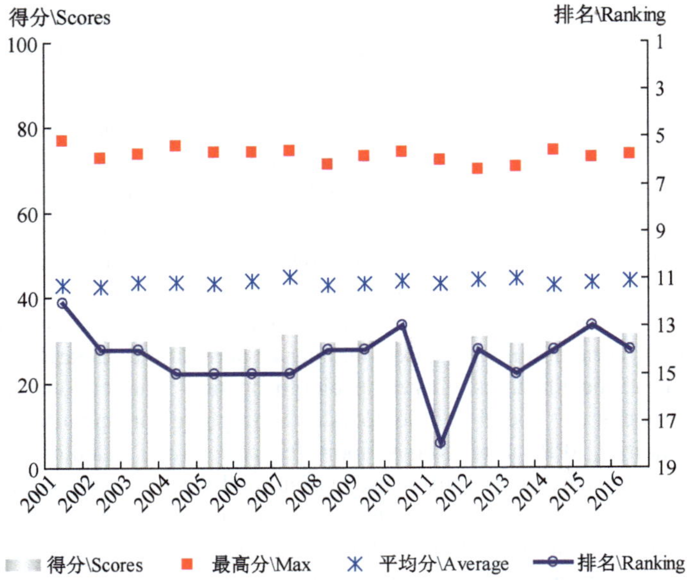

Fig. 7 Russia's NASTIC Scores and Ranking (2001–2016)

(4) Quaternary indicators (location): In 2016, Russia has 8 strong indicators: agricultural foreign direct investment, international cooperation in agricultural research, public curiosity, ease to start businesses, equity for rural education, intensity of fertilizer application, agricultural water consumption and new plant varieties cultivated by domestic residents. Russia has 7 advantageous indicators: penetration rate of ICT, ranking of agriculture-related disciplines for

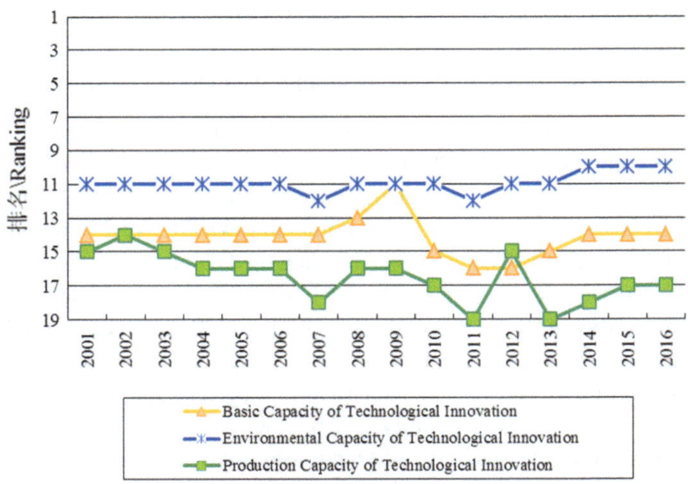

Fig. 8 Ranking of Secondary Indicators of Russia's NASTIC (2001–2016)

Table 8 Russia's NASTIC: score and ranking for secondary indicators

Item year	NASTIFC		NASTIEC		NASTIOC		NASTIC	
	Score	Ranking	Score	Ranking	Score	Ranking	Score	Ranking
2001	16.87	14	41.84	11	31.34	15	30.02	12
2002	17.44	14	39.57	11	32.93	14	29.98	14
2003	18.70	14	41.15	11	30.28	15	30.04	14
2004	18.98	14	40.45	11	26.75	16	28.72	15
2005	18.67	14	39.42	11	24.21	16	27.43	15
2006	20.55	14	39.54	11	24.51	16	28.20	15
2007	24.41	14	40.35	12	29.89	18	31.55	15
2008	21.37	13	41.67	11	25.71	16	29.58	14
2009	23.26	11	41.23	11	25.17	16	29.89	14
2010	21.75	15	40.01	11	27.58	17	29.78	13
2011	21.49	16	39.70	12	14.95	19	25.38	18
2012	21.22	16	40.63	11	30.77	15	30.87	14
2013	21.60	15	41.52	11	24.62	19	29.25	15
2014	21.91	14	44.20	10	22.58	18	29.56	14
2015	22.64	14	44.14	10	25.25	17	30.68	13
2016	22.60	14	44.88	10	27.39	17	31.63	14
Score changes	5.74		3.04		−3.95		1.61	
Ranking changes		0		1		−2		−2
(Dis)advantageous		Neutral		Advantageous		Disadvantage		Neutral

universities and research institutions, R&D for agriculture-related businesses; quantity of agricultural researchers; consumption of agriculture-related energy, agricultural production index and average annual growth of agriculture; Russia remains neutral in 17 indicators: GDP, fiscal revenue, proportion of irrigated area, educational attainment for farmers, investment in public education, investment in agricultural R&D, industry-university collaboration, free and democratic environment for S&T innovation, support for innovation in agricultural S&T, supply of agricultural credit, quality of education system, availability of technology training, intensity of agrochemical application, emission of agriculture-related GHGs, agricultural patents owned by national residents, labor productivity of agriculture and net export of agricultural products. Russia is at a disadvantageous position in 12 indicators: agricultural mechanization, transportation infrastructure, health, cluster development, political stability, IPR protection, appropriateness of agricultural policies, sophistication of market mechanism, production of biofuels, theses of agricultural S&T, productivity of agricultural land and Engel co-efficient in rural areas.

Quaternary indicators (ranking): Among the 17 indicators, 17 show an upward trend: GDP, fiscal revenue, penetration rate of ICT, health, investment in public education, investment in agricultural R&D, agricultural foreign direct investment, industry-university collaboration, cluster development, public curiosity, ease to start businesses, sophistication of market mechanism, availability of technology training, consumption of agriculture-related energy, emission of agriculture-related GHGs, agricultural production index, and productivity of agricultural land. 11 indicators remain unchanged in their ranking: R&D of agricultural businesses, education attainment for farmers, international cooperation in agricultural research, political stability, appropriateness of agricultural policy, equity for rural education, intercity of fertilizer application, agricultural water consumption, new plant varieties cultivated by domestic residents, labor productivity of agriculture and Engel-coefficient in rural areas. 16 indicators show a downward trend: proportion of irrigated area, agricultural mechanization, free and democratic environment for S&T innovation, IPR protection, support for innovation in agricultural S&T, supply of agricultural credit, quality of education system, intensity of agrichemical application, production of biofuels, theses of agricultural S&T, agricultural patents owned by national residents, average annual growth of agriculture and net export of agricultural products.

2.2.5 South Africa's National Agricultural S&T Innovation Capability (NASTIC): Assessment and Analysis

This report will analyze South Africa's 2001–2016 NASTIC scores and South Africa's ranking among the G20.

South Africa's NASTIC scores and ranking among the G20 are as illustrated in Fig. 9.

(1) Overall ranking: in 2016, South Africa ranked the 13th among the G20, 3 places down as compared with in 2011. The ranking shows a downward trend with temporary fluctuation.
(2) Score: In 2016, South Africa scored 34.16 points in NASTIC, 39.20 points lower than the G20 highest and 9.65 points lower than the G20 average. As compared with 2001, South Africa's score decreased by 1.85 points; the gap with the G20 highest of the year narrowed by 1.83 points and the gap with the G20 average widened by 2.83 points.
(3) Ranking in terms of secondary indicators: All 3 secondary indicators (NASTIFC, NASTIOC, NASTIEC) showed a downward trend (Fig. 10; Table 9).
(4) Quaternary indicators (location): In 2016, South Africa has 9 strong indicators: investment in public education, investment in agricultural R&D, international cooperation in agricultural research, support for innovation in agricultural S&T, supply of agricultural credit, sophistication of market mechanism, intensity of agrichemical application, emission of agriculture-related GHGs, and production of biofuels. South Africa has 12 advantageous indicators: ranking of agriculture-related disciplines for universities and research institutions, R&D for agriculture-related businesses, agricultural foreign direct investment, industry-university collaboration, IPR protection, ease to start businesses, availability of

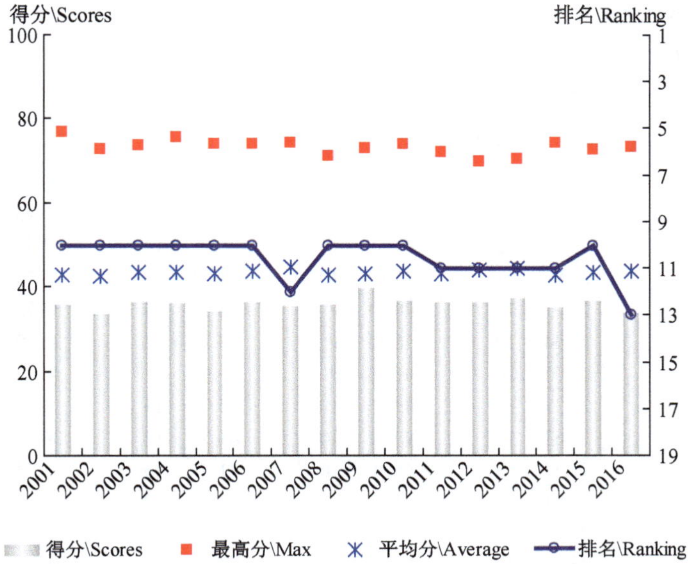

Fig. 9 South Africa's NASTIC Scores and Ranking (2001–2016)

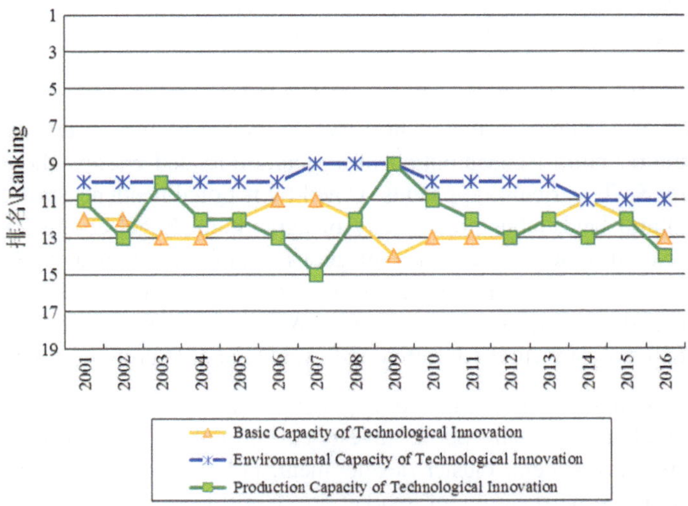

Fig. 10 Ranking of Secondary Indicators of South Africa's NASTIC (2001–2016)

technology training, intensity of agrichemical application, agricultural water consumption, new plant varieties cultivated by domestic residents, agricultural production index and net export of agricultural products. South Africa remains neutral in 14 indicators: proportion of irrigated area, penetration rate of ICT, transportation infrastructure, quantity of agriculture-related research personnel,

Table 9 South Africa's NASTIC: score and ranking for secondary indicators

Item year	NASTIFC		NASTIEC		NASTIOC		NASTIC	
	Score	Ranking	Score	Ranking	Score	Ranking	Score	Ranking
2001	20.86	12	45.11	10	42.03	11	36.00	10
2002	21.67	12	45.08	10	34.74	13	33.83	10
2003	22.11	13	43.55	10	44.38	10	36.68	10
2004	22.67	13	45.04	10	41.58	12	36.43	10
2005	23.39	12	43.53	10	36.65	12	34.53	10
2006	24.87	11	43.83	10	41.44	13	36.72	10
2007	25.91	11	46.44	9	34.91	15	35.75	12
2008	21.94	12	48.14	9	38.18	12	36.09	10
2009	22.05	14	47.16	9	50.91	9	40.04	10
2010	22.37	13	43.26	10	45.03	11	36.89	10
2011	22.62	13	46.01	10	41.58	12	36.74	11
2012	23.47	13	45.40	10	41.31	13	36.73	11
2013	23.81	12	44.28	10	44.75	12	37.61	11
2014	24.58	11	42.99	11	38.53	13	35.37	11
2015	23.03	12	42.49	11	44.89	12	36.80	10
2016	23.27	13	43.00	11	36.20	14	34.16	13
Score changes	2.40		−2.11		−5.84		−1.85	
Ranking changes		−1		−1		−3		−3
(Dis)advantageous		Neutral		Neutral		Neutral		Neutral

cluster development, public curiosity, political stability, appropriateness of agricultural policy, consumption of agriculture-related energy, theses of agricultural S&T, agricultural patents owned by national residents, productivity of agricultural land, labor productivity of agriculture, Engel co-efficient in rural areas. South Africa is at a disadvantageous position in 9 indicators: GDP, fiscal revenue, agricultural mechanization, education attainment for farmers, health, free and democratic environment for S&T innovation, quality of education system, equity for rural education and average annual growth of agriculture.

Quaternary indicators (ranking): Among the 17 indicators, 12 show an upward trend: agricultural mechanization, investment in public education, agricultural foreign direct investment, cluster development, political stability, supply of agricultural credit, sophistication of market mechanism, availability of technology training, intensity of agrichemical application, new plant varieties cultivated by domestic residents, agricultural production index and productivity of agricultural land. 15 indicators remain unchanged in their ranking: R&D of agricultural businesses, quantity of agriculture-related research personnel, health, investment in agricultural R&D, industry-university collaboration, international cooperation in agricultural research, free and democratic environment for S&T innovation, public curiosity, IPR protection, intensity of agrichemical application, agricultural water consumption, consumption of agriculture-related energy, production of biofuels, net export of agricultural products and Engel co-efficient in rural areas. 17 indicators show a downward trend: GDP, fiscal revenue, proportion of irrigated area, penetration rate of ICT, transportation infrastructure, ranking of agriculture-related disciplines for universities and research institutions, education attainment for farmers, appropriateness of agricultural policy, support for innovation in agricultural S&T, ease to start businesses, quality of education system, equity for rural education, emission of agriculture-related GHGs, theses of agricultural S&T, agricultural patents owned by national residents, labor productivity of agriculture and average annual growth of agriculture.

Notes: In this report, strong indicators are those that rank from the 1st to the 5th; advantageous indicators are those that rank from the 6th to the 10th; neutral indicators are those that rank from the 11th to the 15th; disadvantageous indicators are those that rank from the 16th to the 19th.

3 Conclusion

In recent years, the BRICS have made remarkable progress in modern agriculture and their score of innovation capability of agricultural S&T show an upward trend. The BRICS are even taking the lead in certain indicators. For instance, in 2010, only South Africa ranked among the top 10 in agricultural production index while in 2016 all BRICS ranked among the top 10. The score gap between BRICS and agricultural powers have been narrowing on a yearly basis. However, BRICS still lag behind substantially in aspects including research, quality of education, modern infrastructure and eco-protection. The BRICS still enjoy huge space for improvement.

Correction to: BRICS Innovative Competitiveness Report 2017

Xinli Zhao, Minrong Li, Maoxing Huang and Alexander Sokolov

Correction to:
X. Zhao et al. (eds.), *BRICS Innovative Competitiveness Report 2017*, Research Series on the Chinese Dream and China's Development Path, https://doi.org/10.1007/978-981-10-8078-4

In the original version of the book, the affiliation of authors "S. Shashnov, M. Kotsemir, A. Grebenyuk, L. Gokhberg , T. Kuznetsova, and A. Pikalova" was incorrect in chapters 2 and 5.

It has now been corrected to "Institute for Statistical Studies and Economics of Knowledge, National Research University Higher School of Economics, Moscow, Russia".

The updated version of these chapters can be found at
https://doi.org/10.1007/978-981-10-8078-4_2
https://doi.org/10.1007/978-981-10-8078-4_5

X. Zhao et al. (eds.), *BRICS Innovative Competitiveness Report 2017*,
Research Series on the Chinese Dream and China's Development Path,
https://doi.org/10.1007/978-981-10-8078-4_13

Appendix A
Indicators and Evaluation System of Overall Innovation Capabilities of BRICS Countries and Their Roles in Evaluation and Forecasting

A.1 Brief Introduction of STI Evaluating Reports

A.1.1 The Global Competitiveness Report of the World Economic Forum

The World Economic Forum (WEF) started to launch the global competitiveness report since 1979. From 1997 to 1999, the overall ranking of competitiveness adopts the following eight secondary indicators: ① the openness of the economy to international finance and trade; ② government budget, tax and management; ③ the level of financial market development; ④ the quality of transportation, communication, energy and service-oriented infrastructure; ⑤ basic research, applied science and technology science; ⑥ corporate management; ⑦ labor market and its mobility; and ⑧ legal and political systems. Since 2000, WEF has made major adjustments to the index system of global competitiveness evaluation by increasing the proportion of the capability in S&T innovation. From 2001 to 2006, the Technology Indicator was used to determine the Innovation Index, and from 2006 to 2007 the Innovation and its Complexity was used for the ranking. The 2016–2017 Global Competitiveness Report, published in 2016, covers a total of 138 economies worldwide with data from the United Nations, World Bank, International Monetary Fund, and World Economic Forum. The evaluation indicators used in the report include: institutional building, infrastructure, macroeconomic environment, health and primary education, higher education and training, commodity market efficiency, labor market efficiency, financial market development, technology readiness, market size, business sophistication and innovation. The twelve indicators reflect the whole picture of the competitiveness of a country, thus becoming the main basis for evaluation.

X. Zhao et al. (eds.), *BRICS Innovative Competitiveness Report 2017*,
Research Series on the Chinese Dream and China's Development Path,
https://doi.org/10.1007/978-981-10-8078-4

A.1.2 The Global Innovation Index of INSEAD

In 2007, INSEAD and the United Nations University worked together to complete the first Global Innovation Index, which has been released annually since then. The Global Innovation Index represents a comprehensive and quantifiable system of indicators that can be used to assess global innovation activities and the innovation capability of economies across the world, thus providing guidance for innovative practices in various countries and regions. The index report not only includes important indicators such as the proportion of R&D input in GDP and the number of patents and trademarks, but also covers diversified indicators such as infrastructure, business environment and human resources. The report aims at achieving depth and width of research as well as providing a new perspective to the global innovative activities. In addition, this index report uses not only objective, quantitative hard indicators and comprehensive indicators, but also subjective, qualitative soft indicators and other research methods to ensure that the research results are accurate and scientific. The 2016 Global Innovation Index, which was jointly released by the World Intellectual Property Organization, Cornell University and INSEAD, ranked 128 countries and economies based on 82 indicators and explored the impact of innovation-oriented policies on economic growth and development.

A.1.3 Report on National Innovative Competitiveness Development of the Chinese Academy of Social Sciences and Fujian Normal University

In 2011, the Chinese Academy of Social Sciences and Fujian Normal University jointly released the yellow book of the Group of Twenty (G20) National Innovative Competitiveness Development Report (2001–2010), presenting the evaluation of innovative competitiveness of the 20 countries from 2001 to 2010 year by year. Subsequently, they released the yellow papers of the Global Innovative Competitiveness Development Report (2001–2012), the G20 National Innovative Competitiveness Development Report (2011–2013), the G20 National Innovative Competitiveness Development Report (2013–2014), the G20 National Innovative Competitiveness Development Report (2015–2016), assessing the national Innovative Competitiveness of 100 countries and the Group of Twenty. The index system includes five secondary indicators (namely, basic innovative competitiveness, innovative environment competitiveness, innovative input competitiveness, innovative output competitiveness, and innovation sustainability competitiveness) and 32 tertiary indicators that are related to science and technology, economy, and education, etc.

A.1.4 Bloomberg Innovation Index

In 2012, Bloomberg compiled the global innovation index TOP 50. Seven key indicators were used for the index, including R&D intensity, productivity, manufacturing capacity, high-tech density, higher education efficiency, concentration of researchers, and patent status. The data used for the indicators were mainly from the World Bank, the World Intellectual Property Organization, the Conference Board, the OECD and the UNESCO. In 2015, Bloomberg analyzed six indicators of 200 plus countries including R&D, manufacturing, high-tech companies, education, researchers and patents and published the ranking of world's top 50 countries in terms of comprehensive innovation capability. In 2016, the Bloomberg Innovation Index quantified and ranked the innovation capability of countries and regions based on seven comprehensive indicators (i.e. R&D intensity, industrial added value, productivity, high-tech density, higher education coverage, researcher concentration and patent activity).

A.1.5 National Innovation Index of the China Academy of Science and Technology Development (CASTED)

Since 2011, the China Academy of Science and Technology Development (CASTED) started to release the annual National Innovation Index Report. The National Innovation Index Report 2013, which was published in 2014, measured the innovation index of 40 countries by the international common benchmarking method on the basis of statistics from 2011 to 2012. It has drawn on the latest research results on "national competitiveness" and "innovation evaluation" at home and abroad, and internationally authoritative evaluation reports done by the World Economic Forum and the International Institute for Management Development. The National Innovation Index Report 2015, published in 2016, re-evaluated the Innovation Index of 40 major countries in the world by five primary indicators, including innovative resources, knowledge creation, corporate innovation, innovation performance and innovation environment.

A.2 Summary of Evaluating Algorithm

A.2.1 Mathematical Model

The evaluating approach adopted for measuring the comprehensive innovative competitiveness in this report can reflect the time-varying nature of the evaluation system. The mathematical model is as follows:

$$NIC_i = \beta_0 + \beta_1 X_{i1} + \beta_2 X_{i2} + \cdots + \beta_k X_{ik} + \frac{N-k}{N} nic_i + \mu$$

Here, k represents the number of changes of the evaluation method: $k > 0$ when a new evaluation method is added, $k < 0$ when a method is deducted, and $k = 0$ when the evaluation system remains unchanged. N represents the number of evaluation methods of the current evaluation system. X_{ik} (i = 1, 2, …, n) is the innovative competitiveness score the i-th country under a changed evaluation method. β_j (j = 1, 2, …, k) is the regression coefficient. μ represents the random interference. nic_i represents the national innovative competitiveness score based on the original evaluation system. NIC_i represents the national innovative competitiveness score based on the current evaluation system.

In order to reduce the results volatility due to the changes in the evaluation system, this report minimizes sum of square of the difference between the innovative competitiveness scores based on the current evaluation system and the original system and explores the optimal function.

A.2.2 *Forecasting Approach and Its Rationality*

The exponential smoothing method and the Holt's linear exponential smoothing model are used for simple numerical prediction. Under the above-mentioned methods, data of various periods are weighted where recent data are given priority over long-term data and the weight of various periods of data degraded exponentially by scale from recent to long-term. This is consistent with the realities of socio-economic changes. At the same time, this method incorporates all historical data in modeling, which avoids the limitation of the moving average method which uses only partial data; and can reveal the change pattern of the phenomenon by downplaying the irregular change factors.

The research group believes that the prediction results are reasonable. The report data are reliable as they come from 5 authoritative reports widely recognized by both international and domestic experts. The report data are scalable as they can bridge the difference where different reports have different focus and thus reflect the general trend. The report data are in line with historical characteristics and the general knowledge about the impact of economic activities on innovative competitiveness, thus enabling both the horizontal international comparison and the vertical historical comparison.

A.3 Meaning of Agricultural S&T Innovation Capabilities in BRICS and the Indicator System

A.3.1 Meaning

This report analyzes research results on the national agricultural innovation system by domestic and foreign research scholars as well as definitions and features of national agricultural S&T innovation. By doing so, the report establishes the national agricultural S&T innovation system as a unified whole with five subsystems of agricultural education and research, innovation application, innovation value realization, innovation intermediary services and innovation supporting conditions. The national agricultural S&T innovation system is composed of various elements with specific functions, and in this report, the elements are: the basic element, the environmental element, and the output element. The three aspects, which exist across the five subsystems, interact with each other and help to enhance the performance of national agricultural S&T innovation system.

National agricultural science and technology innovation system forms the foundation of organizing and operating agricultural innovation activities, and the function and performance of the system is realized by the innovation capability in this area. The hierarchical structure of the national agricultural innovation system determines the complexity of the formation mechanism of the innovation capability, which this report believes follows such a path as "organizational structure → resources and elements → innovation performance." Based on this analysis, we hold that the national agricultural science and technology innovation capability (NASTIC) is composed of the national agricultural science and technology innovation foundation capability (NASTIFC), the national agricultural science and technology innovation environment capability (NASTIEC), and the national agricultural science and technology innovation output capability (NASTIOC), and the function model is shown in Equation (A.1).

$$\text{NASTIC} = \mathbf{F}(\text{NASTIFC}, \text{NASTIEC}, \text{NASTIOC}) \tag{A.1}$$

(1) The innovative foundation capability not only embodies the foundation and level of national economic and social development, but also reflects how much emphasis a country puts on promoting agricultural innovation ability and the investment into the development in this regard. It is the basis of evaluating the national agricultural science and technology innovation capability. (2) The national agricultural science and technology innovation environment capability is the necessary condition for developing the agricultural innovation capability. Good innovation environment can not only effectively gather innovative resources, but also cultivate a strongly competitive agricultural innovation group and promote the industrialization of the innovation results, thereby improving performance and accumulating innovation and growth capability. (3) The national agricultural science and technology innovation output capability is the achievement and realization

of the innovation activities. The quantity and quality of R&D output reflect the smoothness and implementation of scientific and technological innovation activities.

A.3.2 The Evaluation System

A.3.2.1 Establishing the Evaluation System

Based on the definition, meaning and internal mechanism of national agricultural science and technology innovation capability, the evaluation system has built the analysis framework and the hierarchical indicators by absorbing the existing fine indicators on scientific and technological competitiveness and innovation ability, and picking out representative, operational, and accessible evaluation indicators; the frequency statistical method and Delphi method are adopted to optimize the system; the correlation analysis and discriminant analysis of the indicators are done to further improve the system; part of the country's index data are used for simulating system operation so as to test the results of the operation. The final index system includes three second-level indicators and 44 third-level indicators, as shown in Table A.1.

This report is done based on statistical data released from 2001 to 2016 by the international authoritative institutions such as the World Bank (WB), the Food and Agriculture Organization of the United Nations (FAO), the United Nations Educational, Scientific and Cultural Organization (UNESCO), the World Economic Forum (WEF), the World Trade Organization (WTO), the World Intellectual Property Organization (WIPO), The International Telecommunication Union (ITU) and others.

A.3.2.2 Establishing the Evaluation Model

This report applies the linear weighted evaluation method to score the national agricultural science and technology innovation capability: the higher the evaluation score, the stronger the country's overall agricultural science and technology innovation capability is, and vice versa. The specific calculation model is:

$$\mathrm{Y_i} = \sum \mathrm{x_{ij} w_{ij}} \tag{A.2}$$

$$\mathrm{Y} = \sum\sum \mathrm{x_{ij} w_{ij}} \tag{A.3}$$

In the above equation, Y is the comprehensive evaluation score of innovation competitiveness, Yi is the evaluation score of the i-th indicator, X_i is the

Table A.1 National agricultural S&T innovative capability indicators

Primary indicator	Secondary indicators	Tertiary indicators
National agricultural science and technology innovative capability (NASTIC)	NASTIFC	GDP, fiscal revenue, proportion of irrigated area, agricultural mechanization, penetration rate of information and communication technology, transportation infrastructure, rank of agricultural discipline in universities and scientific research institutions, R&D level of agricultural enterprises, number of agricultural researchers, the education attainment and health status of farmers, public education input, agricultural R&D expenditure, FDI in agriculture, industry-university cooperation, cluster development, international agricultural research cooperation
	NASTIEC	Free and democratic environment for scientific and technological innovation, public curiosity, political stability, intellectual property protection, appropriateness of agricultural policy, support for agricultural science and technology innovation, supply of agricultural credit, ease to start a business, sophistication of market mechanism, the quality of education system, equity in rural education, availability of technology training, intensity of fertilizer application, intensity of agrichemical application, agriculture-related water consumption, agriculture-related energy consumption, agriculture-related GHGs emission, biofuel production
	NASTIOC	The number of papers in agricultural science and technology, the number of agricultural patents in the country, new plant varieties cultivated by domestic residents, the agricultural production index, the agricultural land productivity, the agricultural labor productivity, the average annual growth of agriculture, the net export of agricultural products, the rural Engel coefficient

dimensionless data value of the i-th basic indicator, and w_{ij} is the weight of the basic indicator. Weight of indicator are determined by the commonly-used average weight method.

Appendix B
Fundamental Documents Specifying the Framework of BRICS STI Cooperation

Theme: BRICS Science, Technology and Innovation Cooperation: A Strategic Partnership for Equitable Growth and Sustainable Development

Cape Town Declaration

Cape Town, South Africa, 10 February 2014

1. In line with the mandate of the eThekwini Declaration and Action Plan of March 2013 adopted at the Fifth BRICS Summit held in South Africa, we the Ministers and their representatives for Science, Technology and Innovation of the Federative Republic of Brazil, the Russian Federation, the Republic of India, the People's Republic of China and the Republic of South Africa, met in Cape Town, South Africa for the First BRICS Science, Technology and Innovation Ministerial Meeting on 10 February 2014, to discuss and coordinate positions of mutual interest and identify future directions of institutionalizing cooperation in science, technology and innovation within the framework of BRICS.
2. We reaffirm the vision to strengthen the BRICS partnership for common development and advance cooperation in a gradual and pragmatic manner,

X. Zhao et al. (eds.), *BRICS Innovative Competitiveness Report 2017*,
Research Series on the Chinese Dream and China's Development Path,
https://doi.org/10.1007/978-981-10-8078-4

reflecting the principles of openness, solidarity and mutual assistance, and give substance to all the calls expressed at previous BRICS Summits to intensify cooperation in the spheres of science, technology and innovation, including the peaceful use of space.

3. We stress the paramount importance of science, technology and innovation for human development. Indeed, while recognizing the role and significance of competitiveness in the rapid technologically changing global environment, we agree that people-centred and public-good driven science, technology and innovation, supporting equitable growth and sustainable development, shall form the basis of our cooperation within the framework of BRICS.
4. In order to support this common vision, we agreed to enter into a BRICS Memorandum of Understanding on Cooperation in Science, Technology and Innovation which shall serve as the strategic intergovernmental framework: (i) to strengthen cooperation in science, technology and innovation; (ii) to address common global and regional socio-economic challenges utilizing shared experiences and complementarities; (iii) to co-generate new knowledge and innovative products, services and processes utilizing appropriate funding and investment instruments; (iv) to promote, where appropriate, joint BRICS partnerships with other strategic actors in the developing world.
5. We agree with the text of the BRICS Memorandum of Understanding on Cooperation in Science, Technology and Innovation and propose that it be signed on the occasion of the Sixth BRICS Summit in Brazil in 2014.
6. We agree under this BRICS STI framework the main areas of cooperation shall include: exchange of information on policies and programmes and promotion of innovation and technology transfer; food security and sustainable agriculture; climate change and natural disaster preparedness and mitigation; new and renewable energy, energy efficiency; nanotechnology; high performance computing; basic research; space research and exploration, aeronautics, astronomy and earth observation; medicine and biotechnology; biomedicine and life sciences (biomedical engineering, bioinformatics, biomaterials); water resources and pollution treatment; high tech zones/science parks and incubators; technology transfer; science popularization; information and communication technology; clean coal technologies; natural gas and non-conventional gases; ocean and polar sciences; and geospatial technologies and its applications.
7. In pursuit of cooperation in the above areas, we agree to build upon existing bilateral synergies and other forms of multi-country frameworks of cooperation amongst the BRICS member countries.
8. With a view to supporting the immediate implementation of the objectives outlined in the BRICS Memorandum of Understanding on Cooperation in Science, Technology and Innovation, we recognize and endorse, as a first step, the establishment of five thematic areas and leadership, namely: (a) climate change and natural disaster mitigation, led by Brazil; (b) water resources and pollution treatment, led by Russia; (c) geospatial technology and its applications, led by India; (d) new and renewal energy, and energy efficiency, led by China; (e) astronomy, led by South Africa.

9. We recognize the sharing and exchange of information on science, technology and innovation policies and strategies and the formulation of joint long-term problem-focused cooperation programmes will constitute the central modalities of this cooperation.
10. We recognize that specific cooperative activities under the BRICS STI framework may necessitate the provision of organizational, legal, financial and staffing support. This relates primarily to stimulating joint investment in the development of high technologies, creating common technology platforms, and the setting up of applied research and innovation centres and laboratories.
11. We recognize the importance and centrality of knowledge and technology transfer as the means of mutually empowering BRICS member countries. In this regard we support efforts to establish BRICS mechanisms that enhance technology and knowledge transfer amongst the member countries.
12. We support the establishment of a dedicated BRICS STI training programme to address human capital challenges in BRICS member countries.
13. We commit to strengthen and improve the governance mechanisms for BRICS STI cooperation, including meetings of STI Ministers, senior officials meetings, as well as the network of national coordinators for cooperation in the spheres of science, technology and innovation.
14. Brazil, Russia, India and China extend warm appreciation and sincere gratitude to the Department of Science and Technology of the Republic of South Africa for hosting the First BRICS Science, Technology and Innovation Ministerial Meeting in Cape Town on 10 February 2014.
15. Russia, India, China and South Africa wish the Brazilian government well in its preparations for the Sixth BRICS Summit where deliberations relating to science, technology and innovation will form part of the agenda.

Done in the English language in five copies, each copy being equally authentic, on 10 February 2014 in Cape Town, South Africa.

II BRICS Science, Technology and Innovation Ministerial Meeting

BRICS 2015
Brasília · Brasil

Brasília Declaration

Brasília, Brazil, 18 March 2015

1. In line with the Fortaleza Declaration and the Action Plan adopted at the 6th BRICS Summit, on 15 July, 2014 held in Brazil, we, the Ministers for Science, Technology and Innovation of the Federative Republic of Brazil, the Russian Federation, the Republic of India, the People's Republic of China and the Republic of South Africa, met in Brasília, Brazil, on 18 March, 2015, for the 2nd BRICS Science, Technology and Innovation Ministerial Meeting.
2. Recalling the theme of the 6th BRICS Summit "Inclusive Growth: Sustainable Solutions", we strongly believe that Science, Technology and Innovation play a central role in promoting inclusive macroeconomics and social policies and in the imperative to address challenges to humankind posed by the need to simultaneously achieve growth, inclusiveness, environmental protection and preservation.
3. We reaffirm that sharing and exchanging information on science, technology and innovation policies and strategies; leveraging contacts and programmes aimed at enhancing collaborative innovation projects among BRICS countries; and the formulation of joint long-term problem-focused cooperation programmes shall constitute the central modalities of this cooperation. In order to facilitate this, appropriate mechanisms of cooperation shall be elaborated and established within the implementation of the BRICS Science, Technology and Innovation initiatives.
4. We welcome the outcomes of the 1st BRICS Workshop on Prevention and Mitigation of Natural Disasters, held in Brasília, on 7–8 May 2014; of the BRICS Seminar on National Systems of Innovation, held in Brasília, on 25–27 March 2014; of the Meeting of BRICS Solid State Lighting Working Group, held in Guangzhou, China, on 7–9 November 2014; and of the International Conference on Water Management and Ecology in the Framework of Russian Federation participation in BRICS, held in Moscow, Russia, on 4 June 2014.
5. Following the instructions of the leaders of BRICS member countries, mentioned in paragraph 67 of the Fortaleza Declaration, we express our satisfaction in signing the Memorandum of Understanding on Cooperation in Science, Technology and Innovation (MoU), which establishes a strategic framework for cooperation in priority areas amongst the BRICS member countries.
6. In order to foster further collaboration and achieve concrete results from the MoU directives, we agree to develop and negotiate a Work Plan 2015–2018, based on the Brazilian proposal, during the Russian presidency of BRICS, to be approved in the next BRICS STI-SOM and signed at the next BRICS STI Ministerial Meeting. The Work Plan will focus on the five priority areas and leadership established previously by each country, namely: (a) prevention and mitigation of natural disasters, to be led by Brazil, (b) water resources and pollution treatment, to be led by Russia, (c) geospatial technology and its applications, to be led by India, (d) new and renewable energy, and energy efficiency, to be led by China, and (e) astronomy, to be led by South Africa. New initiatives agreed by the BRICS countries will also be included in the Work Plan.

7. We take note of the following announcements: South Africa will convene the 1st Meeting of the BRICS Working Group on Astronomy shortly after this Ministerial; Russia will host International Scientific and Experimental Conference on Water: Technologies, Materials in Industry and Energy Processes in July 2015, in Ufa; China will host the 2nd Meeting of the BRICS SSL Working Group in November 2015; India will host the BRICS Working Group on Geospatial Technology Application for Development in March 2016. We also welcome the Brazilian–Russian proposal, discussed on the occasion of the 4th STI-SOM, to start negotiations among BRICS countries with a view to establishing biomedicine and life sciences as a new priority area for cooperation.
8. The Work Plan will ensure the development of science, technology and innovation cooperation through the launch of a BRICS Research and Innovation Initiative, which shall cover actions including: (a) cooperation in the framework of major research infrastructures; (b) coordination of existing large-scale national programmes of BRICS countries; (c) setting up a Framework Programme for funding multilateral joint project for research, technology commercialization and innovation; and (d) establishment of a joint Research and Innovation Networking Platform.
9. We support the creation of a BRICS Young Scientists Forum proposed by India, which intends to establish a platform for young students of science, engineering and applied disciplines as well as for those pursuing research careers in the age group of 22–35 years to gather for: (a) addressing the needs for advancement of skills, research competencies, career, talent and next generation scientific leadership; (b) sharing scientific research results and experiences; (c) discussing novel ideas in emerging frontline fields of S&T; (d) analyzing trends and features of globally important scientific issues; (e) suggesting measures to enhance trans-continental mobility in their scientific research careers.
10. To increase the competitiveness of the BRICS economies on the global arena, we commit to supporting the BRICS Economic Partnership Strategy, currently under negotiation, which includes Science, Technology and Innovation as a priority. Long-term cooperation in these areas will help bridge the scientific and technological gap between BRICS and developed economies and provide a new quality of growth based on economic complementarity.
11. We encourage increased participation of business, academia and other relevant stakeholders for science, technology and innovation development among BRICS countries.
12. We welcome the holding of the 4th BRICS Science, Technology and Innovation Senior Officials Meeting in Brasília, on 17 March 2015, and instruct the Senior Officials to organize the 5th BRICS STI-SOM prior to the 3rd Ministerial Meeting.
13. Russia, India, China and South Africa extend their warm appreciation to Brazil for hosting the 2nd BRICS Science, Technology and Innovation Ministerial

Meeting and the 4th BRICS Science, Technology and Innovation Senior Officials Meeting.

14. Brazil, India, China and South Africa convey their appreciation to the Russian Federation for its offer to host the 3rd BRICS Science, Technology and Innovation Ministerial Meeting and the 5th BRICS Science, Technology and Innovation Senior Officials Meeting in 2015 and extend their full support to that end.

III BRICS Science, Technology and Innovation Ministerial Meeting

Theme: BRICS Science, Technology and Innovation Partnership – a Driver of Global Development

Moscow Declaration

Moscow, the Russian Federation, 28 October 2015

1. In line with the Ufa Declaration and Action Plan adopted at the Seventh BRICS Summit on 9 July 2015 held in Russia we, the Ministers and their representatives for Science, Technology and Innovation of the Federative Republic of Brazil, the Russian Federation, the Republic of India, the People's Republic of China and the Republic of South Africa, met in Moscow, the Russian Federation, on the 28th of October 2015, for the III BRICS Science, Technology and Innovation Ministerial Meeting to build further collaboration based on the Memorandum of Understanding on Cooperation in Science, Technology and Innovation (MoU) provisions.
2. Recalling the theme of the Seventh BRICS Summit "BRICS Partnership—a Powerful factor of Global Development", we affirm our willingness to follow the Strategy for BRICS Economic Partnership in addressing common global and regional socio-economic challenges utilizing such drivers as science, technology and innovation (STI).
3. Welcoming the outcomes of the First Meeting of the BRICS STI Funding Parties on the establishment of the BRICS Research and Innovation Initiative (hereinafter—BRICS R&I Initiative) held on 6–7 July 2015, Moscow, Russia, and highlighting the immense research and technological potential in the BRICS member countries and importance of the development of BRICS R&I Initiative (paragraph 62 of the Ufa Declaration) we agree on the following

mechanisms and levels of collaboration: (i) cooperation within large research infrastructures, including mega-science projects; (ii) coordination of the existing large-scale national programmes of the BRICS countries; (iii) development and implementation of a BRICS Framework Programme for funding multilateral joint research projects, technology commercialization and innovation; (iv) establishment of BRICS Research and Innovation Networking Platform.

4. We welcome the establishment of the Working Group on BRICS large research infrastructures, the Working Group on BRICS funding multilateral joint research projects, technology commercialization and innovation.
5. We agree on our commitment to develop and implement the BRICS Framework Programme on multilateral research funding through joint calls. Also we propose to use the possibilities of the New Development Bank (Agreement of the New Development Bank signed during the VI BRICS Summit in Fortaleza) as an additional funding instrument to foster further collaboration.
6. The cooperation focused on the five thematic leadership areas established previously by each country in the Brasilia Declaration, namely: (a) prevention and mitigation of natural disasters, led by Brazil, (b) water resources and pollution treatment, led by Russia, (c) geospatial technology and its applications, led by India, (d) new and renewable energy, and energy efficiency, led by China, and (e) astronomy, led by South Africa, and the activities within these five areas will be implemented by use of the BRICS Research and Innovation Networking Platform developing direct communication channel between stakeholders.
7. To address common societal challenges and to advance BRICS leadership and cooperation on a global level we welcome the new initiatives:
    - Creation of BRICS Young Scientists Forum (India as coordinating country);
    - Cooperation on Biotechnology and Biomedicine including Human Health and Neuroscience (Russia and Brazil as coordinating countries);
    - Cooperation on Information Technologies and High Performance Computing (China and South Africa as coordinating countries);
    - Cooperation on Ocean and Polar Science and Technology (Brazil and Russia as coordinating countries);
    - Cooperation on Material science including Nanotechnology (India and Russia as coordinating countries);
    - Cooperation on Photonics (India and Russia as coordinating countries).
8. Encouraging increased participation of business, academia and other relevant stakeholders for STI development among BRICS countries (paragraph **11**, Brasilia Declaration) we acknowledge the independent initiatives to establish the BRICS Network University aimed at developing master's and Ph.D. programmes along with joint research projects in knowledge fields priorities corresponding with the main areas of cooperation stated in the Article 3 of the MoU and the BRICS University League.
9. We welcome the creation of a BRICS Young Scientists Forum and establishing the BRICS Young Scientist Forum Secretariat in India coordinated by the

Department of Science and Technology with commitment and support from all BRICS countries. We also welcome hosting of the BRICS Young Scientist Conclave in 2016 in India and creation of dedicated website for BRICS Young Scientist Forum.

10. We also support creation of BRICS Research and Innovation Networking Platform.
11. We take note of the following announcements: India and Brazil host the BRICS thematic Session on Prevention and Mitigation of Natural Disasters during the 6th Annual Conference of the International Society for Integrated Disaster Risk Management in October 2015; China hosts the 2nd Meeting of the BRICS Solid-state lightning (SSL) Working Group in November 2015; South Africa hosts the first meeting of the BRICS Astronomy Working Group in December 2015 at the Science Forum South Africa; India hosts the BRICS Working Group on Geospatial Technology Application for Development in March 2016; Russia initiates 2nd Meeting of the Group of STI Funding Parties in January 2016.
12. We endorse the BRICS Science, Technology and Innovation Work Plan 2015–2018 and reaffirm our commitment to implement it (annexed).
13. Brazil, India, China and South Africa convey their appreciation to the Russian Federation for hosting the III BRICS STI Ministerial meeting in Moscow.
14. Russia, Brazil, China and South Africa convey their appreciation to India for its offer to host the IV BRICS STI Ministerial meeting and the VI BRICS STI SOM in 2016 and extend their full support to that end.

IV BRICS Science, Technology and Innovation Ministerial Meeting

Theme: BRICS Science, Technology and Innovation Partnership – Building Responsive Inclusive Collective Solutions

Jaipur Declaration

Jaipur, the Republic of India, 8 October 2016

1. Preparatory to and in line with the proposed Goa Declaration and Action Plan to be adopted at the Eighth BRICS Summit on 15–16 October 2016 in Goa, India, we, the Ministers and their representatives for Science, Technology and Innovation of the Federative Republic of Brazil, the Russian Federation, the

Republic of India, the People's Republic of China and the Republic of South Africa, met in Jaipur, the Republic of India, on the 8th of October 2016, for the 4th BRICS Science, Technology and Innovation Ministerial Meeting to build further collaboration based on the BRICS Memorandum of Understanding on Cooperation in Science, Technology and Innovation (MoU).

2. Taking into consideration the theme of the Eighth BRICS Summit—**Building Responsive Inclusive Collective Solutions**; we reaffirm our commitment to implement the Strategy for BRICS Economic Partnership adopted at the BRICS Ufa Summit which emphasized utilizing Science, Technology and Innovation (STI) as key drivers to address global and regional socio-economic challenges.
3. Welcoming the collective achievements of BRICS partners in the realization of initiatives established in accordance with the BRICS Science, Technology and Innovation Work Plan 2015–2018 (Work Plan 2015–2018) and Moscow Declaration adopted on 28 October 2015, we reaffirm our commitment to implement the Work Plan 2015–2018. We will intensify, diversify and institutionalize STI cooperation as outlined in the BRICS MoU on Cooperation in Science, Technology and Innovation through the mechanism of the BRICS Research and Innovation Initiative.
4. Welcoming the outcomes of the Second Meeting of the BRICS STI Funding Parties on the Development of the BRICS Research and Innovation Initiative and First Meeting of BRICS STI Funding Working Group held in Beijing on 19–21 January 2016, we welcome the signing of the Arrangement of the BRICS STI Framework Program and the Implementation Plan (hereinafter—BRICS Arrangements). These Arrangements will be instrumental in implementation of BRICS countries' joint initiative on multilateral interdisciplinary research & innovation funding under the BRICS STI Framework Program as evident from the launching of the 1st BRICS Pilot Call 2016 in mutually agreed priority areas. We take note of the huge response of BRICS scientists to work together in the BRICS multilateral research projects.
5. We take note of the conclusions of the First Photonics Conference of BRICS countries held on May 30–31, 2016, Moscow. We welcome the establishment of a BRICS Working Group on Photonics.
6. We welcome the establishment of BRICS Geospatial Working Group and its 1st Meeting held in India on 3 March 2016.
7. We welcome the hosting of 1st BRICS Young Scientist Conclave by India during 26–30 September, 2016, under the framework of the BRICS Young Scientist Forum being coordinated by India, as mandated by BRICS Leaders during 7th BRICS Summit. We take note of the recommendations of the BRICS Young Scientists Conclave.
8. We welcome India's proposal to host the BRICS Young Scientist Forum-Conclave on a rotation basis in the BRICS Chair country to keep the momentum for engaging youth of BRICS countries and explore mechanisms for implementation.

9. We welcome India's proposal to establish the **BRICS Innovative Idea Prize for Young Scientists** within the framework of the BRICS Young Scientist Forum.
10. We welcome the establishment of the BRICS Working Group on Astronomy and its meetings held in South Africa and Russia.
11. We take note of the outcomes of the BRICS thematic session on Prevention and Mitigation of Natural Disasters outlined during 6th Annual Conference of the International Society for Integrated Disaster Risk Management hosted by India in October 2015 in New Delhi; and of the BRICS Special Session on Natural disaster risk prevention and Mitigation in Coastal Areas jointly organized by Russia and Brazil in Saint Petersburg on 26 August, 2016
12. We take note of the 1st Meeting of the BRICS Working Group on Ocean and Polar Science and Technology held in Beijing on 26–28 September 2016 coordinated by Brazil.
13. We take note of the 2nd BRICS Water Forum hosted by Russia during 29–30 September 2016.
14. We take note of the 2nd Meeting of BRICS Working Group on Solid State Lighting hosted by China in November, 2015.
15. We agree to launch next the BRICS Framework Program call for research and innovation proposals in May 2017.
16. We agree on the Spearly establishment of the BRICS Working Group on Research Infrastructure, and Mega-Science to reinforce the BRICS Global Research Advanced Infrastructure Network (BRICS-GRAIN). We recommend exploring the possibility of supporting such initiatives through New Development Bank as well as other similar organizations.
17. We encourage synergies of the BRICS Research and Innovation Initiative with the BRICS Network University.
18. We welcome India's proposal to establish a BRICS Science and Technology driven Entrepreneurship and Innovation Partnership. We agree to start consultations and discussions to implement this initiative.
19. Pursuant to the BRICS Work Plan 2015–2018, we take note on the progress made during 2015–2016 and endorse the Action Plan 2016–2017 as updated.
20. Brazil, Russia, China and South Africa convey their appreciation to India for hosting the 4th BRICS STI Ministerial Meeting and 6th BRICS STI SOM in Jaipur.
21. India, Brazil, Russia and South Africa welcome the offer of China to host the 5th BRICS STI Ministerial Meeting and the 7th BRICS STI SOM in 2017.

Done at Jaipur on October 8, 2016.

V BRICS Science, Technology and Innovation Ministerial Meeting

Theme: Leading through Innovation & Deepening Cooperation

BRICS 2017 CHINA

The 5th BRICS Science, Technology & Innovation (STI) Ministerial Meeting

Hangzhou Declaration

Hangzhou, China, 18 July 2017

1. In line with the *BRICS Memorandum of Understanding on Cooperation in Science, Technology and Innovation* signed in March 2015 and the *Goa Declaration* adopted at the BRICS Summit held in India on October 16, 2016, we, the Ministers for Science, Technology and Innovation of the Federative Republic of Brazil, the Russian Federation, the Republic of India, the People's Republic of China and the Republic of South Africa met in Hangzhou, China, on 18 July, 2017, for the 5th BRICS Science, Technology & Innovation (STI) Ministerial Meeting.
2. Recalling the theme of the BRICS XIAMEN SUMMIT "BRICS: Stronger Partnership for a Brighter Future", we will continue to strengthen pragmatic cooperation in science, technology and innovation (STI) among the BRICS countries, create new cooperation opportunities, expand partnerships, and jointly tackle global challenges.
3. Based on the theme of the 5th BRICS Science, Technology & Innovation Ministerial Meeting "Leading through Innovation & Deepening Cooperation", we reaffirm the importance of innovation dialogues leading to outcomes and STI cooperation for promoting innovation-driven development and supporting the robust and sustainable growth of the world economy. We will continue to strengthen STI cooperation and implement relevant BRICS research and innovation initiatives mainly by means of exchanges in innovation policies and strategies and drafting of long-term cooperation plans to address common developmental challenges faced by all BRICS countries.

4. In order to promote innovation and leverage the central role of science and technology in enhancing socio-economic development and driving global sustainable development, we agree to adopt the *BRICS Action Plan for Innovation Cooperation*. We agree to promote entrepreneurship and build platforms in BRICS countries and mainly collaborate in technology cooperation, technology transfer and translation, science and technology parks, youth innovation and entrepreneurship and in fostering strategic and long term university-industry partnerships so as to build sound ecosystems for innovation and entrepreneurship.
5. Building on the positive experience and spin-off of the 1st BRICS Young Scientist Conclave under the framework of the BRICS Young Scientist Forum held in India last year, we welcome the convening of the 2nd BRICS Young Scientist Forum in Hangzhou, China. We recognize the potential of the Young Scientist Forum to develop into a powerful networking platform for BRICS young scientists and entrepreneurs and become an important arena to stimulate new academic ideas and train young professionals for the BRICS. We therefore fully support South Africa's decision to host the 3rd Young Scientist Forum in the lead up to the 6th BRICS STI Ministerial Meeting in South Africa in 2018. We encourage representatives of the BRICS thematic working groups to support participation of youth and invite themes for BRICS Young Scientist Forum.
6. We welcome the approval of the first set of BRICS R&D projects in priority areas. We recognize the importance of the BRICS STI Framework Programme as a mechanism for pooling innovation resources and strengths, and driving development in major areas and key technologies. We welcome the decision to launch the 2nd BRICS STI Call 2017 in six priority areas with Russia continuing as the Call Secretariat. We support the restated commitment of BRICS Science and Technology Ministries and their relevant funding agencies to continue jointly funding such multilateral R&D projects.
7. Acknowledging the importance of supporting cutting-edge high-impact research, we will encourage researchers from BRICS countries to publish the results of their research in international high-impact journals and participate as external foreign reviewers in the review of research proposals submitted to the funders in other BRICS countries, ensuring the quality of scientific review system within BRICS.
8. Recognizing the need for setting concerted priorities for S&T cooperation, we promote to support joint activities on identified priorities for S&T cooperation of BRICS countries based on foresight and monitoring of global S&T development.
9. We welcome India's initiative to coordinate the 1st meeting of BRICS Science and Technology Driven Entrepreneurship and Innovation Partnership in April 9th, 2017; and endorse the Term of Reference of the BRICS Working Group on Science Technology, Innovation and Entrepreneurship Partnership (STIEP).
10. We welcome China hosting the 1st Working Group Meeting and Innovation Collaboration Forum on Information and Communication Technology and

High Performance Computing in April, 2017 which presented several cooperation proposals including working together in relevant flagship projects.

11. We welcome Russia hosting the 1st Meeting of the BRICS Working Group on Research Infrastructure and Mega-Science Projects to strengthen cooperation on the BRICS Global Research Advanced Infrastructure Network and mega-science projects.
12. We welcome the convening of the 3rd BRICS STI Funding Working Group Meeting in South Africa in May, 2017 for discussion and negotiation on the approval of the first set of projects to be funded and the second call for proposals, and the outcomes of various thematic working group meetings or workshops.
13. Based on the BRICS STI Work Plan 2015–2018, we recognize the progress of BRICS STI cooperation since 2015 and adopt the updated BRICS STI Action Plan 2017–2018.
14. Acknowledging the importance of supporting STI investment and the need to establish inter-BRICS investment instruments, we support explore the possibilities of driving BRICS cooperation on innovation and entrepreneurship through the National Development Banks, New Development Bank and other existing financing platforms.
15. Brazil, Russia, India and South Africa extend their warm appreciation to the Ministry of Science and Technology of China for hosting the 5th BRICS STI Ministerial Meeting and the 7th BRICS STI Senior Officials Meeting.
16. Brazil, Russia, India, and China convey their appreciation to South Africa for its offer to host the 6th BRICS STI Ministerial Meeting and the 8th BRICS STI Senior Officials Meeting and extend their full support to that end.

Postscript

Science and technology is the cornerstone of a prosperous country, innovation is the spirit of a nation in its progress. Historical experience shows that those who can gain advantage in science, technology and innovation (STI) will gain the initiative of development. At present, the world economy is going through a treacherous recovery amidst deep adjustment and a new round of global technological and industrial revolutions is gaining momentum. In this world of dynamic and intensive innovation and in this age of new competition, all countries are making vigorous effort to promote innovation and seize opportunities. National innovative competitiveness which is underpinned by STI has become a focus of attention for all countries. Emerging economies represented by BRICS countries are taking up an increasingly important role in the global arena. In particular, in the wake of the global financial crisis in 2008, BRICS countries have attracted much world attention with its speedy recovery and development, making a sharp contrast to the troubling situation in developed economies. The prospect of sustained development in BRICS countries is very much determined by their comprehensive innovation capability with STI at its core.

In 2017, China takes over the BRICS presidency, and will host the ninth BRICS Summit in Xiamen, Fujian province in September. The Chinese Ministry of Science and Technology is responsible for organizing the fifth BRICS STI Ministerial Meeting and Senior Officials Meeting, and other related side events in July in Hangzhou. The China Science and Technology Exchange Center (CSTEC), which is the Chinese liaison office for BRICS STI cooperation, will support the organization of the ministerial meeting, the senior officials meeting and other side events.

To support the work relating to BRICS STI cooperation under the Chinese presidency in 2017, CSTEC, as entrusted by the Ministry of Science and Technology, established a High-level Expert Group consisting of leading professionals from the Central Party School, the Chinese Academy of Sciences, Tsinghua University, Renmin University, Fujian Normal University and other organizations to research the comprehensive innovative competitiveness of BRICS countries based on the latest statistics, analyzed the status quo, problems and potential of BRICS STI cooperation, and conducted country and thematic studies on BRICS countries' performance in STI. Under the guidance of the High-level Expert Group

X. Zhao et al. (eds.), *BRICS Innovative Competitiveness Report 2017*,
Research Series on the Chinese Dream and China's Development Path,
https://doi.org/10.1007/978-981-10-8078-4

and with the participation of science and technology sections of Chinese embassies in other BRICS countries as well as the Russian Higher School of Economics and other BRICS think tanks for research on STI, we have complied the *BRICS Innovative Competitiveness Report 2017* (Chinese and English editions) based on the results of initial research in order to provide input for the decisions at the BRICS STI Ministerial Meeting and the drafting of other related documents.

This report contains four parts, with a total of 12 reports.

The first part has two reports. The first is the analysis report on the overall evaluation and forecast of innovative competitiveness of BRICS countries and the current status and strategic priority of BRICS STI cooperation; the second is the research report on the priority areas of BRICS STI cooperation based on the win-win strategy. The two reports have evaluated the comprehensive innovative competitiveness of the BRICS countries since 2001 and predicted the innovative competitiveness of the five countries in the future. They have also assessed the current state and effect of China's STI cooperation with other BRICS countries, and analyzed the priority areas of BRICS STI cooperation, providing a valuable reference for BRICS countries to make key decisions on accelerating the building of national innovative competitiveness. This part of the report is drafted by Zhao Xinli, Wang Dan, Xiao Yi, Dong Quanchao, Huo Hongwei, Ma Zongwen, Xin Bingqing and Russian experts Alexander Sokolov, Sergey Shashnov, Maxim Kotsemir and Anna Grebenyuk.

The second part is country reports, which evaluates and analyzes the innovative competitiveness of individual BRICS countries and studies STI cooperation within the BRICS framework. Each country report starts with an analysis on its cooperation with China and within the BRICS framework, presents the features, strategies and differences of the innovative competitiveness of respective BRICS countries, and highlights the competitiveness advantage and weaknesses of each country. The science and technology sections of Chinese embassies in BRICS countries are also involved in the drafting of this part, providing the most up-to-date and authoritative information included in the country reports. The Brazil part is drafted by Wang Lei, Gao Changlin, Dong Quanchao and Shi Tao; the Russian section includes one report drafted by Leonid Gokhberg, Tatiana Kuznetsova, Anna Pikalova and Alexander Sokolov, and another report drafted by Chen Qiang, Zheng Shimin, Xiao Yi and Yang Yefeng; the India part is drafted by Shan Zuhua, Bi Liangliang and Xin Bingqing; the China part is drafted by Huo Hongwei, Wang Zhongcheng and Li Wenjing, and the South Africa part is drafted by Wang Zhongyang, Zhang Dong and Ma Zongwen.

The third part is thematic reports, which features an in-depth analysis on hotspot issues related to STI. The experts of the High-level Expert Group are also involved in the drafting of this part. They have analyzed the current status and challenge of STI development in the country and offered targeted proposals for cooperation. Specifically, the reports in this part include four subjects: one on the innovation of digital economy drafted by Huang Maoxing, Tang Jie and Huang Xinhuan, one of energy technology innovation drafted by Zhang Shirong, Zhang Pei and Gao Mingyuan, one on technology innovation for promoting financial inclusion drafted by Zhao Xijun, Wei Boyang and Shao Mengzhu, and one on agricultural innovative competitiveness drafted by Guo Xiangyu, Wang Dan and Zhao Xinli. The detailed information on the level and potential of STI development in each BRICS country included in the report provides useful reference to help better understand the innovative competitiveness of BRICS countries.

The fourth part is the appendix, which introduces the relevant evaluation indicators and compiles the relevant documents of BRICS STI cooperation.

The report comprises near 200,000 words. Research Fellow Zhao Xinli, the Chair of the High-level Expert Group on BRICS STI Cooperation 2017, Deputy Director (DG level) of CSTEC and Academician of the International Eurasian Academy of Sciences, Professor Li Minrong, former Party Group Secretary of Fujian Provincial Administration of Press and Publication (Fujian Provincial Intellectual Property Office) and Director of China Institute of Science and Technology Evaluation, and Professor Huang Maoxing, Dean of the School of Economics of Fujian Normal University and Director of the Fujian Normal University Sub-center of the National Research Center of Comprehensive Economic Competitiveness are the chief editors of the book. The experts of the High-level Expert Group and their colleagues, the science and technology sections of Chinese embassies in BRICS countries, Russian Higher Schools of Economics and other BRICS think tanks on STI have jointly contributed to this project. Li Zhiqiang, Zheng Wei, Pan Hua, Chen Yuheng, Dai Le and Xia Huanhuan of CSTEC have translated the report.

The book has also made direct or indirect citations or references to the works of other researchers. We express our deepest appreciation to the authors.

Due to time constraint and limited knowledge and experience of the drafting team, mistakes and errors are hardly avoidable. Your valuable opinions will be most appreciated.

Editors.

MIX
Papier aus verantwortungsvollen Quellen
Paper from responsible sources
FSC® C105338

If you have any concerns about our products,
you can contact us on
ProductSafety@springernature.com

In case Publisher is established outside the EU,
the EU authorized representative is:
Springer Nature Customer Service Center GmbH
Europaplatz 3, 69115 Heidelberg, Germany

Printed by Libri Plureos GmbH
in Hamburg, Germany